THREATS FROM CAR TRAFFIC TO THE QUALITY OF URBAN LIFE:
PROBLEMS, CAUSES, AND SOLUTIONS

THREATS FROM CAR TRAFFIC TO THE QUALITY OF URBAN LIFE:
PROBLEMS, CAUSES, AND SOLUTIONS

Edited by

TOMMY GÄRLING
Göteborg University, Sweden

and

LINDA STEG
University of Groningen, The Netherlands

Emerald

United Kingdom – North America – Japan
India – Malaysia – China

Emerald Group Publishing Limited
Howard House, Wagon Lane, Bingley BD16 1WA, UK

British Library Cataloguing in Publication Data
A catalogue record for this book is available from the British Library

ISBN: 978-0-08-044853-4

INVESTOR IN PEOPLE

PREFACE

When thinking about current growth trends in motorized traffic and in particular private car use, there are many reasons to be worried about the future, even the near future. The ever-increasing documentation of negative effects on the environment, most importantly the transport sector's adverse effects on global climate change, is one reason for serious concern. Not enough seems to be done. Some have argued that the political system is non-linear – when the situation becomes *really* severe, politicians will react forcefully. This may or may not be an over-optimistic view.

There must be reasons why the public does not react strongly. The private car is instrumental for many important and desirable activities that people have time to engage in. And they gain even more time from using the car, at least as long as it remains a fast mode of daily travel. However, we know that this is no longer always the case. That people continue to use the car may therefore appear strange. Apparently, other factors account for this: freedom of choice, resistance to change a habit, affective attachment to the car, and the pleasure to drive. A diluted responsibility for undertaking required changes is an additional important factor.

In particular, in urban areas the negative effects of private car use are felt. Noise pollution, air pollution, pedestrian traffic accidents, infringement on land use resulting in the destruction of historic, cultural, and restorative qualities are among the most severe negative threats to the quality of urban life. A primary cause is the immense growth in urban populations, car ownership, and car use.

How can urban-life quality be restored? In any solution private car use must most likely be restrained, although not banished. Is increasing the price a solution? Regulation? Information and education?

We were lucky to manage to recruit scholars as authors of the chapters in this book, who are experts on various aspects of (i) what the threats are from car traffic, (ii) which the determinants of car use are, and (iii) what possible policy measures for curtailing car use can be implemented. This guaranteed a broad coverage of both positive and negative aspects of private car use in urban areas. We hope readers coming from one of the many disciplines represented by the authors of chapters in this book will appreciate this broad coverage. At the same time, we are particularly pleased that all chapters take a behavioural perspective on the problems as well as their solutions. This is needed as a contrast to other perspectives that tend to dominate. After all, it is ordinary people who are both drivers benefiting from the car (excluding the benefits to the car producers) *and* are exposed to the negative effects. We hope that this message will get through to policy makers in the transport sector.

We would like to thank all authors for their work and the following persons who were willing to thoroughly review chapter drafts and did so in a timely manner: Staffan Hygge, Lena Nilsson, Dan Strömberg, Bert Van Wee, Erik Verhoef, Bertil Vilhelmson, and Emile Quinet.

Tommy Gärling
Linda Steg

October, 2006

CONTRIBUTORS

Gary L. Allen
Department of Psychology, University of South Carolina, Columbia, SC, USA

Kay W. Axhausen
Institute of Transport Planning, Swiss Federal Institute of Technology, Zürich, Switzerland

Anke Blöbaum
Workgroup of Cognition and Environmental Psychology, Ruhr-University Bochum, Bochum, Germany

Karel Brookhuis
Department of Psychology, Faculty of Behavioural and Social Sciences, University of Groningen, Groningen, The Netherlands; Delft University of Technology, Faculty of Technology, Policy and Management

Dick de Waard
Department of Psychology, Faculty of Behavioural and Social Sciences, University of Groningen, Groningen, The Netherlands; Delft University of Technology, Faculty of Technology, Policy and Management

Philippe Domergue
Conseil Supérieur du Service Public Ferroviaire (CSSPF), Paris, France

Satoshi Fujii
Department of Civil Engineering, Tokyo Institute of Technology, Tokyo, Japan

Tommy Gärling
Department of Psychology, Göteborg University, Göteborg, Sweden

Birgitta Gatersleben
Department of Psychology, University of Surrey, Guildford, UK

Robert Gifford
Department of Psychology, University of Victoria, Victoria, BC, Canada

Reginald G. Golledge
Department of Geography, University of California, Santa Barbara, CA, USA

Phil Goodwin
Centre for Transport and Society, University of the West of England, Bristol, UK

Terry Hartig
Institute for Housing and Urban Research, Uppsala University, Gävle, Sweden

Cecilia Jakobsson
Department of Psychology, Göteborg University, Göteborg, Sweden

Jeff Kenworthy
Institute for Sustainability and Technology Policy, Murdoch University, Murdoch, WA, Australia

Peter Loukopoulos
Institute for Human-Environment Systems, Swiss Federal Institute of Technology Zurich, Zürich, Switzerland

Ellen Matthies
Workgroup of Cognition and Environmental Psychology, Ruhr-University Bochum, Bochum, Germany

Henk M.E. Miedema
Department of Environment and Health, Netherlands Organisation for Applied Scientific Research (TNO), Delft, The Netherlands

Peter Newman
Institute for Sustainability and Technology Policy, Murdoch University, Murdoch, WA, Australia

Emile Quinet
Ecole Nationale des Ponts et Chaussées, Paris, France

Geertje Schuitema
Department of Psychology, Faculty of Behavioural and Social Sciences, University of Groningen, Groningen, The Netherlands

Linda Steg
Department of Psychology, Faculty of Behavioural and Social Sciences, University of Groningen, Groningen, The Netherlands

Stephen Stradling
Transport Research Institute, Napier University, Edinburgh, UK

John Thøgersen
Department of Marketing and Statistics, Aarhus School of Business, Aarhus University, Aarhus, Denmark

Barry Ubbels
NEA Transport Research and Training (member of Panteia), Rijswijk, The Netherlands

Bert Van Wee
Delft University of Technology, Delft, The Netherlands

Erik Verhoef
Department of Spatial Economics, Free University of Amsterdam, Amsterdam, The Netherlands

Bertil Vilhelmson
Department of Human and Economic Geography, School of Economics and Law, Göteborg University, Göteborg, Sweden

Charles Vlek
Department of Psychology, Faculty of Behavioural and Social Sciences, University of Groningen, Groningen, The Netherlands

Contents

Tommy Gärling and Linda Steg (Editors)

1

INTRODUCTION

Linda Steg, Department of Psychology, Faculty of Behavioural and Social Sciences, University of Groningen, Groningen, The Netherlands
Tommy Gärling, Department of Psychology, Göteborg University, Göteborg, Sweden

ABSTRACT

This introduction briefly overviews the following chapters in the book. The chapters focus on a wide range of behavioural issues related to (i) what the threats are to the urban quality of life from car traffic (and how urban life quality may be defined and measured); (ii) which are the determinants of car use (instrumental, affective/symbolic, or habit) including the possible role played by an ecological orientation; and (iii) how the problems of car use may effectively be reduced through policies forcing or encouraging changes in car use.

BACKGROUND

Motorised traffic is a major contributor to environmental problems at a global scale. In urban areas quality of life is threatened by the steady growth of motorised traffic. Private car use is a major source of these problems. It is widely acknowledged that these problems cannot be effectively controlled by means of new technology aimed at reducing the negative impacts per vehicle. Changes in volumes of car traffic are needed as well (OECD, 1996; Gärling *et al.*, 2002; Steg and Gifford, 2005). Thus, policies must target the demand for car use.

To effectively reduce the problems resulting from motorised traffic, the nature of these problems must be understood. Moreover, knowledge is needed regarding which behaviours contribute to these problems; which factors affect such behaviours; and how the relevant behaviours (and underlying determinants) may be changed to reduce the problems. Given the nature of these problems and the many different factors affecting travel behaviour, and more specifically, car use, a multidisciplinary perspective is warranted to address the urgent issues.

Private car use is a major source of threat to urban quality of life. Therefore, this volume focuses on private car use. In the past decades, scholars from different disciplines have conducted numerous relevant studies on problems resulting from car use, factors influencing the level of car use, and ways to reduce car use to manage these problems. These studies have typically been conducted from a unidisciplinary perspective. Insights from such unidisciplinary studies need to be combined and integrated to understand the complexity of the problems of car use and possible solutions for it.

AIMS

This book aims to provide a comprehensive overview of research on problems resulting from car use, factors influencing car use, and effective strategies to manage these problems by reducing the level of car use. These issues are discussed from a behavioural science perspective this book integrates insights from different disciplines. The book consists of three sections, in which the following three main questions are being addressed: (i) What are the threats to the urban quality of life from car traffic; (ii) Which are the determinants of car use; and (iii) How can the problems of car use effectively be reduced via behavioural changes of individual car users?

OVERVIEW OF CHAPTERS

The first part of the volume is devoted to problems resulting from car use. A detailed description is given regarding negative impacts, such as air pollution, traffic noise, destruction of natural areas and aesthetic qualities, and congestion. In Chapter 2, Van Wee reviews environmental effects of urban traffic and related health effects. He discusses emission of toxic and harmful substances related to climate change, acidification, and air pollution (e.g., CO_2, NO_x, CO), as well as the so-called livability effects related to running and parked vehicles. Furthermore, possible ways of reducing these negative environmental effects are discussed. In Chapter 3, Gifford and Steg discuss quality-of-life effects of (reductions in) car use. They argue that sustainable transportation implies finding a balance between collective qualities and individual quality of life. Approaches to measuring quality of life are discussed, as well as implications for informing policy.

Adverse effects of traffic noise are discussed in Chapter 4 by Miedema, who describes various effects of noise annoyance focussing on reduced attention, increased arousal, and affective reactions, such as fear. A distinction is made between instantaneous and chronic effects, and their relationships with impacts on health are discussed. Furthermore, Miedema discusses acoustic and non-acoustic factors that influence (effects of) noise annoyance. In Chapter 5, Allen and Golledge review research on car drivers' navigation related to spatial structure. They demonstrate how cognitive structures of spatial knowledge and physical structures of the built environment organise and constrain travellers' spatial behaviour. Furthermore, they discuss implications of these influences for motorised traffic in urban areas, with an emphasis on causes and potential remedial actions. Chapter 6 by Hartig focuses on positive and negative effects of car traffic on opportunities for restoration. He argues that, on the one hand, motorised traffic inhibits restoration, for instance, because of loss of green space and depreciation of restorative quality of housing with the construction of new road infrastructure. On the other hand, motorised traffic may serve restoration objectives, by enabling people to travel to recreational settings of greater restorative quality or by allowing a restorative interlude between daily demands.

In Chapter 7, Domergue and Quinet describe external costs of traffic and transport. They list effects that can be evaluated and effects that can be monetarised. Next, they elaborate on methods to evaluate and monetarise impacts of car traffic, which enable decision-makers to make optimal trade-offs between environmental protection and other social objectives. They demonstrate how these methods may be used to provide input to decision-making processes.

The second part of the volume focuses on factors influencing mode choice, and more specifically, choice of private car. Historical trends in car ownership and use as well as possible future developments are described. Next, societal and individual factors affecting car use are discussed. In particular, the chapters elaborate on relevant individual factors. In Chapter 8, Vilhelmson illustrates that socio-spatial organisation of society has stimulated high levels of mobility, and an increased car dependency. He explores the development, structure, and distribution of various car-dependent urban activities. Moreover, he refers to individual as well as external factors that stimulate car dependency, and in particular the time-space organisation of society.

In Chapter 9, Axhausen reviews the macroscopic dynamics of the travel and communication market. Based on this, he proposes dynamic frameworks for travel behaviour, and elaborates a research agenda for travel behaviour and mobility tool ownership. Chapter 10 by Stradling looks at how car ownership, car use, and prospects for modal shifts from car to more sustainable modes vary across different segments of a population. Stradling develops a theoretical overview of factors influencing travel and transport choices, distinguishing car dependent places, trips, and persons. He demonstrates the prevalence of different types and levels of car-dependence, and the prospects for modal shifts for groups differing in car dependency. Based on this, policy suggestions are given for how to reduce car dependency.

The subsequent chapters focus on individual factors affecting car use. The instrumental values of car use as well as its social and affective values are discussed. In Chapter 11, Jakobsson focuses on instrumental reasons for car use. She highlights the role of external factors as well as internal motives for private car use. The argument is made that many people fail to carry out intentions to reduce car use, because instrumental motives are in conflict with the intended change in behaviour, and she concludes that people must be presented with good (instrumental) reasons to reduce their car use, as well as the possibility to form a conscious plan for how to overcome various barriers to do so. Affective and symbolic values of car use are discussed by Gatersleben in Chapter 12. She reviews studies demonstrating the significance of non-instrumental aspects indicating that cars have more than instrumental value. Gatersleben concludes that in informing travel demand management (TDM), it is necessary to know more about what drives people to use their car rather than more sustainable modes of transport.

Habits and routines also play a large role. Chapter 13 by Fujii and Gärling focuses on the development and impact of car use habits. They argue that habitual car users often fail to reduce their car use, even when they intend to do so. Their chapter answers questions like, how car-use habits are developed, how habits affect travel choice, and how car-use habits may be broken. Moreover, they evaluate various methods for measuring car-use habits.

Chapter 14 by Matthies and Blöbaum describes relationships between ecological norm orientations and car use. They demonstrate that norms to cut back on driving in order to protect the environment are particularly distinctive in some western European countries, such as Germany, Switzerland, Sweden, and The Netherlands. However, these norms do often not result in actual reductions in car use. After a review of studies that have examined relationships between ecological norms and behaviour, Matthies and Blöbaum propose the conditions under which ecological norms may influence mode choices.

The third part of the volume describes ways to reduce the problems caused by car use through behavioural changes. The effectiveness, political feasibility, and acceptability of various strategies of behavioural change are critically discussed. In Chapter 15, Loukopoulos provides a classification of TDMstrategies detailing their important properties. He demonstrates that TDM measures have a much broader aim than they originally had, that is, they are no longer aimed at reducing car use only, but are aimed at reducing negative impacts related to car use as well. Next, Loukopoulos discusses several attributes or dimensions of TDM strategies, and indicates how these dimensions may affect policy effectiveness, acceptability, and feasibility.

The following chapters provide an in-depth discussion of four general TDM strategies: urban planning, prohibition, transport pricing, and marketing. These strategies differ in how they trigger behaviour change (Steg, 2003). The first three strategies are aimed at changing social conditions and structures that inhibit car use or facilitate the use of sustainable modes of transport. Urban planning is aimed at facilitating or inhibiting certain types of travel behaviour by changing physical structures and infrastructure. Prohibition is based on enforcing behaviour changes via laws, regulations, and standards adopted by the government. Transport pricing makes car use less attractive by increasing prices of car use, or by making the use of sustainable modes of transport relatively cheaper, thereby increasing their relative attractiveness. In contrast, the fourth strategy, social marketing, is aimed at promoting the use of sustainable transport modes by influencing people's perceptions, beliefs, attitudes, values, and norms.

In Chapter 16 Newman and Kenworthy illustrate how urban transport is shaped by the physical structure and infrastructure of a city. Based on this, they suggest how physical planning may change travel behaviour. Newman and Kenworthy argue that sustainable transport options should in particular be competitive with cars with respect to speed, and they propose various planning measures that may help achieve this aim. In Chapter 17, Gärling and Loukopoulos focus on prohibition of car traffic. They discuss why voluntary measures are not effective in reducing car use, and why coercive measures need to be implemented. They demonstrate that coercive measures may have negative side effects. Furthermore, Gärling and Loukopoulos describe the circumstances under which coercive measures may or may not be acceptable to the public and to the policy-makers, respectively.

Chapters 18 and 19 focus on transport pricing strategies. Transport pricing strategies have been extensively studied by many scholars. They are believed to be effective in reducing car use. However, transport pricing is not easily implemented because of lack of public and political support. In Chapter 18, Ubbels and Verhoef provide an overview of the economic theory of transport pricing. They distinguish several transport pricing objectives, of which economic efficiency should take precedence, according to economics. They demonstrate that efficient prices can often not be set in existing markets. Therefore, regulators have to resort to so-called "second-best" pricing, that is optimal prices, given the prevalent constraints, which are more difficult to define and implement than are the "first-best" prices (i.e., efficient prices under ideal circumstances). Steg and Schuitema provide a psychological perspective on transport pricing in Chapter 19. They discuss various factors influencing the effectiveness and acceptability of transport pricing from a theoretical perspective, and provide empirical evidence for their propositions. The role of individual as well as policy characteristics is discussed. Steg and Schuitema also analyse the relationship between effectiveness and acceptability.

In Chapter 20, Thøgersen discusses social marketing strategies aimed at promoting the use of non-motorised modes of transportation. He provides a concise definition of social marketing in this context, and reviews its effectiveness based on practical experiences as well as research. From this, Thøgersen outlines policy recommendations and makes suggestions for further research to better understand under which conditions social marketing may be effective.

Chapter 21 illuminates the value of driver support systems in cars to reduce the problems caused by car traffic. In this chapter, Brookhuis and De Waard distinguish driver support systems that operate in advisory, semi-automatic or automatic mode, and illustrate that these may have different effects for traffic safety and the environment. They discuss possible desired and adverse effects of driver support systems on traffic safety, efficiency, and the environment. Moreover, Brookhuis and De Waard outline issues related to acceptability of driver support systems.

A general overview of the effectiveness of transport policies is provided in Chapter 22 by Goodwin, who reviews a vast body of empirical evidence with a bearing on what policies are likely to be effective in reducing car travel. Prospective research results as well as actual effects of policy initiatives are critically discussed. Goodwin concludes that, given the political will and a suitably consistent policy support to reduce unintended effects, it is not very difficult to adopt measures that reduce car use from 10% to 30%.

The different perspectives on problems resulting from car use, their causes and possible solutions are integrated in the concluding chapter by Vlek (Chapter 23). Vlek discusses 12 key issues for research and policy-making to effectively manage the problems of car use. He argues that management of these problems requires (i) diagnosis of the problems resulting from car use and behavioural factors underlying these problems; (ii) policy-decision making about which behaviour changes are needed to reduce the relevant problems; (iii) implementation of practical intervention strategies that are effective, acceptable, and feasible; and (iv) evaluation of effectiveness of these strategies. For each of these phases, specific focal issues are elaborated. Vlek argues that freedom in markets (i.e., freedom to travel) must be pitted against required governmental control and direction in order to protect human well-being and environmental quality. He emphasises the importance of a multidisciplinary research, and stresses that effective solutions of the problems caused by car use will be reached only when a coherent mix of policy measures that address the main factors underlying the growth of car use is implemented.

Overall, the volume provides a comprehensive overview of the problems resulting from car use, the causes of these problems, and effective and acceptable ways to manage the problems through behavioural changes.

REFERENCES

Gärling, T., D. Eek, P. Loukopoulos, S. Fujii, O. Johansson-Stenman, R. Kitamura, and R. Pendyala (2002). A Conceptual Analysis of the Impact of Travel Demand Management on Private Car Use, *Transport Policy*, 9, 59–70.

OECD, (1996). Towards Sustainable Transportation, OECD Publications, Paris.

Steg, L. (2003). Factors Influencing the Acceptability and effectiveness of transport pricing. In: *Acceptability of Transport Pricing Strategies* (J. Schade and B. Schlag, eds), Elsevier, Amsterdam, 187–202.

Steg, L. and R. Gifford (2005). Sustainable Transport and Quality of Life, *Journal of Transport Geography*, **13**, 59–69.

PART 1:
THREATS TO THE QUALITY OF URBAN LIFE FROM CAR TRAFFIC

2

Environmental Effects of Urban Traffic

Bert Van Wee, Delft University of Technology, Delft, The Netherlands

ABSTRACT

This chapter starts with a general overview of environmental effects related to traffic followed by an overview of urban environmental effects. Effects relate to emissions of CO_2 (climate change), NO_x (acidification, air quality), PM, CO and HC (air pollution) as well as to noise production. It also gives an overview of the related health impacts. Attention is then paid to the health benefits of walking and cycling. Finally, this chapter focuses on possible measures to reduce negative environmental impacts.

INTRODUCTION

Over the last four decades, the impact of transport on the environment has changed from a new subject to one that is now generally recognized as very important and that should play a key role in transport policies. In the late 1960s resistance of both the public and policymakers to large-scale reconstructions of central urban areas to provide more space for traffic (roads, parking places) started to increase, resulting in many plans being abolished. In the early 1970s many Western countries introduced regulations for the maximum emissions of pollutants per vehicle kilometre for cars and other road vehicles, to reduce negative health impacts. Later regulations for noise emissions were also implemented. The attention paid to environmental issues increased rapidly due to reports such as the first report of the Club of Rome (Meadows, 1972) and Our Common Future (WCED, 1987). In the mid-1990s the World Bank introduced a general framework for the evaluation of all kinds of developments and policies that consisted of three blocks: impacts on the economy, social impact and environmental impacts (Serageldin and Steer, 1994). This framework has been adopted worldwide for evaluation purposes, including the evaluation of transport policies. This illustrates that environmental impacts have now become a well-established impact category in the area of transport.

This chapter aims to give an overview of what the impacts of transport on the environment are, with emphasis on urban road transport. It also aims to give an overview of the determinants of these impacts as well as an overview of the policy measures to reduce these impacts. It is not the aim of this chapter to give a full overview of all the environmental impacts of transport, their causes and cures and policy options, but rather to give an introduction to this field. For further reading, we recommend Hensher and Button (2003) and the journal *Transportation Research Part D – Environment.*

The second section gives an overview of the environmental impacts of traffic; the third section presents the contribution of transport to environmental impacts. The fourth section describes a conceptual model for the impact of transport on the environment. The health impacts of transport, probably the most important effects at the urban scale, are presented in the fifth section. The sixth section elaborates the positive impacts of walking and cycling. The seventh section gives an overview of policy measures to reduce the environmental impacts of urban traffic. In the eighth section the most important conclusions of this chapter are summarized.

OVERVIEW OF THE ENVIRONMENTAL IMPACT OF TRAFFIC

With respect to the spatial scale of traffic and environmental effects, a distinction can be made between the urban and non-urban (higher than urban) scale, resulting in a 2 × 2 table: urban traffic contributes to urban and non-urban environmental problems, as does non-urban traffic. Below we categorize the effects based on spatial scales, starting with urban (local) environmental problems, followed by non-urban problems.

Urban/Local Environmental Problems

Local air pollution is one of the most important environmental effects. The emissions of road vehicles contribute heavily to especially peak levels of concentrations of pollutants, causing health effects and odour nuisance, but also dirt on windows and other parts of buildings, garden furniture and laundry. For health, the emissions of PM (particulate matter – small particulates), NO_2, VOC and CO are relatively important. Note that traffic is not the only emitter of these pollutants. The overall concentration at a certain location near a road consists of a regional or national background concentration, a local background concentration that is higher at a greater distance from the urban fringe (i.e. in the central parts of a city) and the contribution of a nearby road. Note that although the regional/national background concentration is the result of the emissions of many sources, traffic also contributes to these background concentrations. Concentrations often exceed the standards for air quality.

Another dominant effect is noise, especially from road vehicles, rail vehicles and aircraft. Unfortunately, owing to methodological inconsistencies, it is hardly or not at all possible to give quantitative numbers on noise nuisance at an international level, such as the EU (EEA, 2000). A little less problematic, though still quite uncertain, are the figures for noise exposure. Berglund *et al.* (1999) give numbers for a selection of Western (mainly European) countries. The percentage of the population exposed to noise levels over 65 dB(A) due to road transport varies from 3% to 30%. These numbers are also very uncertain. To illustrate this: whereas in Sweden this percentage is only 3%, in the also sparsely populated neighbour country Norway the figure is 12%, the numbers for neighbours Germany and the Netherlands are also 12% and 3%, respectively. Although differences in policy measures or in the land-use and transport system might contribute to the differences in percentages, it is very likely that the differences between these neighbour countries are in practice smaller than these numbers suggest. In almost all countries the share of the population exposed to high levels of noise from aircraft and rail are lower than from road.

A third local environmental problem is the impact of acidification on buildings: emissions of NO_x, SO_2 and NH_3 contribute to acidification. Traffic does not contribute to NH_3 (which is mainly emitted by agriculture), but it does contribute to SO_2 and especially NO_x. Owing to acidifying emissions statues and some parts of buildings are affected, the results of which include a reduction in the quality or even loss of cultural heritage.

The health effects of air pollution, effects on buildings and noise are mainly urban problems, not only because in Western countries most people live in urban areas, but also because emission levels (expressed as per square kilometre emissions) are higher in such areas.

The environmental problems described so far are generally recognized in overviews of environmental problems as well as in policy documents. Since the mid-1990s there has been a growing awareness, both in research and policymaking, that even if vehicles were completely quiet, did not emit pollutants and only used renewable clean energy, some problems would still remain, problems that are often included in what is called 'livability'. One can think of streets with parked vehicles on both sides, making it impossible for children to play on the street, too much traffic having the same effect, or barrier effects making it difficult or impossible for children or old people to cross streets.

Non-Urban Environmental Problems

Apart from the urban problems as described above three other environmental problems related to transport (including urban transport) are of importance. The first is climate change, mainly as a result of the combustion of fossil fuels causing CO_2 emissions. The second problem is the effects of acidification on nature, agriculture and the landscape; the final is large-scale air pollution, not only resulting from the emissions themselves but also from complex chemical reactions taking place in the atmosphere, resulting in ozone formation.

The Impact of Transport on the Environment Expressed in Monetary Terms

How big is the impact of transport on the environment? Economists have developed methods to calculate the effects. Of course, there is still a lot of discussion about whether these effects can be expressed in monetary terms at all. It is beyond the scope of this article to elaborate on this discussion, but the outcomes of quantifications are relevant. As an example we present some data from the Netherlands: environmental costs are in the range of 3–8 billion euros per year (based on Haab and McConnell, 2002), which is in the same order of magnitude as the costs of accidents (4–8 billion euros; based on Bleijenberg *et al.*, 1994; Haab and McConnell, 2002; SWOV, 2005) and much higher than the costs of congestion (1–2.5 billion euros; based on Koopmans and Kroes, 2003). This example shows that, whereas in policy discussions and the media the attention paid to congestion is much larger than the attention paid to the environment, the external costs of the environmental impacts of transport very likely exceed those of congestion. Note that not all environmental costs (as well as costs of accidents and congestion) are urban.

THE SHARE OF TRANSPORT IN ENVIRONMENTAL PROBLEMS

Transport is a major contributor to many environmental problems. Table 1 presents the share of transport in total emissions of SO_2, NO_x, CO, HC and CO_2 in the EU (EU15 and EU25). Table 1 shows that the transport sector emits about 60 per cent of NO_x and CO

Table 1: Share of Transport in Emissions in the EU, 2000

	EU25	*EU15*
SO_2	4	5
NO_x	60	62
CO	55	59
HC (excluding CH_4)	33	33
CO_2	23	25

Source: Eurostat (http://europa.eu.int/newcronos/).

emissions, about one-third of HC emissions, and a quarter of CO_2 emissions. The share of SO_2 seems to be small but emissions of ships at sea are excluded, because these do not occur within a country's boundaries and are therefore not included in national statistics.

Table 1 does not fully express transport's share in the health impacts of these pollutants since, on average, the distance between road traffic and the people exposed is much shorter than for other sources of pollution, such as power plants. Traffic emissions therefore have a greater health impact per kilogram than average emissions (Dorland and Jansen, 1997; Eyre *et al.*, 1997; Newton, 1997). This is especially the case for road transport, more than for non-road transport. The importance of the distance between the source and the receptor is visualized by the concentrations of pollutants as a function of the distance between a location and the road: the greater the distance, the lower the concentration. As an illustration Figure 1 visualizes the contribution of road traffic–related emissions PM10 and PM2.5 in μg/m^3 as a function of the distance from the road, using a motorway location in the Netherlands. Bennett *et al.* (2002) have introduced the concept of the 'intake factor', being the intake of a pollutant by individuals and expressed in mass, divided by the mass of emissions. Evans *et al.* (2002) give an overview of studies on this subject and refer to a study by Smith (1993a, 1993b) concluding that the intake of emissions of particulates of vehicles is 10 times higher than those of a power plant. For further examples of this concept, see Marshall *et al.* (2003, 2005). To conclude, the share in emissions of different sectors is only a rough indicator of the share in effects. The impact of distance on effects should be included to give insights into the environmental effects of emissions, resulting in a relatively high share of road transport emissions.

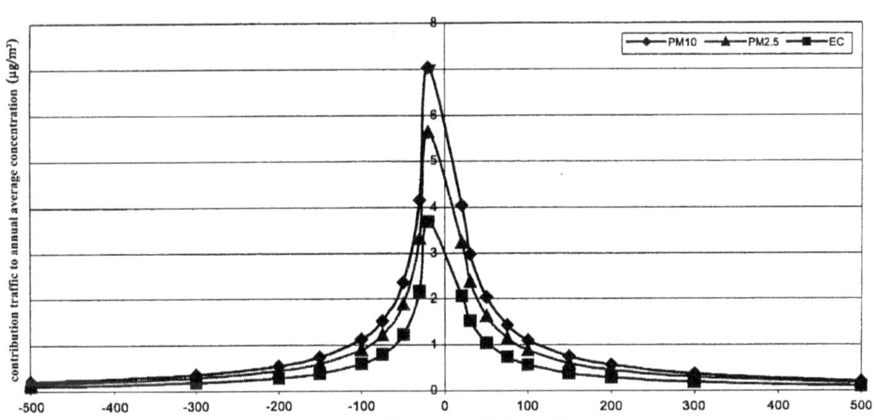

Source: Janssen *et al.* (2002).

Figure 1: Contribution of Road-Traffic-Related Emissions as a Function of the Distance to the Road

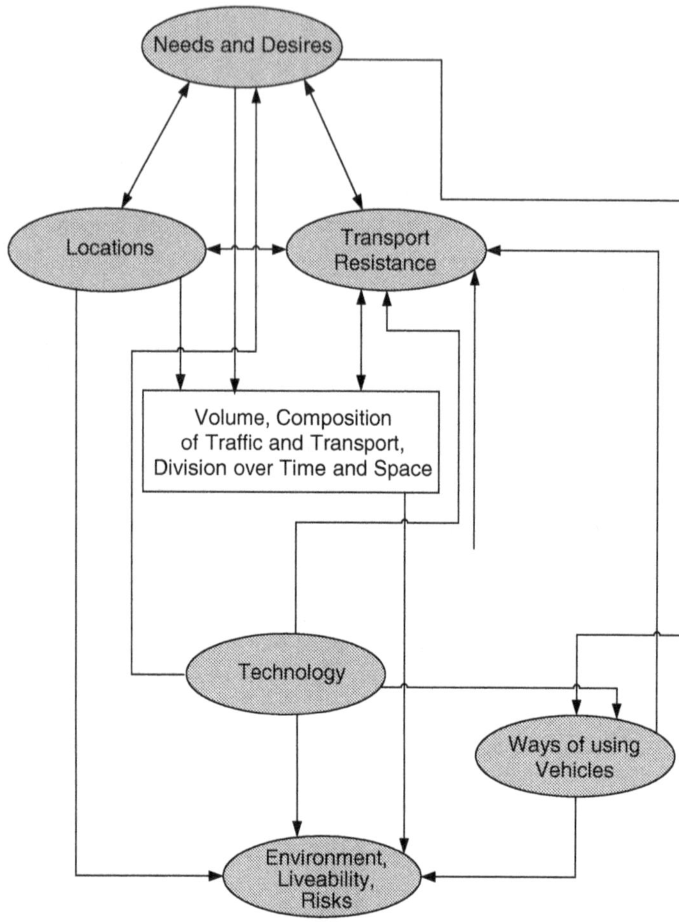

Figure 2: A Conceptual Framework for Factors Having an Impact on Transport Volumes and the Impact of Transport Volumes on the Environment, Accessibility and Safety

A CONCEPTUAL MODEL OF THE IMPACT OF TRANSPORT ON THE ENVIRONMENT

Figure 2 gives a conceptual framework for factors having an impact on the environmental pressure of transport. Environmental effects highly depend on transport volumes.

Transport volumes and travel patterns (such as transport mode and vehicle type) result from

- The wants, needs, preferences and choice options of people and firms;
- The locations of activities such as living, working and shopping;
- Transport resistance, often expressed in time, money costs and other factors, that we refer to as 'trouble' and that include risks, reliability of the transport system and effort.

These transport volumes and travel patterns vary according to socio-economic factors such as age and household type.

Technology and the driving behaviour of people (expressed by, for example, speed and acceleration/deceleration behaviour) also have an impact on the environment. Driving behaviour is influenced by peoples' preferences. Driving fast does not only reduce travel times but people may also like it. The distribution of traffic over space and over time also has an impact on the environment. The distribution over space includes the breakdown between traffic within and outside the build-up area and by road class. For example, traffic on a road along which hardly any houses are sited causes less noise nuisance compared to a traffic on a road along which many houses are located at close distances. Concentrations of pollutants on the pavements are higher if the pavement is located near a busy road. In terms of the impact of traffic on noise nuisance over time a breakdown into hours of the day is very relevant since night traffic causes much more noise nuisance than daytime traffic. However, night traffic hardly causes congestion.

HEALTH IMPACTS OF ROAD TRAFFIC

The health impacts of concentrations of pollutants are often considered to be the most important environmental impact category at the urban scale[1]. We therefore elaborate on health impacts in more detail in this section. We first present a conceptual model for these impacts, followed by the results of empirical research into these effects.

A Conceptual Model

This section presents a conceptual model for the relationship between emissions, concentrations, exposure and health effects. The model is derived from literature on transport emissions, dispersion of pollutants and public health (see, for example, Van Pul *et al.*, 1996; Whitelegg *et al.*, 1993, for the relations between emissions and concentrations, and several references below for the health impact of exposure to concentrations).

---

[1] For health impacts of traffic noise we refer to Chapter 4 by Miedema.

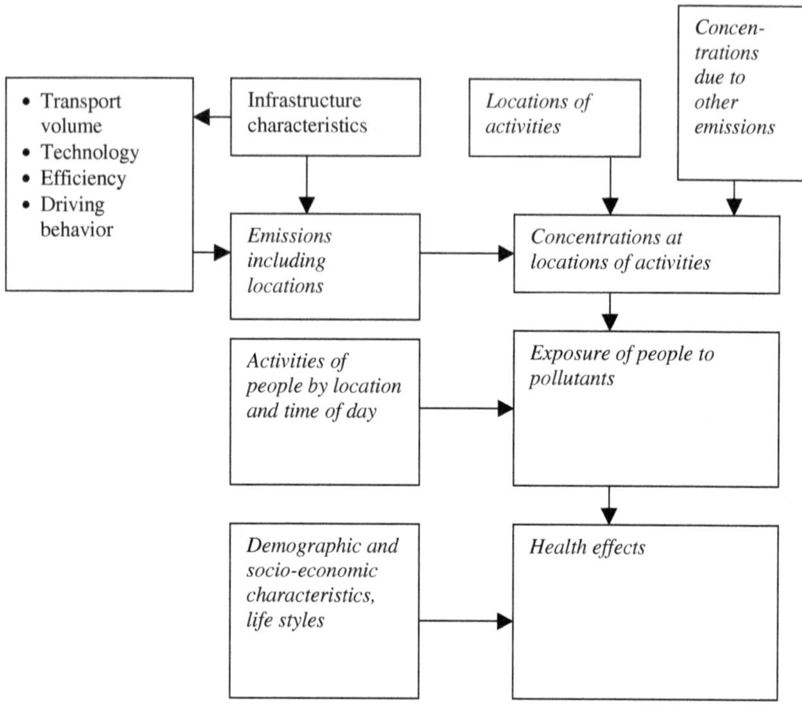

Figure 3: A Conceptual Model for the Health Impacts of Transport Emissions

Figure 3 shows the health effects for people that result from exposure to concentrations of pollutants. However, apart from this exposure, peoples' characteristics are also relevant. For example, people at an advanced age are more vulnerable to high concentrations of pollutants than young people. The peak in mortality in periods of high concentrations of pollutants in Western cities mainly relates to older people. Lifestyle is also important. For example, the additional negative impacts of exposure to high concentrations of pollutants might differ between smokers and non-smokers. The exposure of people depends on the concentrations of pollutants at locations where people carry out their activities, and the time people spend at these locations, including the characteristics of these activities. For example, being physically active in a high-concentration location causes more health impacts than doing office work. The concentrations of pollutants at locations depend on the emissions of road vehicles and the dispersion of these pollutants. The distance between the locations of emissions (roads) and the locations of activities is particularly important, but also the average wind speed and wind direction matter, as well as the objects between the locations of emissions and activities. The emissions of vehicles depend on transport and traffic volumes (such as the number of vehicle kilometres), the technology of the vehicles and fuels used, and driving behaviour. For example, the speed and acceleration characteristics matter: driving at high speeds and aggressively causes higher overall emission levels (El-Shawarby *et al.*, 2005).

Exposure of Road Users

The following sections give an overview of the results of empirical studies, starting with the exposure of road users to concentrations of pollutants. Table 2 gives an overview of the main conclusions with respect to the exposure of road users to concentrations of pollutants. Table 2 shows that road users are exposed to concentrations that are higher than average. This, of course, is not surprising because they stay at the very same location where vehicles emit pollutants. Cyclists and pedestrians are exposed to lower concentrations than car drivers. Note that cyclists and pedestrians inhale more than twice as much air as car drivers, due to their physical activity. They therefore inhale a little

Table 2: Exposure of Road Users to Pollutants

Main Conclusion	*Further Information*	*References*
Concentrations in vehicles are higher than in the ambient air	In vehicles the concentrations are between 1.5 and over 10 times higher The difference between concentrations is greater for CO, benzene and NO compared to PM2.5 and PM10	Van Wijnen and Van der Zee (1998) Van Wijnen *et al.* (1995) Van Bruggen *et al.* (1991) Chan *et al.* (1991a) Den Tonkelaar and Van der Tuin (1983) Zagury *et al.* (2000) Adams *et al.* (2001) Praml and Schierl (2000) BeVan *et al.* (1991)
Concentrations to which cyclists and pedestrians are exposed are higher than in the ambient air	As 'in vehicles' (first row in this table)	Van Wijnen *et al.* (1998) Van Bruggen *et al.* (1991)
	Concentrations for cyclists and pedestrians are lower compared to 'in vehicle concentrations'	Chan *et al.* (1991a) Chan *et al.* (1991b) Rank *et al.* (2001)
Concentrations are lower on rural roads than on urban roads	The difference is especially great for CO and benzene, and much less pronounced for NO_2 and PM	Van Wijnen *et al.* (1995) Chan *et al.* (1991a) Adams *et al.* (2001) Praml and Schierl (2000) BeVan *et al.* (1991)
Concentrations depend on wind speed, travelling speed and temperature	An increase in these variables decreases concentrations	Avol *et al.* (1998) Den Tonkelaar and Van der Tuin (1983) Van Wijnen *et al.* (1995) Van Bruggen *et al.* (1991) BeVan *et al.* (1991)

Table 3: Exposure of People Living Near Busy Roads

Main Conclusion	*Further Information*	*References*
Concentrations at locations near urban roads are higher than at locations further away from the road	This conclusion was found for PM, NO_2, soot, benzene and polycyclic aromatic HC (PAH), the impact of distance differs between pollutants	Zagury *et al.* (2000) Bloemen *et al.* (1993) Fischer *et al.* (2000) Janssen *et al.* (1997) Roemer and Wijnen (2001a) Kinney *et al.* (2000) Pfeffer (1994) Janssen *et al.* (2001) Roorda-Knape *et al.* (1998) Kirby *et al.* (1998)
This effect is found both for urban roads and for motorways		Janssen *et al.* (2001) Roorda-Knape *et al.* (1998)
Concentrations are higher in narrow streets with high buildings		Kirby *et al.* (1998)

more benzene and CO, and significantly more NO_2 than car drivers. As examples of the studies as presented in Table 2, Van Wijnen *et al.* (1995) and Van Bruggen *et al.* (1991) conclude that the CO-concentration in a car in Amsterdam was about four times higher than at a city background measuring point. The exposure for cyclists was 1.5 times higher than the background concentration and the exposure for pedestrians was comparable with that of cyclists.

Exposure of People Living Near Busy Urban Roads and Motorways

Table 3 gives an overview of the main conclusions with respect to the exposure of people living near busy roads. For example, Kirby *et al.* (1998) carried out a large number of NO_2 measurements over two years at 80 different locations in Cambridge, England. The NO_2 concentrations were about twice as high near busy roads as at background locations in the city. The same holds for narrow streets. Janssen *et al.* (2001) measured traffic-related air pollution at 24 schools located within 400 m of a motorway. Both in the ambient air and in the classrooms, the concentrations of soot and PM2.5 were significantly higher when there was more heavy traffic passed on the motorway, and significantly lower the further the school was located from the motorway.

Health Effects of Traffic-Related Air Pollution for People Living or Working Near Busy Roads

Table 4 gives an overview of the main conclusions with respect to the health effects of traffic-related air pollution for people living or working near busy roads. For example,

Table 4: Health Effects of Traffic-Related Air Pollution for People Living or Working Near Busy Roads

Main Conclusion	*Further Information*	*References*
For people living near a motorway or busy urban road the risk of death due to cardiovascular diseases and respiratory disorders is about twice as high as average. Some studies found an effect on lung cancer	On days with a high level of air pollution more people die Some studies do not focus on the source of pollutions. Not only traffic contributes to the exposure of people	Hoek *et al.* (1997) Dockery *et al.* (1993) Pope *et al.* (1995) Abbey *et al.* (1999) Pope *et al.* (2002) Nyberg *et al.* (2000) Roemer and Van Wijnen (2001b)
	Soot and PM2.5 (or smaller) seem to be the most important pollutants	Janssen *et al.* (2001) Laden *et al.* (2000) Katsouyanni *et al.* (2001) Hoek *et al.* (2002)
People living near busy roads have a greater chance of respiratory disorders and allergy	Several studies use self-reporting as a research method, resulting in the risk of over-reporting	
Children living near busy roads have relatively more problems with the lung function	Mainly related to heavy traffic (diesel)	Brunekreef *et al.* (1997) De Hartog *et al.* (1997) Van Vliet *et al.* (1997) Janssen *et al.* (2001) Pope *et al.* (2002)
People working near busy roads face more negative health impacts	Impacts were found on chromosomal aberrations and lung cancer	Ruchirawa *et al.* (2002) Knudsen *et al.* (1999) Hansen *et al.* (1998) Raaschou-Nielsen *et al.* (1995) Evans *et al.* (1988) Karita *et al.* (2001)

Dockery *et al.* (1993) found 14% higher mortality rates per 10 μg PM2.5 per m^3, combined with a 4 per cent higher sulphate concentration. A relationship between mortality and PM10 concentrations was also found: an increase of 10 μg per m^3 results in an increase of 4.3% in mortality for adults over 30 years of age. Hoek *et al.* (2002) found a doubling of the risk of dying from heart and lung diseases for people living less than 100 m from a motorway or less than 50 m from a busy urban road.

In the Netherlands a study was carried out that included 24 schools located less than 400 m from a motorway in the Netherlands. A comprehensive measurement programme was carried out. The participating children were tested not only for respiratory symptoms and lung function but also for irritability of the respiratory system and allergies. The study shows a relationship between the intensity of the freight traffic and the respiratory complaints. As in previous research the researchers did not find a relationship with the intensity of passenger traffic. They did not find a significant relationship between distance to the motorway and irritability of the lung function either

(Aarts *et al.*, 1999; Janssen *et al.*, 2001). The intensity of heavy vehicles is of importance because of the diesel these vehicles use. Health impacts are strongly related to soot and PM-emissions, and diesel vehicles have much higher emission factors for these pollutants than petrol vehicles, although in many western countries the difference is now decreasing due to recent and forthcoming regulations. Note that at the time of the research the share of diesel in the car fleet was much lower than at present. In most EU countries diesel car sales have increased. Sales of diesel cars in France, Spain, Austria, Belgium and Luxembourg already exceed those of gasoline cars (Ricardo, 2002). Nowadays diesel cars might have a bigger share in traffic-related health impacts than some years ago. On the other hand, owing to EU regulations, the emission factors for diesel cars (as well as for petrol cars) are nowadays much lower than, for example, 20 years ago.

An example of a study into the health effects of people working near a busy road is the study of Hansen *et al.* (1998), who found that professional drivers active in Copenhagen were diagnosed with lung cancer more often than control persons who were not exposed to a similar degree of traffic-related air pollution. The risk was 1.6 times higher among taxi-drivers and 1.3 times higher among bus- and lorry-drivers. The difference between lorry-drivers, taxi-drivers and partly also bus-drivers might be explained by the share of kilometres – lorry-drivers drive outside the built-up area, where concentrations are lower than in cities.

HEALTH EFFECTS OF WALKING AND CYCLING

Literature on transport and the environment often only focuses on the negative impacts of driving, and in much fewer cases the impact of parked vehicles. Depending on the definition or demarcation of transport's impact on the environment, some authors pay attention to the health impacts related to being a participant of the transport system. In Western countries many people are much less active than is healthy for them. This often starts at school age, especially if parents bring their children to and from school by car, not allowing children to walk or cycle. They sometimes have good reasons for that, such as the risks of children traveling alone or the fact that bringing their children to school or picking them up from school is combined with the commute, not allowing parents to walk or cycle between home and school. Many adults commute by car and use their car for most of their other trips as well. If they do not exercise (sports, jogging, cycling), these adults are rarely physically active, and certainly not active enough from a health perspective. Although there is still a lot of debate, several studies conclude that some urban forms result in a larger share of slow transport modes (see Meurs and Haaijer, 2001, for an example of empirical research; see Handy, 1996, for an overview). Generally speaking, these studies show that an urban form with a high density of buildings with mixed use, an attractive infrastructure for cycling and walking, and a nice environment for walking and cycling results in a higher share of these modes and a positive impact on health. Some authors have explicitly linked this impact on travel behaviour to an impact on health (Dora, 1999; Lumsdon and Mitchel, 1999; Morisson *et al.*, 2004; Badland and Schofield, 2005; see Handy, 2005, for an overview of theory and empirical evidence). The importance of the possible health impacts of transport policy measures is illustrated by a Norwegian study on the costs and benefits of cycling infrastructure, which showed that the cycle infrastructure examined had much higher benefits for society than costs. The health benefits vary between cities and towns, ranging from 55% to 75% of all benefits (Saelensminde, 2004).

POLICY MEASURES

Having a general overview of the environmental impacts of transport an important question is: what can policymakers do to reduce these impacts? This section gives an overview of possible policy measures. The environmental impact of transport depends on a number of determinants. The first is the overall volume of transport, expressed as passenger kilometres (persons) or tonne kilometres (goods). The second category is the modal split (for passengers this is car-driver, car passenger, train, bus/tram/ metro, walking, cycling, aircraft, ship) and the third is the technology used. A fourth category is the efficiency of using vehicles (for lorries this is the load factor, and for cars, trains and buses it is the occupancy rate). The fifth is the manner in which vehicles are used (speed, acceleration, deceleration). Governments have several types of policy instruments to influence these determinants, including regulations, prices, land-use planning, infrastructure planning, marketing and the provision of information and communication. Table 5 shows the relationship between the type of policy instruments and the determinants.

Table 5 shows that regulations and pricing measures potentially have the widest range of impacts, followed by land-use and infrastructure measures. We now give examples of the dominant effects as presented in Table 1. *Restrictions* can reduce overall volume, e.g. due to limiting access to certain areas. If access is limited to certain vehicles only

Table 5: Dominant Relationships between Determinants for Environmental Impact of Transport and Policy Instruments

	Volume	*Modal Split*	*Technology*	*Efficiency of Using Vehicles*	*Use of Vehicles/ Driving Behaviour*
Restrictions	*	*	*	*	*
Prices	*	*	*	*	*
Land-use planning	*	*			*
Marketing		*			
Information and communication	*	*		*	*

Source: After Blok and Van Wee (1994).

(e.g. motorized transport, which is often the case in central urban areas), it can affect the modal split. Most technological changes in vehicles that reduce emissions result from regulations. Restrictions on which lorries are allowed in certain areas may affect the load factors (and therefore the efficiency of using freight vehicles). The CAFÉ regulations in the USA have reduced the energy use of cars. Speed limits also reduce energy use and some emissions. Higher fuel prices may reduce overall transport volume, and may lead to a shift to other modes. Higher taxes for vehicles without a three-way catalytic converter in the years before 1993 (the introducing of the Euro 1 standards) increased the sales of 'clean' cars in Germany and Holland and therefore led to technological changes in the car fleet. Higher fuel prices increase the share of car use (efficiency) and lead to lower speeds on motorways to save fuel. More compact land-use and mixed use may reduce overall transport volume, and increase the share of slow modes. It might also change the distribution of car kilometres over road classes and therefore affect driving behaviour. Marketing public transport might increase its share. The provision of information and communication might change people's travel behaviour. For example, route information might reduce trip distance; public transport information might increase its share. Information provided, thanks to Information and Communication Technology (ICT) possibilities, might increase the efficiency of freight transport by making it possible to pick up a load on a return trip. Information provided by on-board units might change people's driving behaviour. Providing more and faster infrastructure increases transport volumes and driving speeds. Investing more in public transport might increase its share.

The main messages we want to give here is firstly that a broad range of policy measures exist to influence the environmental impact of urban transport and secondly that policy measures may have a much wider impact than on the environment only. These complex relationships have an important impact on decision-making with respect to policymaking: it is seldom possible to draw policy recommendations based on research into mobility on environmental effects alone. Policy options should be evaluated using a much wider evaluation framework and a set of indicators. Indicators must include the impact on emissions, the impact on concentrations or the environment and the impact on health, including those related to the positive effects of cycling and walking, accessibility impacts (including congestion), safety impacts, in terms of land-use measures also the residential preferences of people and location preferences of firms, and the robustness: can the land-use and transport system be maintained for many decades under different scenarios for energy, preferences, demography, etc.

The policy discussion on the environmental impacts of transport is a broad one. Here we will limit ourselves to a few major issues that often lead to 'wrong' conclusions. The first issue is the cost-effectiveness of transport measures. In the case of policy-related studies that should give insights into how to meet certain environmental standards, particularly the maximum levels of emissions of certain pollutants, it is common to pay attention to the cost-effectiveness of possible future policy options. Such studies for example look at the costs of the reduction of NO_x or PM emission by 1 tonne. However, on average the distance between urban road traffic and people exposed is much shorter than for other sources such as power plants (see the third section). Therefore, per kilogram, road traffic emissions have a greater health impact than average. We think that the cost-effectiveness of measures to reduce emissions is a poor indicator of the cost-effectiveness of health impacts. Distance needs to be taken into account.

The second issue we want to address is modal shift policy. Many countries, as well as the EU, have proposed a shift to public transport and slow modes, from cars and lorries, based on average figures of emissions per kilometre. Though we certainly would not like to suggest that modal shift policies can never be fruitful or are never to be recommended, some warnings may be relevant. Firstly, average figures might not be applicable for specific situations. If one wants to compare long-distance goods transport by rail and road, urban and short distance road trips should be excluded, making the average performance of road for long distances better than for all road goods transport. Secondly, owing to the limited overlap in markets between cars and public transport, making public transport cheaper or faster primarily results in an increase in its use, but only a minority of the extra use originates from previous car users. Depending on the specific situation, making public transport cheaper or faster may result in an increase in emissions. See Van Wee *et al.* (2005) for an extensive elaboration on this discussion.

The third issue is the issue of self-selection. In both travel behaviour research as well as in research into the impacts of transport (e.g. on health), it is nowadays common to include important determinants like income, age, sex and household structure. The phenomenon of self-selection refers to differences within the so-called homogeneous (with respect to income, age, etc.) population groups. *Within* a certain group of people within the same age, income, etc. category differences exist. For example, some people prefer to travel by car whereas others with seemingly the same characteristics prefer to travel by public transport and slow modes (Kitamura *et al.*, 1997; Bagley and Mokhtarian, 2002). Self-selection refers to the fact that people make choices according to their preferences, attitudes and lifestyles. Differences between people in their preferences, attitudes and life-styles do exist, even within people with the same income level, age, class, etc. Including self-selection probably results in a lower impact of infrastructure and land-use measures than assumed by models using cross-section based data that include income, age, sex and other relevant variables. However, self-selection may also exist with respect to residential choice. For example, people that hate traffic noise probably are less often than average to live near roads, railway lines or airports. The same might be true for people with asthma or other problems with their lungs. If so, this self-selection results in an underestimation of the negative impacts of transport on noise nuisance or health when a new infrastructure is built in or near a quiet area: in this area the share of sensitive people might be higher than at current locations near infrastructure.

CONCLUSIONS

The main conclusions of this chapter are presented below.

During the past four decades the environmental impacts of transport have become a well-established impact category in both policy and assessment methods.

Urban traffic contributes to both local environmental problems, such as local air pollution and noise, as well as to environmental problems at a higher spatial scale, such as acidification and climate change. On the other hand, local problems can be caused by local traffic, but also by non-local traffic, examples of the last category being noise due to aircraft or motorways and air pollution due to regional, national or international traffic.

At the EU level the share of transport in the emissions of CO, CH, NO_x and CO_2 lies roughly between 25% and 60%, making transport probably the most important sector causing environmental problems. The share in emissions underestimates the share in health effects, since in general the distance between road transport and the locations where people stay is shorter than on average, resulting in a relatively high health impact per kilogram of emissions.

The environmental impacts of transport depend on transport volume (e.g. vehicle kilometres), technology, the way vehicles are used (speed, acceleration), the distribution of the use of vehicles over space and time, and the locations of exposed people, nature and buildings. For health impacts the socio-demographic characteristics of the population are relevant, as are activity patterns over time.

People in road vehicles, cyclists and pedestrians are exposed to high concentrations of pollutants causing health impacts. Concentrations near roads are higher than at a greater distance from road, the related health impacts are higher for people living in houses near roads than for those who live at a greater distance from heavily trafficked roads. The distance decay effect varies strongly between pollutants. On days with a high level of air pollution more people die.

Walking and cycling has a positive health effect. Some land-use concepts contribute to a higher share of slow modes in travel behaviour, mixed use, high densities, attractive infrastructure for cycling and walking, and a nice environment for walking and cycling being the most important land-use characteristics.

Policy measures to reduce the environmental impact of transport include restrictions, pricing measures, land-use planning, infrastructure measures, marketing, communication and information provision. These measures may have an impact on transport volume, modal split, technology, the efficiency of the use of vehicles (load factors, occupancy rates) as well as the way vehicles are used (speed, aggressive driving). Because several policy measures have a greater impact than on the environmental one, decisions should be based on a framework that not only includes the environmental impacts, but also other relevant impacts (such as safety impacts and the preferences of people with respect to their residential location).

In the policy debate of transport's impacts on the environment the relatively short distance between road traffic and people exposed should be taken into consideration in cross-sectoral cost-effectiveness comparisons of measures to reduce emissions.

Modal shift measures may certainly contribute to a decrease in the environmental effects of transport, but the effect is often overestimated.

Self-selection may be an important factor for the impact of transport on the environment, but is generally ignored in both research and policymaking.

REFERENCES

Aarts, F., P. Van Vliet, N.A.H. Janssen, H. Harssema, and B. Brunekreef (1999). Nader onderzoek naar effecten Van verkeersgerelateerde luchtverontreiniging op de luchtwegen Van kinderen wonend nabij snelwegen (luchtweg 2). Wageningen: Wageningen Universiteit afd Gezondheidsleer, EOH 1999-486.

Abbey, D.E., N. Nishino, W.F. McDonnell, R.J. Burchette, S.F. Knutsen, W.L. Beeson, and J.X. Yang (1999). Long-Term Inhalable Particles and Other Air Pollutants Related to Mortality in Nonsmokers, *American Journal of Respiratory and Critical Medicine*, **159** (2), 373–382.

Adams, H.S., M.J. Nieuwenhuijsen, R.N. Colvile, M.A. McMullen, P. Khandelwal, and P. Fine (2001). Particle (PM2.5) Personal Exposure Levels in Transport Microenvironments, London, UK, *Science of the Total Environment*, **279** (1–3), 29–44.

Avol, E.L., W.C. Navidi, and S.D. Colome (1998). Modeling Ozone Levels in and around Southern California Homes, *Environmental Science & Technology*, **32** (4), 463–468.

Badland, H. and G. Schofield (2005). Transport, Urban Design, and Physical Activity: An Evidence-Based Update, *Transportation Research Part D*, **10**, 177–196.

Bagley, M.N. and P. Mokhtarian (2002). The Impact of Residential Neighborhood Type on Travel Behaviour: A Structural Equations Modeling Approach, *Annals of Regional Science*, **36**, 279–297.

Bennett, D.H., T.E. McKone, J.S. Evans, W.W. Nazaroff, M.D. Margni, O. Jolliet, and K.R. Smit (2002). Defining Intake Fraction. *Environmental Science & Technology* (May), **36**, 3–7.

Berglund, B., T. Lindvall, and D.H. Schwela (eds) (1999). *Guidelines for Community Noise*, WHO, Geneva.

BeVan, M.A.J., C.J. Proctor, J. Baker-Rogers, and N.D. Warren (1991). Exposure to Carbon Monoxide, Respirable Suspended Particulates and Volatile Compounds While Commuting by Bicycle, *Environmental Science & Technology*, **25**, 788–791.

Bleijenberg, A.N., W.J. Van den Berg, and G. De Wit (1994). *Maatschappelijke kosten van het verkeer. Literatuuroverzicht*, CE, Delft.

Bloemen, H.J.T., T.T.M. Balvers, H.J. Van Scheindelen, and E. Lebret (1993). *Benzeen onderzoek Zuid Kennemerland*, RIVM, Bilthoven.

Blok, P. and B. Van Wee (1994). Het Verkeersvraagstuk. In: *Basisboek Milieu-Economie* (F. Dietz, W. Hafkamp, and J. Van der Straaten (Retd.), eds), Boom, Amsterdam/ Meppel.

Brunekreef, B., N.A. Janssen, J. De Hartog, H. Harssema, M. Knape, and P. Van Vliet (1997). Air Pollution from Truck Traffic and Lung Function in Children Living Near Motorways, *Epidemiology*, **8** (3), 298–303.

Chan, C.C., H. Ozkaynak, J.D. Spengler, and L. Sheldon (1991a). Driver Exposure to Volatile Organic Compounds, CO, Ozone and NO2 under Different Driving Conditions, *Environmental Science and Technology*, **25**, 964–972.

Chan, C.C., J.D. Spengler, H. Azkaynak, and M. Lefkopoulou (1991b). Commuter Exposures to VOCs in Boston, Massachusetts, *Journal of the Air & Waste Management Association*, **41** (12), 1594–1600.

De Hartog, J.J., P.H. Van Vliet, B. Brunekreef, M.C. Knape, N.A. Janssen, and H. Harssema (1997). Relationship between Air Pollution Due to Traffic, Decreased Lung Function and Airway Symptoms in Children, *Nederland Tijdschrift voor Geneeskunde*, **41** (38), 1814–1818.

Den Tonkelaar, W.A.M. and J. Van der Tuin (1983). *Luchtverontreiniging in auto's. Onderzoek naar de expositie Van inzittenden Van personenauto's*. IMG-TNO, Delft.

Dockery, D.W., C.A. Pope III, X. Xiping, J.D. Sprengler, J.H. Ware, M.A. Fay, B.G. Ferries Jr, and F.E. Speizer (1993). An Association between Air Pollution and Mortality in Six US Cities, *New England Journal of Medicine*, **329** (24), 1753–1759.

Dockery, D.W., C.A. Pope III, X. Xu, J.D. Sprengler, J.H. Ware, M.A. Fay, B.G. Ferries Jr, and F.E. Speizer (1993). An Association between Air Pollution and Mortality in Six U.S. Cities. *New England Journal of Medicine*, **329** (24), 1753–1759.

Dora, C. (1999). A Different Route to Health: Implications of Transport Policies, *British Medical Journal*, **318**, 1686–1689.

Dorland, C. and H.M.A. Jansen (1997). *Externe Transport – The Netherlands. Dutch Case Studies on Transport Externalities*, Free University, Institute for Environmental Studies (IVM), Amsterdam.

El-Shawarby, I., K. Ahn, and H. Rakha (2005). Comparative Field Evaluation of Vehicle Cruise Speed and Acceleration Level Impacts on Hot Stabilized Emissions, *Transportation Research Part D*, **10** (1), 13–30.

European Environment Agency (2000). Are We Moving in the Right Direction? Indicators on Transport and Environmental Integration. In the *EU: TERM2000*. EEA, Kopenhagen.

Eyre, N.J., E. Ozdemiroglu, D.W. Pearce, and P. Steele (1997). Fuel and Location Effects on the Damage Costs of Transport Emissions, *Journal of Transport Economics and Policy*, **31** (1), 5–24.

Evans, J.S., S.K. Wolff, K. Phonboon, J.I. Levy, and K.R. Smith (2002). Exposure Efficiency: An Idea Whose Time Has Come? *Chemosphere*, **49**, 1075–1091.

Evans, R.G., K. Webb, S. Homan, and S.M. Ayres (1988). Cross-Sectional and Longitudinal Changes in Pulmonary Function Associated with Automobile Pollution among Bridge and Tunnel Officers, *American Journal of Industrial Medicine*, **14** (1), 25–36.

Fischer, P.H., G. Hoek, H. Van Reeuwijk, D.J. Briggs, E. Lebret, J.H. Van Wijnen, S. Kingham, and P.E. Elliott (2000). Traffic-Related Differences in Outdoor and Indoor Concentrations of Particles and Volatile Organic Components in Amsterdam, *Atmospheric Environment*, **34**, 3713–3722.

Haab, T.C. and E. McConnel (2002). *Valuing Environmental and Natural Resources: The Econometrics of Non-Market Valuation*, Edward Elgar Publishing, Cheltenham.

Handy, S. (1996). Methodologies for Exploring the Link between Urban Form and Travel Behaviour, *Transportation Research D*, **1** (2), 151–165.

Handy, S. (2005). Critical Assessment of the Literature on the Relationships among Transportation, Land Use, and Physical Activity Resource paper for TRB Special Report 282: Does the built environment influence physical activity? Examining the evidence.

Hansen, J., O. Raaschou-Nielsen, and J.H. Olsen (1998). Increased Risk of Lung Cancer among Different Types of Professional Drivers in Denmark, *Occupational and Environmental Medicine*, **55** (2), 115–118.

Hensher, D. and K. Button (eds) (2003). *Handbook of Transport and the Environment*. Elsevier, Amsterdam.

Hoek, G., B. Forsberg, M. Borowska, S. Hlawiczka, E. Vaskövi, H. Welinder, M. Branis, I. Benes, F. Kotesovec, L.O. Hagen, J. Cyrys, M. Jantunen, R. Roemer, and B. Brunekreef (1997). Wintertime Concentrations of PM10 and Black Smoke from 28 European Regions Studied in the Framework of the PEACE Study, *Atmospheric Environment*, **31**, 3609–3622.

Hoek G., B. Brunekreef, and S. Goldbohm (2002). Association between Mortality and Indicators of Traffic Related Air Pollution in the Netherlands: A Cohort Study. *The Lancet*, **360**, 1203–1209.

Janssen, N.A.H., B. Brunekreef, G. Hoek, and P. Keuken (2002). *Traffic-Related Air Pollution and Health*, Utrecht/Apeldoorn: Utrecht University, Institute for Risk Assessment Sciences, Environmental and Occupational Health Division/TNO Milieu, Energie en Procesinnovatie.

Janssen, N.A.H., D.F.M. Van Mansom, K. Van der Jagt, H. Harssema, and G. Hoek (1997). Mass Concentration and Elemental Composition of Airborne Particulate Matter at Street and Background Locations, *Atmospheric Environment*, **31**, 1185–1193.

Janssen, N.A.H., P.H.N. Van Vliet, F. Aarts, H. Harssema, and B. Brunekreef (2001). Assessment of Exposure to Traffic Related Air Pollution of Children Attending Schools Near Motorways, *Atmospheric Environment*, **35**, 3875–3884.

Karita, K., E. Yano, W. Jinsart, D. Boudoung, and K. Tamura (2001). Respiratory Symptoms and Pulmonary Function among Traffic Police in Bangkok, Thailand, *Archives of Environmental Health*, **56** (5), 467–470.

Katsouyanni, K., G. Touloumi, E. Samoli, A. Gryparis, A. Le Tertre, Y. Monopolis, G. Rossi, D. Zmirou, F. Ballester, A. Boumghar, H.R. Anderson, B. Wojtyniak, A. Paldy, R. Braunstein, J. Pekkanen, C. Schindler, and J. Schwartz (2001). Confounding and Effect Modification in the Short-Term Effects of Ambient Particles on Total Mortality: Results from 29 European Cities within the APHEA2 Project, *Epidemiology*, **12** (5), 521–531.

Kinney, P.L., M. Aggarwal, M.E. Northridge, N.A. Janssen, and P. Shepard (2000). Airborne Concentrations of PM(2.5) and Diesel Exhaust Particles on Harlem Sidewalks: A Community-Based Pilot Study, *Environmental Health Perspectives*, **108** (3), 213–218.

Kirby, C., A. Greig, and T. Drye (1998). Temporal and Spatial Variations in Nitrogen Dioxide Concentrations across and Urban Landscape: Cambridge, UK, *Atmospheric Environment*, **52**, 65–82.

Kitamura, R., P.L. Mokhtarian, and L. Laidet (1997). A Micro-Analysis of Land Use and Travel in Five Neighbourhoods in the San Francisco Bay Area, *Transportation*, **24**, 125–158.

Knudsen, L.E., H. Norppa, M.O. Gamborg, P.S. Nielsen, H. Okkels, H. Soll-Johanning, E. Raffn, H. Järventaus, and H. Autrup (1999). Chromosomal Aberrations in Humans Induced by Urban Air Pollution: Influence of DNA Repair and Polymorphisms of Glutathione S-transferase M1 and N-Acetyltransferase 2, *Cancer Epidemiology Biomarkers & Prevention*, **8** (4 Pt 1), 303–310.

Koopmans, C. and E. Kroes (2003). The Real Cost of Congestion in The Netherlands, Paper presented at the European Transport Conference, Strasbourg, France, 8–10 October.

Laden, F., L.M. Neas, D.W. Dockery, and J. Schwartz (2000). Association of Fine Particulate Matter from Different Sources with Daily Mortality in Six U.S. Cities. *Environmental Health Perspectives*, **108** (10), 941–947.

Lumsdon, L. and E. Mitchel (1999). Walking, Transport and Health: Do We Have the Right Prescription? *Health Promotion International*, **14** (3), 271–280.

Meadows, D.L. (1972). The Limits to Growth – A Report for the Club of Rome Project, Universe Books, New York.

Marshall, J.D., W.J. Riley, Th.E. McKone, and W. Nazaroff (2003). Intake Fraction of Primary Pollutants: Motor Vehicle Emissions in the South Coast Air Basin, *Atmospheric Environment*, **37**, 3455–3468.

Marshall, J.D., T.E. Mc Kone, E. Deaking, and W.W. Nazaroff (2005). Inhalation of Motor Vehicle Emissions: Effects of Urban Population and Land Area, *Atmospheric Environment*, **39**, 283–295.

Meurs, H. and R. Haaijer (2001). Spatial Structure and Mobility, *Transportation Research part D*, **6** (6), 429–446.

Morisson, D.S., M. Petticrew, and H. Thomson (2004). What Are the Most Effective Ways of Improving Population Health through Transport Interventions? Evidence from Systematic Reviews, *Journal of Epidemiology and Community Health*, **57**, 327–333.

Newton, P.N. (ed) (1997). *Reshaping Cities for a More Sustainable Future – Exploring the Link between Urban Form, Air Quality, Energy and Greenhouse Gas Emissions*. Research Monograph 6, Melbourne: Australian Housing and Research Institute (AHURI). http://www.ea.gov.au/atmosphere/airquality/urban-air/urban air docs.html.

Nyberg, F., P. Gustavsson, L. Jarup, T. Bellander, N. Berglind, R. Jakobsson, and G. Pershagen (2000). Urban Air Pollution and Lung Cancer in Stockholm, *Epidemiology*, **11** (5), 487–495.

Pfeffer, H.U. (1994). Ambient Air Concentrations of Pollutants at Traffic-Related Sites in Urban Areas of North-Rhine-Westphalia, Germany, *Science of the Total Environment*, **146/147**, 263–273.

Pope III, C.A., R.T. Burnett, M.J. Thun, E.E Calle, D. Krewski, K. Ito, and G.D. Thurston (2002). Lung Cancer, Cardiopulmonary Mortality and Long-Term Exposure to Fine Particulate Air Pollution. *Journal of the American Medical Association*, **287** (9), 1132–1141.

Pope III, C.A., M.J. Thun, M.M. Namboodiri, D.W. Dockery, J.S. Evans, F.E. Speizer, and C.W. Heath (1995). Particulate Air Pollution as a Predictor of Mortality in a Prospective Study of U.S. Adults, *American Journal of Respiratory and Critical Care Med*, **151** (3 Pt 1), 669–674.

Praml, G. and R. Schierl (2000). Dust Exposure in Munich Public Transportation: A Comprehensive 4-Year Survey in Buses and Trams, *International Archives of Occupational and Environmental Health*, 73 (3), 209–214.

Raaschou-Nielsen, O., M.L. Nielsen, and J. Gehl (1995). Traffic-Related Air Pollution: Exposure and Health Effects in Copenhagen Street Cleaners and Cemetery Workers, *Archives of Environmental Health*, **50** (3), 207–213.

Rank, J., J. Folke, and P.H. Jespersen (2001). Differences in Cyclists and Car Drivers Exposure to Air Pollution from Traffic in the City of Copenhagen, *Science of the Total Environment*, **279** (1–3), 131–136.

Ricardo (2002). *Ricardo Expects New Technology to Continue Fuelling Diesel Sales*, http://www.ricardo.com/mediaCentre/pressRelease/2002_diesel_report.doc (retrieved 5-2-2004).

Roemer, W.H. and J.H. Van Wijnen (2001a). Daily Mortality and Air Pollution along Busy Streets in Amsterdam, 1987–1998, *Epidemiology*, **12** (6), 649–653.

Roemer, W.H. and J.H. Van Wijnen (2001b). Differences among Black Smoke, PM(10), and PM(1.0) Levels at Urban Measurement Sites, *Environmental Health Perspectives*, **109** (2), 151–154.

Roorda-Knape, M.C., N.A.H. Janssen, J.J. De Hartog, P.H.N. Van Vliet, H. Harssema, and B. Brunekreef (1998). Air Pollution from Traffic in City Districts Near Major Motorways, *Atmospheric Environment*, **32**, 1921–1930.

Ruchirawa, M., C. Mahidol, C. Tangjarukij, *et al.* (2002). Exposure to Genotoxins Present in Ambient Air in Bangkok, Thailand – Particle Associated Polycyclic Aromatic Hydrocarbons and Biomarkers, *Science of the Total Environment*, **287** (1–2), 121–132.

Saelensminde, K. (2004). Cost–Benefit Analyses of Walking and Cycling Track Networks Taking into Account Insecurity, Health Effects and External Costs of Motorized Traffic, *Transportation Research Part A*, **38** (8), 593–606.

Serageldin, I. and A. Steer (1994). *Making Development Sustainable: From Concepts to Action*, The World Bank, Washington DC.

Smith, K.R. (1993a). Fuel Combustion, Air Pollution, and Health: The Situation in Developing Countries, *Annual Review of Energy and the Environment*, **18**, 529–566.

Smith, K.R. (1993b). Taking the True Measure of Air Pollution, *EPA Journal*, **19** (4), 6–8.

Stichting Wetenschappelijk Onderzoek Verkeersveiligheid (SWOV) (2005). Factsheet Kosten Van verkeersonveiligheid. Leidschendam: SWOV (http://www.swov.nl/nl/publicaties/ index.htm).

Van Bruggen, M., H.W.A. Jans, A.P. Verhoeff, and J.H. Van Wijnen (1991). *Oriënterend onderzoek naar de blootstelling Van weggebruikers aan enkele buitenluchtcontaminanten in het stadsverkeer*, GGD Amsterdam, Amsterdam.

Van Pul, W.A.J., E.D.G. Van Zandvoort, F.A.A.M. De Leeuw, and R.J.C.F. Sluyter (1996). Urban Air Quality Assessment Model UAQUM, RIVM Report 722401010, Bilthoven, National Institute of Public Health and the Environment.

Van Vliet, P., M. Knape, J. De Hartog, N. Janssen, H. Harssema, and B. Brunekreef (1997). Motor Vehicle Exhaust and Chronic Respiratory Symptoms in Children Living Near Freeways, *Environmental Research*, **74** (2), 122–132.

Van Wee, B., P. Janse, and R. Van den Brink (2005). Comparing Environmental Performance of Land Transport Modes, *Transport Reviews*, **25** (1), 3–24.

Van Wijnen, J.H. and S.C. Van der Zee (1998). Traffic-Related Air Pollutants: Exposure of Road Users and Populations Living Near Busy Roads, *Reviews on Environmental Health*, **13** (1–2), 1–25.

Van Wijnen, J.H., A.P. Verhoeff, H.W. Jans, and M. Van Bruggen (1995). The Exposure of Cyclists, Car Drivers and Pedestrians to Traffic-Related Air Pollutants, *International Archives of Occupational and Environmental Health*, **67** (3), 187–193.

Whitelegg, J., A. Gatrell, and P. Naumann (1993). *Traffic and Health*, Environmental Epidemiology Research Unit, University of Lancaster, Lancaster.

World Commission on Environment and Development (WCED) (1987). *Our Common Future*, Oxford University Press, Oxford.

Zagury, E., Y. Le Moullec, and I. Momas, I. (2000). Exposure of Paris Taxi Drivers to Automobile Air Pollutants within Their Vehicles, *Occupational and Environmental Medicine*, **57** (6), 406–410.

3

THE IMPACT OF AUTOMOBILE TRAFFIC ON QUALITY OF LIFE

Robert Gifford, Department of Psychology, University of Victoria, Victoria, BC, Canada
Linda Steg, Department of Psychology, Faculty of Behavioural and Social Sciences, University of Groningen, Groningen, The Netherlands

ABSTRACT

We consider how quality of life is affected by the continuing increase in the use and density of automobiles. We propose that sustainable transportation implies finding a proper balance between collective and individual interests, because a fully sustainable transport system is difficult to envision if most citizens believe it will significantly reduce their quality of life. Approaches to measuring quality of life, its relation to sustainable transport alternatives, and the potential implications for informing transport policy are considered.

INTRODUCTION

Automobile use has greatly increased during the last few decades. The number of passenger kilometres by private car per capita increased by 90 per cent in western Europe and 13 per cent in the US between 1970 and 1990. Drivers may be unaware how much they drive. When asked to prospectively estimate how many trips they would take during the next week, Swedish drivers took 80 per cent more trips than they expected to take (Jakobsson, 2004). How do these objective increases in the use of automobiles, and the subjective lack of awareness about it, affect quality of life (QoL)?

Automobile use threatens environmental quality, urban QoL, and the accessibility of important destinations. Many observers have argued that the current transportation system is not sustainable (e.g., OECD, 1996; see also Chapter 2 by Van Wee). Increased automobile use may affect people and environments in ways other than sustainability in a purely economic sense, including negative effects on such QoL elements as health. For example, each added daily hour spent in a car adds six per cent to one's chances of being obese (Frank *et al.*, 2004).

This chapter focuses on private transport, especially car use, because about 80 per cent of all vehicles are used for passenger transport. It introduces an instrument for assessing the QoL effects of differing transport plans and for enabling the examination of the degree to which sustainable transport policies are acceptable to the public. Psychological factors that affect individual QoL judgments and the acceptability of transport plans are reviewed. Finally, some conclusions and the practical value of instruments for assessing sustainable transport are offered.

DEFINING AND MEASURING QUALITY OF LIFE

QoL is a multidimensional construct that may be defined as the extent to which important values and needs are fulfilled (e.g., Diener, 1995; Diener *et al.*, 1999). QoL refers to well-being, conceptualized either in terms of objective living conditions or as a person's own assessment of his or her well-being in life, or both. This chapter primarily focuses on the well-being aspects of QoL as experienced, that is, individuals' cognitive and affective evaluations of their lives (Diener, 2000).

A group of researchers centred at the University of Groningen has recently developed an instrument to measure a comprehensive list of experiential QoL indicators, and linked them to various sustainability issues (see Slotegraaf and Vlek, 1996; Skolnik, 1997; Vlek *et al.*, 1998, 1999; Gatersleben, 2000; Poortinga *et al.*, 2001, 2004; Steg *et al.*, 2002; De Groot and Steg, 2006a, 2006b). This list appears to represent a wide range of non-overlapping dimensions that are important to community members, at least in the Netherlands.

Table 1 presents the most recent version of this QoL instrument. The mean importance rating of each QoL indicator is included, based on data from a questionnaire study of 455 Dutch respondents in 1999 (see Poortinga *et al.*, 2001, 2004, for more details). The table shows that most QoL indicators naturally are considered to be very important by these respondents, but some are valued more than others. Health, partner and family, social justice, freedom, and safety are valued more highly (at least were by Dutch people in 1999) than material beauty, spirituality and religion, status and recognition, and challenge and excitement.

QUALITY OF LIFE RELATED TO SUSTAINABLE TRANSPORT

A sufficient QoL is an important precondition for sustainable development. Although no single definition of sustainable development is universally accepted, several fairly

Table 1: Description and Importance Ratings of 22 QoL Indicators

Indicator	*Description*	*M*
Health	Being in good health. Having access to adequate health care	4.9
Partner and family	Having an intimate relationship. Having a stable family life and good family relationships	4.7
Social justice	Having equal opportunities and the same possibilities and rights as others. Being treated in a just manner	4.7
Freedom	Freedom and control over the course of one's life, to be able to decide for yourself, what you will do, when and how	4.5
Safety	Being safe at home and in the streets. Being able to avoid accidents and being protected against criminality	4.5
Education	Having the opportunity to get a good education and to develop one's general knowledge	4.3
Identity/self-respect	Having sufficient self-respect and being able to develop one's own identity	4.2
Privacy	Having the opportunity to be yourself, to do your own things and to have a place of your own	4.2
Environmental quality	Having access to clean air, water, and soil. Having and maintaining good environmental quality	4.2
Social relations	Having good relationships with friends, colleagues, and neighbours. Being able to maintain contacts and to make new ones	4.2
Work	Having or being able to find a job and being able to fulfil it as pleasantly as possible	4.2
Security	Feeling attended to and cared for by others	4.1
Nature/biodiversity	Being able to enjoy natural landscapes, parks, and forests. Assurance of the continued existence of plants and animals and maintenance of biodiversity	4.1
Leisure time	Having enough time after work and household work and being able to spend this time satisfactorily	4.0
Money/income	Having enough money to buy and to do the things that are necessary and pleasing	3.6
Comfort	Having a comfortable and easy daily life	3.5
Aesthetic beauty	Being able to enjoy the beauty of nature and culture	3.5
Change/variation	Having a varied life. Experiencing as many things as possible	3.3
Challenge/excitement	Taking up challenges and experiencing pleasant and exciting things	3.2
Status/recognition	Being appreciated and respected by others	3.0
Spirituality/religion	Being able to live a life with the emphasis on spirituality and/or with your own religious persuasion	2.9
Material beauty	Having nice possessions in and around the house	2.6

Source: Adapted from Poortinga *et al.* (2004).

convergent definitions are available (Beatley, 1995). Sustainable development implies a balance between current and future environmental, social, and economic considerations (e.g., WCED, 1987; Ruckelhaus, 1989; OECD, 1996; Litman, 2003). Exactly how these considerations should be balanced is less clear. Although various attempts have been made to define, in particular, indicators of sustainable transport (see below), a set that adequately reflects environmental, social, and economic qualities has not yet been identified.

Ideally, theory-based conceptions and operationalizations of sustainable transport indicators should be developed, first by defining sustainable transport, and then by deriving significant performance indicators that enable the measurement of sustainable transport. At present, many performance indicators have been derived from current practices (e.g., in transport plans and policies) and stakeholder perceptions of sustainable transport. Indicator development often has not been based on an explicit definition or vision of sustainable transport (Gilbert and Tanguay, 2000).

The sustainability of transportation systems might be considered by examining objective positive and negative features and externalities of traffic and transport as they are apparent now or in the near future. Various attempts have been made to list these, such as energy use, CO_2 emissions, emissions of toxic and harmful substances, land use, disruption and fragmentation of natural areas, waste, traffic safety, noise pollution, health consequences of transport, accident costs, the contribution of the transport sector to economic welfare, and accessibility (e.g., Gilbert and Tanguay, 2000; Gudmundsson, 2001; Litman, 2003). Other objective indicators have been defined that are based on the quality of the current transport system, including commuting speed, congestion delay, variety and quality of transport options available in a community, accessibility of activities (for drivers and non-drivers), and the proportion of household expenditures devoted to transport (e.g., Litman, 2003).

Discussions on sustainable development focus on environmental sustainability. Indicators are also needed to examine effects of environmentally sustainable (and unsustainable) transportation systems on QoL. Social indicators should reflect effects on QoL (e.g., OECD, 1976, 1982). In one recent study that examined the effect of potential transport scenarios on social indicators (Geurs and Van Wee, 2000), environmentally sustainable transport criteria, such as emissions of CO_2, NO_x, VOS, particles, noise, and land use, were defined. Then, three environmentally sustainable transport scenarios that would meet these criteria were developed, following a backcasting method: a high-technology scenario (only technological changes), a mobility-change scenario (only behaviour changes aimed to reduce car dependency), and a combination scenario (technological and behavioural changes). Next, the policy measures that would be needed to reach these environmentally sustainable transport systems were identified. Finally, the possible social consequences of the combination scenario were compared to those of a business-as-usual scenario. The social effects were qualitatively assessed by experts, and thus might be considered a hybrid of objective and experiential forms of QoL assessment.

The authors concluded that the social consequences of environmentally sustainable transport scenarios appear to be less drastic than is often assumed. However, Geurs and Van Wee (2002) focused on a relatively small number of social indicators (safety, health, perceived environmental qualities, and community relationships), when numerous other social indicators, such as equity, freedom, convenience, and comfort, may also be affected by future transport scenarios. Of particular interest to this chapter, Geurs and Van Wee assert that there is a need for the timely implementation of measurement instruments.

Prescriptive studies like these are important for examining whether and how sustainable transportation systems are feasible. They clarify what a sustainable future might look like. However, an important subsequent question is: How does the public evaluate such sustainable futures? Is a sustainable transport system widely acceptable? The answers will depend, among other things, on the extent to which members of the public believe that these futures will result in an increase or at least not a reduction in their QoL. Improvements in the collective QoL, as embodied in sustainable transport, may conflict with individual short-term interests, especially when individuals must adapt their lifestyles in order to reach sustainability goals. Thus, collective and individual interests may be at odds, or at least appear so to some citizens. A sustainable transport system may actually improve QoL by increasing the amount a driver walks or rides a bike, for example, but many drivers may not be ready to see the value of these alternatives.

Nevertheless, sustainable transport issues often seem to contrast individual with collective goals. To achieve a sustainable transport system, drivers may well have to drive less. For many, driving a car is more attractive than other modes of transport, because of its convenience, independence, flexibility, comfort, speed, perceived safety, and privacy. The car also provides more status and pleasure than other modes of transport; it is a means of self-expression, and enables one to control a powerful machine (e.g., Reser, 1980; Steg, 2003a, 2005).

Thus, improved QoL for citizens in general may imply that drivers forfeit some of the individual advantages of car use, which may (at least initially) be perceived as a threat to their individual QoL. Among those who may especially be affected are those who live far from public transport, those who routinely transport several passengers, such as parents with children, and the elderly, many of whom believe they need to drive to maintain their independence, lifestyle, social ties, and access to shopping and services (Rudinger *et al.*, 2006). These citizens, and others, are inclined to drive, that is, to act in their own interest, because these interests are experienced as important and immediate, whereas the collective problems of steadily increasing traffic are visible only in the long term. Moreover, individuals *qua* individuals cannot control the problems caused by car use. Thus, to them, foregoing the advantages of driving does not seem sensible, at least in the usual everyday life in which long-term outcomes have low salience.

However, various factors may encourage citizens to act in the common interest, even though doing so may not have immediate positive consequences for themselves. Some of these factors include problem awareness (e.g., Nordlund and Garvill, 2003), perceived responsibility for the problem, trust in others' contributions, consideration of future consequences (Joireman *et al.*, 2004), and personal norms (see Gifford, 2002; Steg, 2003c, for overviews).

From these considerations, one may conclude that not only should the sustainability of different transport scenarios for society as a whole be examined, but also the extent to which such scenarios affect individual QoL and whether such scenarios are acceptable to individual members of society. More specifically, knowing which elements of different sustainability scenarios have high or low public acceptance would be important. Obviously, one can hardly expect sustainable transport to be implemented by governments if most citizens believe that it will significantly reduce their QoL. Indeed, the Brundtland Commission Report's own definition of sustainability implies the importance of QoL: "meeting the needs of the present without compromising the ability of future generations to meet their own needs" (WCED, 1987, p. 43). Thus, sustainable transport must also be concerned with human needs at the individual level as well as at the societal level. The effects of strategies aimed at creating sustainable transport should also be assessed in terms of human needs and values.

A reasonable assertion that follows from this is that policymakers should give special attention to possible effects on the most important QoL indicators when they design and implement sustainable transport policies. Apart from this obvious conclusion, policymakers are likely to hear from the public when they fail to do this: citizens can be expected to more strongly oppose measures that negatively affect these QoL indicators. Policymakers must search for ways to achieve sustainable transport that would affect these QoL indicators in less negative, or even in positive, ways. They might also consider possible ways to compensate any expected negative effects. The assessment of the needs of individuals must be done by consulting individuals, partly because individuals will support (vote for) or not (vote against) those who propose policies based on their needs as they experience them. Therefore, an instrument for measuring these QoL indicators is essential.

QUALITY OF LIFE, THE ENVIRONMENT, AND TRANSPORTATION

QoL often is viewed by the average citizen as primarily related to the quality of relationships, their own health, the health of significant others, finances, work, and social life (e.g., Bowling and Windsor, 2001). Environmental effects on QoL are not often considered, either by members of the public or by researchers, although a poor environment can scarcely avoid compromising a person's QoL. Perhaps, the relatively good quality of environments in the industrialized world, where virtually all research is conducted, causes citizens and researchers to overlook its importance. However, some attention has been given to QoL in relation to the quality of the environment (Eyles, 1990; Rogerson, 1995; Cairns, 1997; Dasgupta, 2001).

QoL effects of transportation scenarios or plans may be assessed by asking lay respondents to indicate the extent to which various sustainable transportation scenarios would affect relevant QoL indicators in positive or negative ways. To obtain a more precise view of these effects, these expected changes may be weighted, based on the importance judgments of the relevant QoL indicators, because changes in important QoL indicators presumably will be more significant for individuals than changes in less-important QoL indicators. Subsequently, the overall expected changes in QoL may be calculated. A multi-attribute QoL scale may then be created by summing the expected changes on the QoL indicators, each multiplied by the importance judgment assigned to it. It should be noted that such an approach does assume an "involved" decision maker, and the average citizen rarely actively evaluates which transportation choice is best. Decision-making by the "uninvolved" citizen may be better explained by a variety of other decision-making models that emphasize habit, mindlessness, or "fast and frugal" criteria (Gigerenzer and Todd, 1999). Drivers typically have many things on their minds in their daily lives besides sustainability, and must be "cognitive misers" (Fiske and Neuberg, 1990) to survive and prosper.

The QoL instrument has been successfully used in several studies of sustainable household consumption (e.g., Vlek *et al.*, 1998; Gatersleben, 2000; Steg *et al.*, 2002). One Dutch study involving transport examined the extent to which sustainable household energy consumption scenarios would affect the judged QoL (Poortinga *et al.*, 2001). Scenarios were presented that systematically varied on three dimensions, one of which concerned the locus of energy saving (home versus transport). The respondents expected the transport scenarios would result in QoL reductions in comfort, work, money, privacy, and freedom. However, improvements in QoL were expected in nature/biodiversity and environmental qualities. Overall QoL appeared not to be affected much, which implies that the expected improvements nearly compensated for the expected reductions in QoL. The respondents expected the most negative consequences from the transport scenario that involved technological as well as behavioural changes that would result in small energy savings.

In another study, participants in five European countries (i.e., Austria, the Czech Republic, Italy, Sweden, and the Netherlands) evaluated the extent to which a doubling of the costs of car use would affect the 22 QoL aspects listed in Table 1. Next, they indicated to what extent this measure would affect their overall QoL as well (De Groot and Steg, 2006a, 2006b). In general, respondents anticipated negative effects for certain QoL indicators like comfort, money and income, leisure time, change and variation, freedom and work, but they believed that other QoL indicators (environmental quality, nature and biodiversity, and safety) would improve. Differences among the five countries were found: respondents from the Netherlands and Sweden were more pessimistic about the QoL consequences of the policy than respondents from the Czech Republic, Italy and, to a lesser degree, Austria. Interestingly, the sum of expected changes in QoL indicators (i.e., the sum of expected positive and negative changes) was correlated with the expected changes in overall QoL (i.e., the overall judgement of expected changes in QoL), which suggests that the QoL instrument is a valid way to elicit judgements of expected changes in QoL of sustainable transport scenarios.

FACTORS THAT INFLUENCE EXPERIENTIAL QUALITY OF LIFE

Based on the studies reviewed above, we believe that the QoL concept is useful for assessing the expected effects of future scenarios. They not only reveal that QoL judgments are affected by various transport plans, but also how they are affected, that is, which QoL indicators would improve and which would deteriorate. The studies reveal that deteriorations in specific QoL indicators may be compensated for by improvements in other dimensions, at least when drivers are actively involved in thinking about transportation choices. Sustainable scenarios typically threaten some experienced QoL indicators such as comfort, freedom, and privacy, whereas QoL indicators that refer to societal indicators such as environmental quality and nature and biodiversity would improve. This once again illustrates the conflict between individual and collective interests, and demonstrates that individual and collective interests must be balanced. However, other experienced QoL indicators, such as health and the need for variety and change may improve once drivers discover the value of walking or riding a bike, so that the trade-off for individuals is not necessarily all loss.

Most studies reported above (except Steg *et al.*, 2002) examined only anticipated changes in QoL, that is, respondents indicated to what extent they expected their QoL to be affected in such cases. These may differ from actual QoL changes that would occur when the proposed changes would be implemented. For example, studies of the acceptability of transport policies have shown that public support may be higher after transport policies have been implemented (Tretvik, 2003; see also Steg, 2003b), even though changes typically are resisted at first because they may have negative consequences. When individuals are unsure of the consequences, they prefer the status quo (Kahneman and Tversky, 1984). Nevertheless, support for changes may increase after the changes occur, when respondents' opinions are better informed because they have more experience with the pros and cons of the measures.

For example, attitudes towards bus riding improved and bus riding increased after a policy change, that is, the implementation of a U-Pass that allowed free bus transport for students after a mandatory addition to their University tuition fees (Heath and Gifford, 2002). In two Paris districts, after 30 km per hour traffic restriction zones were implemented, residents' evaluation of their sonic environment and their reported QoL improved (Rozec, 2003). The two zones had sonic environments with fewer traffic characteristics and more human sounds, including conversation and other elements of the soundscapes, compared to other zones. Individuals may also become more convinced of the advantages of the new policies because they perceive that collective problems are being solved.

Similar processes may play a role when people are asked to assess what changes in QoL they would expect from future (transport) scenarios. Therefore, the QoL concept should also be used to monitor QoL over time and to examine the extent to which changes in society or in transport affect judged QoL. This highlights the importance of the way in which future scenarios are presented. To ensure that respondents provide well-considered judgments of expected QoL effects of transport plans, the plans should be described in a plausible and imaginable way. Clear description of proposed changes in the transport system is important for helping respondents to think through the possible consequences of the plans for themselves. The public should also be involved in the development of sustainable transport plans. This should result in better and more acceptable sustainable transport plans.

A few other factors should be considered when examining QoL effects of sustainable transport scenarios. Diener *et al.* (1999) found that general subjective well-being has not changed much over the last several decades, even though incomes and consumption levels have increased significantly (see also Veenhoven, 2004). Individuals seem to adapt to positive as well as to negative changes in their lives; they change their expectations and goals (e.g., Meyers, 1992; Suh *et al.*, 1996; Diener, 2000). QoL is judged in comparison to some social or personal standard (Ormel *et al.*, 1997; Diener, 2000). People seem to feel more satisfied with their lives when they believe that they are better off than others are, when they are better off than yesterday, or when they are closer to their aspirations. The standards used to judge QoL do change over time, that is, individuals appear to adapt their comparative standards to changes in their circumstances (which might improve or deteriorate).

All this implies that further increases in consumption levels, including transport, will not necessarily enhance QoL, and that reductions in consumption and transport levels may not necessarily reduce QoL. Although individuals may initially experience a reduction in QoL, they may well adapt soon after the changes (Diener, 2000). Thus, the conviction that a sustainable transportation system is not feasible because environmentally sound transportation systems will seriously threaten QoL may not be correct, and should at least be tested.

Theories of QoL and human well-being typically assume that a general set of indicators for QoL can be defined that does not differ over time or between cultures (e.g., Maslow,

1954; Rokeach, 1973; Schwartz and Bilsky, 1987, 1990; Max-Neef, 1991; Schwartz, 1992; see Vlek *et al.*, 1999, for a review). However, the way individuals prefer to fulfil their needs and values does change over time and differs between cultures. Moreover, the relative importance of various QoL indicators (or needs and values) differs between groups (see Inglehart, 1990; Gatersleben and Vlek, 1998; Gatersleben, 2000; Poortinga *et al.*, 2001). For example, Dutch respondents with greater environmental concern evaluate environmental quality and personal freedom as more important, and material wealth as less important than do respondents with less environmental concern. Dutch women value personal freedom and maturity more than men do, and unmarried persons evaluate family, health, and safety as less important than couples and families do (Poortinga *et al.*, 2001). The relative importance of environmental values also depends on the context. For example, the relative importance of one's environmental values varies with the presumed immediacy of the environmental impact or economic gain, self-interest, and social norm (Heath and Gifford, 2006). The view that values are situation-independent does not appear to be valid.

Obviously, current and future sustainable forms of transport may affect various groups in society differently, and group differences may exist in what is considered to be sustainable (or livable) transport (see also Button, 1982; Adams, 1999). Consequently, the interests of various groups should be balanced, and it may be necessary to compensate groups that are disproportionately affected by current and future transport systems. Also, the relative importance of QoL indicators may vary over time (see Inglehart, 1990; Gatersleben, 2000). This implies that the multi-attributive evaluation of QoL effects of sustainable transport scenarios may be time-dependent. Which QoL aspects should be considered is known, but the relative importance of various QoL aspects, and consequently, overall (multi-attributive) QoL effects should be monitored regularly. This will also reveal to what extent actual QoL effects differ from anticipated effects. Based on these considerations, policies for developing sustainability may need to be adapted.

TRANSPORT AND THE ULTIMATE QUALITY OF LIFE INDICATORS: ACCIDENTS, INJURIES, AND DEATH

Experiential QoL is an important construct, particularly given the connection between public opinion and policy-making. However, certain objective statistics also bear on the experiential nature of the issue. Those pertaining to accidents, injuries, and death perhaps have the most dramatic negative impact on both the objective and experienced QoL of survivors, their families, and friends. Serious accidents change the life of victims, friends, family, and even guilty parties, such as the drunk driver who fatally injures someone else. In the United States, 44 per cent of all accidental deaths occur as a result of motor vehicle accidents. In round numbers, about 45,000 people have been killed every year in the US over the last 30 years (retrieved February 10, 2006, from www.benbest.com/lifeext/causes.html). This is about the same number of American lives lost during the entire Vietnam War. Although this number has been declining in recent years (from about 52,000 to about 42,000), the US thus experiences approximately a Vietnam-War level of lost lives every year. Furthermore, the 45,000 figure does not include the much larger number of citizens who are injured, nor the emotional loss to the hundreds of thousands of others associated with the death or severe injury to a family member, friend, or employment associate. Losses in Europe are smaller, but are in rough proportion; for example, traffic deaths in Germany have fallen from about 21,000 to 7,000 per year recently.

But how does this relate to sustainable transport and experienced QoL? Assuming that buses are, in general, more sustainable than automobiles, greater use of buses probably would reduce the enormous accident toll, because buses are safer. Per passenger mile, automobiles are 25 times more likely to lead to death than buses (retrieved February 10, 2006, from www.benbest.com/lifeext/causes.html). Incidentally, motorcycles are 35 times more likely to lead to death, on a per-mile basis, than automobiles, which suggests that they are over 800 times as dangerous as buses. Data are difficult to come by, but one might guess that metropolitan subway trains, as another sustainable form of transport, are even safer than buses, which must share the road with cars. Turning to injuries, motor vehicle accidents very often leave their victim with cases of posttraumatic stress disorder, which of course severely affects their QoL (e.g., Cagnetta and Cicognani, 1999; Gudmundsdottir *et al.*, 2004). Thus, automobile accidents have a large impact on society, and a huge impact on the QoL of those who are directly affected.

SIGNIFICANCE FOR POLICYMAKING

As noted earlier, policymakers should take into account the extent to which their policies will affect judged QoL. Transport policies will be less acceptable, and consequently, less feasible and less effective, if they have significant negative impacts on QoL. Policymakers may be reluctant to implement policies that lack public support. Moreover, restrictions on freedom of choice may evoke psychological reactance (Brehm, 1966). As a consequence, restrictive policies may be less effective, or even have effects opposite from what was intended (Tertoolen *et al.*, 1998). If specific transport policies aimed at reducing car use are believed to threaten freedom of choice, drivers might be motivated to continue driving to the extent to which possible negative consequences will be bearable. Policies that restrict driving should emphasize the potential benefits of not driving, such as improved health from walking or riding bikes to the value of a change in routine and increased variety of daily experience.

Sustainable transport may imply different things in different regions and cultures, and consequently, specific sustainable transport plans may be evaluated differently in these regions and cultures. For example, North American society is more strongly tuned towards the regular use of cars than many European societies. Of course, inter-city distances in Canada and parts of the US are much greater than those in European countries such as the Netherlands. Also, the public transport system in the Netherlands is sophisticated compared to that in many parts of North America. Thus, car dependency (i.e., the level of car use, car-oriented land use and quality of travel alternatives; Newman and Kenworthy, 1999) is much higher in North America compared to the Netherlands.

This implies that reductions in car use may have more significant consequences for the QoL of North Americans than for the Dutch. Differences may also emerge when comparing regions within a country. For example, as demonstrated by Rozec (2003), reductions in traffic volume may significantly enhance the QoL of people in densely populated areas (e.g., fewer traffic jams, less noise, better urban QoL), but may reduce the QoL of rural dwellers (e.g., some key locations and activities may be much more difficult to access).

CONCLUSIONS

Although no common definition of sustainable transport exists, most observers would agree that sustainable transport implies balancing current and future economic, social, and environmental qualities. Current traffic and transport trends do not appear sustainable in the long term, yet a standardized set of sustainable transport indicators has not yet been identified. The negative environmental, social, and economic externalities of current transport systems often outweigh their social and economic values. Sustainable transport mainly is investigated by examining the sustainability of current transport systems, such as the positive and negative values and externalities related to energy and land use, waste, traffic safety, traffic noise, health consequences, accident costs, accessibility, and economic wealth. Governments and international bodies such as the Organisation for Economic Co-operation and Development (OECD) often apply this approach. Sustainability indicators are defined and operationalised as sustainable transport policy goals, and whether the transport system is moving towards sustainability is monitored. In some cases future projections are also made.

In addition, the effects of various transport plans on sustainability are being assessed. This implies a need to consider a broader range of sustainability indicators, because changes in current transport systems may affect other sectors that also contribute to unsustainable development, such as employment levels or health care costs. Various methods and models have been developed to assess economic, social, and environmental consequences of transport plans. However, at present, only a few social indicators are being considered, because of the lack of knowledge and valid methods, tools, and techniques for assessing relevant social impacts.

Obviously, an important question concerns how the public evaluates such sustainable futures, and whether transitions to sustainable transport systems are acceptable to the public. These transitions may not be acceptable to everyone, because sustainable transport may conflict with some individual short-term interests, especially when individual car users are asked to significantly adapt their lifestyles and transport behaviour. Therefore, the extent to which transitions to sustainable transport would affect individual QoL should be examined, and also the extent to which such transitions will be acceptable to the public.

This chapter proposes a compensatory method for assessing the QoL effects of transitions to transportation systems that systematically differ in the extent to which they are sustainable. QoL is a multidimensional construct defined as the extent to which important values and needs are fulfilled. Subjective evaluations of QoL, that is, cognitive evaluations of citizens' lives as a whole, are considered. A list of 22 QoL indicators is introduced for assessing QoL effects of transport policy plans; the list represents a wide range of dimensions that are important to consumers (and thus travellers).

The effects on QoL of possible transportation scenarios may be assessed by asking respondents to indicate to what extent various transportation scenarios would affect relevant QoL indicators, and how important each indicator is to their lives. The overall expected changes in QoL may be calculated by summing the expected changes on the QoL indicators, possibly after multiplying the importance assigned to each indicator. Several empirical studies revealed that the QoL concept is useful for assessing actual and expected QoL effects of various sustainable transport scenarios. However, such studies, whether theory- or policy-oriented, require that respondents be awake to the issues involved; many everyday transport users are not actively involved in deciding which transport method is best for them or for society.

Such studies reveal not only whether overall QoL is or would be affected by transport plans, but also how QoL would be affected, that is, which QoL indicators would improve or deteriorate under different sustainable transport scenarios. Once the results are known, their implementation may require massive awareness campaigns so that the recommended changes are received by drivers who are actively cognizant of the issues involved. Research in consumer psychology shows that campaigns aimed at involved consumers do not work as well with uninvolved consumers, for whom a different kind of publicity campaign is required (e.g., Greenwald and Leavitt, 1984; Solomon *et al.*, 2005).

The proposed list of experiential QoL indicators enables examination of which groups' QoL would be affected most strongly. Based on this, politicians and policymakers should be able to decide whether and how specific groups should be compensated, and which indicators with potential salutogenic outcomes might be emphasized, to better inform the public about expected (positive and negative) effects of the proposed sustainable policies.

This would greatly improve the current situation, in which decisions are largely based on the preferences of special-interest groups. At present, significant minorities that wield sufficient political power often are able to obstruct particular solutions or compromises, which leaves governments with options that are unacceptable to others, or watered down so much that their effectiveness becomes questionable.

The objective and experiential approaches described above are not contradictory; they complement each other. QoL effects of both kinds must be considered when designing and implementing sustainable transport plans, because they are crucial for the acceptability, and consequently, the feasibility and effectiveness, of such plans. Sustainable transport plans will be strongly opposed when citizens start believing that the plans will significantly reduce their QoL.

To improve the chances for sustainable development, the basis of the expectations that sustainable transport will reduce QoL must be investigated. If the expectations are realistic, policymakers should consider other ways to achieve sustainable transport that would affect QoL less negatively, or even positively. The extent to which possible negative effects could be compensated, for instance, by implementing additional policies, should be examined. However, it may also be that such expectations are based on misperceptions or a lack of knowledge (e.g., the public is unaware of environmental problems caused by automobile traffic). In this case, the public should be informed and educated to the need for, and possible consequences of, sustainable transport.

The list of indicators introduced in this chapter may be used to collect community members' assessments of the changes in QoL that would be anticipated if possible future scenarios were to be enacted. Many psychological processes will influence these judgments. For example, well-considered judgments about the expected QoL effects of transport plans may not be obtained if respondents do not think enough about the advantages and disadvantages of sustainable transport compared to a business-as-usual scenario. This might be facilitated by providing citizens or study respondents with clear descriptions or visualizations of plausible changes in transport, and by describing what each one implies for them personally. Members of the public must be involved in the development of sustainable transport scenarios.

Changes understandably are met with initial resistance, as long as individuals are unconvinced of the salutogenic consequences. Individuals generally judge their expected QoL in comparison to some standard, for instance, the QoL of others, their current QoL, or their aspirations. These standards are adapted in response to changes in their circumstances. This implies that changes in transport may initially negatively influence QoL, but if individuals adapt fairly quickly, significant reductions in QoL may not occur in the long term. Thus, support for sustainable transport plans may become stronger after they have been implemented. Finally, because the relative importance of QoL indicators may vary over time, the expected and actual changes in QoL of sustainable transport scenarios should be monitored continuously, and policies should be adapted when necessary.

Although much important work has been done to understand sustainable transport, many questions still remain. First, the methods used for assessing sustainable transport and for assessing QoL effects of sustainable transport scenarios need to be further developed. For example, methods must be developed to examine how valid judgments can best be collected, and how psychological processes that may affect QoL evaluations can best be understood. Second, whether the results of studies like the ones reported here may be generalized to transport behaviour in everyday life need to be examined. As noted earlier, a multiattribute model may be especially appropriate when citizen involvement is high, but other models are necessary when involvement is low, as it often is. The everyday preferences of many citizens might be better predicted by fast-and-frugal or noncompensatory models. Third, whether the present list of QoL indicators is comprehensive should be investigated. Although the list in Table 1 is believed to be complete, additions and changes may be needed. Fourth, relations among QoL indicators should be examined more thoroughly. For example, some QoL indicators refer to goals (e.g., comfort, status, affection), but others refer to resources (e.g., money, time, or health) that may be used to fulfil these goals (see Ormel *et al.*, 1997). Thus, the indicators may be found to consist of factors such as goals versus resources, or may form other clusters, which could simplify the structure of the full list, and therefore reduce the complexity of the relations between transport system proposals and QoL.

The development of sustainable transport scenarios should be combined with assessments of QoL effects of those scenarios. On the one hand, individuals may assess the QoL effects of transport plans that fulfil general sustainability criteria. On the other, scientists may assess the sustainability of transport plans that optimize QoL of current as well as future generations.

REFERENCES

Adams, J. (1999). The Social Implications of Hypermobility. In: *OECD, Project on Sustainable Transport (EST). The Economic and Social Implications of Sustainable Transportation. Proceedings of the Ottawa Workshop, 20–21 October 1998*, OECD Publications, Paris. http://www.olis.oecd.org/olis/1999doc.nsf/63c71d2d4054d0fdc125685d0053aee4/c125685b002f5004c125686b005cb510/$FILE/00071363.PDF#page=95

Beatley, T. (1995). The Many Meanings of Sustainability, *Journal of Planning Literature*, 9, 339–342.

Bowling, A. and J. Windsor (2001). A Population Survey of Dimensions of Quality of Life, *Journal of Happiness Studies*, 2, 55–82.

Brehm, J.W. (1966). *A Theory of Psychological Reactance*, Academic, New York.

Button, K.J. (1982). *Transport Economics*, Heinemann, London.

Cagnetta, E. and E. Cicognani (1999). Surviving a Serious Traffic Accident: Adaptation Processes and Quality of Life, *Journal of Health Psychology*, **4**, 551–563.

Cairns, J. (1997). Environmental Monitoring for Sustainable Use of the Planet, *Population and Environment: A Journal of Interdisciplinary Studies*, **18**, 463–471.

Dasgupta, P. (2001). *Human Well-Being and the Natural Environment*, Cambridge, New York.

De Groot, J. and L. Steg. (2006a). Impact of Transport Pricing on Quality of Life, Acceptability, and Intentions to Reduce Car Use: An Explorative Study in Five European Countries, *Journal of Transport Geography*, **14**(6), 463–470.

De Groot, J. and L. Steg (2006b). The Role of Value Orientations in Evaluating Quality of Life Consequences of a Transport Pricing Policy, *Transportation Research Part D*, **11** (2), 160–165.

Diener, E. (1995). A Value-Based Index for Measuring National Quality of Life, *Social Indicators Research*, **36**, 107–127.

Diener, E. (2000). Subjective Well-Being: The Science of Happiness and a Proposal for a National Index, *American Psychologist*, **55**, 34–43.

Diener, E., E.M. Suh., R.E. Lucas, and H.L. Smith (1999). Subjective Well-Being: Three Decades of Process, *Psychological Bulletin*, **125**, 276–302.

Eyles, J. (1990). Objectifying the Subjective: The Measurement of Environmental Quality, *Social Indicators Research*, **22**, 139–153.

Fiske, S.T. and S.L. Neuberg (1990). A Continuum of Impression Formation, from Category-Based to Individuating Processes: Influences of Information and Motivation on Attention and Interpretation, *Advances in Experimental Psychology*, **23**, 1–74.

Frank, L.D., M.A. Andresen, and T.L. Schmidt (2004). Obesity Relationships with Community Design, Physical Activity, and Time Spent in Cars, *American Journal of Preventive Medicine*, **27**, 87–96.

Gatersleben, B.C.M. (2000). *Sustainable Household Metabolism and Quality of Life: Examining the Perceived Social Sustainability of Environmentally Sustainable Household Consumption Patterns*, Doctoral Dissertation, Faculty of Psychological, Pedagogical and Sociological Sciences, University of Groningen, The Netherlands.

Gatersleben, B. and C. Vlek (1998). Household Consumption, Quality-of-Life and Environmental Impacts: A Psychological Perspective and Empirical Study. In: *Green households? Domestic Consumers, Environment and Sustainability* (K.J. Noorman and A.J.M. Schoot Uiterkamp, eds.), Earthscan, London, 141–183.

Geurs, K.T. and G.P. Van Wee (2000). *Environmentally Sustainable Transport: Implementation and Impacts for the Netherlands for 2030*, RIVM, Bilthoven, The Netherlands.

Gifford, R. (2002). *Environmental Psychology. Principles and practice*, Optimal Books, Colville, WA.

Gigerenzer, G. and P.M. Todd (1999). *Simple Heuristics that Make us Smart*, Oxford, New York.

Gilbert, R. and H. Tanguay (2000). *Sustainable Transportation Performance Indicators Project. Brief Review of Some Relevant Worldwide Activity and Development of an Initial Long List of Indicators*, The Centre for Sustainable Transportation, Toronto, Ontario.

Greenwald, A. and C. Leavitt (1984). Audience Involvement in Advertising: Four Levels, *Journal of Consumer Research*, **11**, 581–592.

Gudmundsdottir, B., J.G. Beck, S.F. Coffey, L. Miller, and S. Palyo (2004). Quality of Life and Post Trauma Symptomatology in Motor Vehicle Accident Survivors: The Mediating Effects of Depression and Anxiety, *Depression and Anxiety*, **20**, 187–189.

Gudmundsson, H. (2001). *Indicators and Performance Measures for Transportation, Environment and Sustainability in North America* (Research Notes No. 148). Roskilde, Denmark: Ministry of Environment and Energy, National Environmental Research Institute. Available at http://www.dmu.dk/1_viden/2_publikationer/3_arbrapporter/rapporter/ AR148.pdf.

Heath, Y. and R. Gifford (2002). Extending the Theory of Planned Behavior: Predicting the Use of Public Transportation, *Journal of Applied Social Psychology*, **32**, 2154–2189.

Heath, Y. and R. Gifford (2006). *Are Value Hierarchies Stable or Situation-Dependent?* Manuscript in preparation.

Inglehart, R. (1990). *Culture Shift in Advanced Industrial Society*, New Jersey University Press, Princeton.

Jakobsson, C. (2004). Accuracy of Household Planning of Car Use: Comparing Prospective to Actual Car Logs, *Transportation Research Part F: Traffic Psychology & Behaviour*, 7, 31–42.

Joireman, J.A., P.A.M. Van Lange, and M. Van Vugt (2004). Who Cares About the Environmental Impact of Cars? Those with an Eye Toward the Future, *Environment and Behavior*, **36**, 187–206.

Kahneman, D. and A. Tversky (1984). Choices, Values and Frames, *American Psychologist*, **39**, 341–350.

Litman, T. (2003). *Sustainable Transportation Indicators*. Victoria, BC: Victoria Transport Policy Institute. Available at www.vtpi.org.

Maslow, A.H. (1954). *Motivation and Personality*, Harper & Row, New York.

Max-Neef, M.A. (1991). *Human Scale Development. Conception, Application and Further Reflections*, Apex Press, New York.

Meyers, D.G. (1992). *The Pursuit of Happiness. Discovering Pathways to Fulfilment, Well-Being and Enduring Personal Joy*, Avon Books, New York.

Newman, P. and J. Kenworthy (1999). *Sustainability and Cities: Overcoming Automobile Dependency*, Island Press, Washington, DC.

Nordlund, A.M. and J. Garvill (2003). Effects of Values, Problem Awareness, and Personal Norm on Willingness to Reduce Personal Car Use, *Journal of Environmental Psychology*, **23**, 339–347.

OECD (1976). *Measuring Social Well-Being. A Progress Report on the Development of Social Indicators*, OECD Publications, Paris.

OECD (1982). *The OECD List of Social Indicators*, OECD Publications, Paris.

OECD (1996). *Towards Sustainable Transportation*, OECD Publications, Paris.

Ormel, J., S. Lindenberg, N. Steverink, and M. Vonkorff (1997). Quality of Life and Social Production Functions: A Framework for Understanding Health Effects, *Social Science and Medicine*, **45**, 1051–1063.

Poortinga, W., L. Steg, and C. Vlek (2004). Values, Environmental Concern and Environmental Behavior: A Study into Household Energy Use, *Environment and Behavior*, **36**, 70–93.

Poortinga, W., L. Steg, C. Vlek, and G. Wiersma (2003). Household Preferences for Energy-Saving Measures. A Conjoint Analysis, *Journal of Economic Psychology*, **24**, 49–64.

Poortinga, W., G. Wiersma, L. Steg, C. Vlek, K.J. Noorman, H. Moll, and T. Schoot Uiterkamp. (2001). *Expected Quality of Life Impacts of Experimental Scenarios for Sustainable Household Energy Use.* Unpublished manuscript, Groningen (NL): Centre for Environmental and Traffic Psychology/Centre for Environmental Studies, University of Groningen.

Reser, J. (1980). Automobile Addiction: Real or Imagined? *Man–Environment Systems*, **10**, 279–287.

Rogerson, R.J. (1995). Environmental and Health-Related Quality of Life: Conceptual and Methodological Similarities, *Social Science and Medicine*, **41**, 1373–1382.

Rokeach, M. (1973). *The Nature of Human Values*, Free Press, New York.

Rozec, V. (2003). The Influence of a Specific Urban Planning on Sonic Environment. In: *People, Places, and Sustainability* (G. Moser, E. Pol, Y. Bernard, M. Bonnes, *et al.*, eds), Hogrefe & Huber, Ashland, OH, 209–219.

Ruckelhaus, W.D. (1989). Toward a Sustainable World, *Scientific American*, **257**(3), 114–120.

Rudinger, G., K. Donaghy, and S. Poppelreuter (2006). Societal Trends, Mobility Behavior, and Sustainable Transport in Europe and North America, *European Journal of Transport and Infrastructure Research*, **6**, 61–76.

Schwartz, S.H. (1992). Universals in the Content and Structure of Values: Theoretical Advances and Empirical Tests in 20 Countries, *Advances in Experimental Social Psychology*, **25**, 1–65.

Schwartz, S.H. and W. Bilsky (1987). Toward a Universal Psychological Structure of Human Values, *Journal of Personality and Social Psychology*, **53**, 550–562.

Schwartz, S.H. and W. Bilsky (1990). Toward a Theory of the Universal Content and Structure of Values: Extensions and Cross-Cultural Replications, *Journal of Personality and Social Psychology*, **58**, 878–891.

Skolnik, M. (1997). *Duurzaam verder? Een onderzoek naar de betekenis van economie en milieu voor de kwaliteit van leven [Sustainable Progress? A Study on the Meaning of Economy and Environment for Quality of Life]*, Masters Thesis, University of Groningen, Department of Psychologie, Groningen, The Netherlands. (in Dutch).

Slotegraaf, G. and C. Vlek (1996). *The Perception of Global Warming and the Acceptability of Energy Saving Policies*. Unpublished Manuscript, University of Groningen, Department of Psychology Groningen, the Netherlands (in Dutch).

Solomon, M.R., J.L. Zaichkowsky, and R. Polegato (2005). *Consumer Behaviour: Buying, Having, and Being*, Pearson Prentice Hall, Toronto.

Steg, L. (2003a). Can Public Transport Compete with the Private Car? *IATSS Research*, 27, 27–35.

Steg, L. (2003b). Factors Influencing the Acceptability and Effectiveness of Transport Pricing. In: *Acceptability of Transport Pricing Strategies* (B. Schlag and J. Schade, eds), Elsevier Science, Oxford.

Steg, L. (2003c). Motives and Behaviour in Social Dilemmas Relevant to the Environment. In: *Human Decision Making and Environmental Perception. Understanding and Assisting Human Decision Making in Real-Life Settings* (L. Hendrickx, W. Jager, and L. Steg, eds), Department of Psychology, University of Groningen, Groningen, 83–102.

Steg, L. (2005). Car Use: Lust and Must. Instrumental, Symbolic and Affective Motives for Car Use, *Transportation Research A*, 39 (2–3), 147–162.

Steg, L., W. Abrahamse, C. Vlek, G. Wiersma, K.J. Noorman, M. Oostra, P. Van Kampen, P. Smeets, S. Graas, H. Grimberg, and, J. Roos (2002). *Energiebesparing door gedragsverandering. De invloed van informatie, feedback en commitment op energiegebruik, psychologische achtergronden en kwaliteit van leven [Reducing Energy Use via Behavioural Change]*. Department of Psychology, University of Groningen, Groningen, the Netherlands (in Dutch).

Suh, E., E. Diener, and F. Fujita (1996). Events and Subjective Well-Being: Only Recent Events matter, *Journal of Personality and Social Psychology*, 70, 1091–1102.

Tertoolen, G., D. Van Kreveld, and E.C.H. Verstraten (1998). Psychological Resistance against Attempts to Reduce Private Car Use. *Transportation Research A*, 32, 171–181.

Tretvik, T. (2003). Urban Road Pricing in Norway: Public Acceptability and Travel Behaviour. In: *Acceptability of Transport Pricing Strategies* (B. Schlag and J. Schade, eds), Elsevier Science, Oxford, 77–92.

Veenhoven, R. (2004). *World Database of Happiness, Catalogue of Happiness in Nations*. Available at: http://www.eur.nl/fsw/research/happiness/

Vlek, C.A.J., A.J., Roojers, and E.M. Steg (1999). *Duurzaam consumeren: meer kwaliteit van leven met minder materiaal? [Sustainable consumption: better quality of life with less material?]*. Department of Psychology, University of Groningen, Groningen, The Netherlands (in Dutch).

Vlek, C., M. Skolnik, and B. Gatersleben (1998). Sustainable Development and Quality of Life: Expected Effects of Prospective Changes in Economic and Environmental Conditions, *Zeitschrift für Experimentalle Psychologie*, 45, 319–333.

World Commission on Environment and Development (WCED) (1987). *Our Common Future*, Oxford University Press, Oxford.

Tommy Gärling and Linda Steg (Editors)

4

ADVERSE EFFECTS OF TRAFFIC NOISE

Henk M.E. Miedema, Department of Environment and Health, Netherlands Organisation for Applied Scientific Research (TNO), Delft, The Netherlands

ABSTRACT

Adverse effects of traffic noise occur in different degrees in a substantial part of the world population, especially in areas with a dense population and dense transportation networks. As a consequence, traffic noise is a global issue. This chapter reviews and systematizes the research on adverse effects of traffic noise in residential areas. Four routes are distinguished through which noise exerts its primary influence: sound masking (communication disturbance), attention (concentration disturbance), arousal (sleep disturbance), and affective/emotional responses (fear/anger). Results presented can be used to design the best practical ways of reducing noise-induced effects through the reduction of noise exposure.

INTRODUCTION

Traffic noise causes adverse effects and it is widespread and increasing. Lambert and Vallet (1994) estimated that in the European Union (EU) during daytime approximately 77 million people (i.e., 22% of the total population in 1994) were exposed to transportation noise levels exceeding 65 dB – a level which many countries consider to be unacceptable. Almost 170 million Europeans (49%) lived in, what the authors call, grey zones, that is, zones that do not ensure acoustic comfort to residents. Depending on the country, road traffic noise annoyed between 20% and 25% of the population. Even though the uncertainty in these estimates is very large, there is no doubt about the high prevalence of noise annoyance in the EU. A survey (Al-harthy and Tamura, 1999) in Muscat City, Oman, illustrates that noise and noise annoyance are not confined to the industrialized societies, but grow very fast in cities in the developing countries. The length of the paved roads in Muscat City increased from not more than 50 km in 1975 to 156 km in the old city and 1213 km in the entire city in 1995. This explains the finding that in 1995 (lack of) "quietness" caused the highest dissatisfaction in a sample of 452 inhabitants. It was higher than the dissatisfaction with the 12 other aspects of the surrounding that were rated, such as "public facilities" and "safety". The above figures illustrate that traffic noise is widespread in the industrialized countries, as well as in urban areas in the developing countries. The growing transportation network with increasing traffic intensities is an important driving force of the increase in the worldwide noise exposure.

Noise is a background stressor interwoven with our daily life (cf. Rotton, 1990). Being able to cope with daily background stressors is important for well-being and health. The adverse effects of traffic noise depend on acoustical characteristics of the noise (e.g., loudness, time pattern) and on aspects of the noise situation that may involve cognitive processing, such as expectation regarding the future development of the noise exposure (will it get better or worse), lack of short-term predictability, and a feeling of lack of control over the source of the noise. As for other stressors, people differ in their appraisal of noise situations and in their coping style. For noise, active approaches (e.g., improve insulation, file a complaint) and passive approaches (e.g., put in perspective, ignore as much as possible) can be distinguished. In addition to coping style, other personal traits, in particular noise sensitivity, influence the response to noise. However, traffic noise is not only a personal matter, but also a societal problem that is beyond the influence of most individuals. When individual coping is not successful and a person has to endure stressful noise, the outcome will be affected by the presence of other stressors, and whether there remains opportunity for resilience.

However, the above-described stress perspective is too narrow for describing the impact of traffic noise (cf. Bell *et al.*, 2001). There are adverse effects of traffic noise that do not cause stress, and that do not elicit any coping behaviour. Recognizing this, a framework is used here that distinguishes four primary interferences caused by traffic noise which may or may not be accompanied by acute stress responses. These primary effects may lead to long-term effects, and chronic stress is likely to be important in some of these long-term effects (e.g., cardiovascular).

The following four routes are distinguished through which noise exerts its primary influence (see Figure 1): *sound masking* (communication disturbance), *attention* (concentration disturbance), *arousal* (sleep disturbance), and *affective/emotional responses* (fear/anger). It appears that these routes are in part functionally separated in the following sense: for each two routes, noise can induce an effect through one of them without having an effect through the other, and vice versa. For example, noise can have effects on the attention without the occurrence of masking, arousal, or an affective/emotional response. The four routes are discussed in separate sections. The effects through these different routes may have secondary effects. Also such effects will be discussed in separate sections on cognitive impairment, cardiovascular disease, and annoyance.

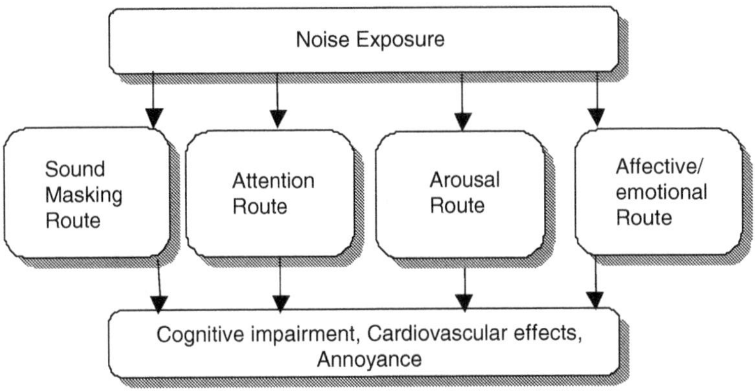

Figure 1: Four Routes (Primary Effects) That Contribute with Different Weights to Cognitive Impairment, Cardiovascular Effects, and Annoyance. The Primary Interferences may be Accompanied by Short-Term Stress Responses and Chronic Stress may Play at Role in Long-Term Effects

Noise annoyance is a sensitive indicator of adverse noise effects and by itself means that noise affects quality of life. Therefore, it is often taken as an indicator of the acoustical climate. For noise annoyance, extensive research has provided relationships that give the expected noise annoyance at a given level of noise exposure. The noise exposure metric used, L_{den} (in dB(A)), is the yearly average of the daytime level (7–19 h), evening level (19–23 h) plus 5 dB(A), and night-time level (23–7 h) plus 10 dB(A) at the most exposed facade of a dwelling. The "(A)" indicates that contributions to noise from different frequencies are weighted according to the sensitivity of the ear for those frequencies. Noise annoyance relationships will be presented for exposure to noise from a single transportation source (air, road, or rail traffic) and for combined exposures. In addition, a section treats the role of acoustical factors not reflected by L_{den} and the role of non-acoustical factors in noise annoyance. The section preceding the discussion illustrates the usage of the annoyance relationships in practice when traffic noise is on the policy agenda. It illustrates how information on health impacts as presented here is used for evaluating existing and possible future noise situations.

THE SOUND MASKING ROUTE: COMMUNICATION DISTURBANCE

Because masking reduces the comprehension of speech, it is an important phenomenon. In addition to the masking of speech, the masking of signals, music, or natural sounds may cause a problem or be unpleasant. The main factors that determine whether noise masks other sound are the overlap in time and the overlap of the frequency spectra of the noise and the other sound. Roughly, a noise totally masks a tone if the noise and the tone occur simultaneously, and the level of the tone is at least 2–6 dB lower than the level of the noise in the frequency band around the tone. However, it is a very large step from knowing such details of the masking of a momentary simple sound to predicting the effect of noise on communication. Metrics used to quantify the influence of masking on intelligibility of speech are the speech interference level (SIL), articulation index (AI), and the speech transmission index (STI). These metrics give an indication of the intelligibility of speech on the basis of the frequency spectrum of the noise, and, depending on the metric, additional factors that influence masking such as the reverberation time of the room and the characteristics of the speaker.

When the traffic noise level is higher, the speaker usually increases the effort and speaks louder, or reduces the distance between speaker and listener. Table 1 indicates how, given a combination of speaker–listener distance and vocal effort, the background noise affects the comprehensibility. For example, the table indicates that with a "relaxed" voice of the speaker, "excellent" comprehension at 1 m distance (one-to-one conversation) requires that the noise level does not exceed 41 dB(A). This holds for a constant noise with a standard road traffic spectrum. To obtain with the same voice, the same quality for a listener at 4 m distance (group around a table), the noise level must be below 29 dB(A). Considering the existing traffic noise levels, in particular in urban environments, the table implies that effects of traffic noise on communication are ubiquitous, especially in cities.

The widely prevalent masking of speech by traffic noise affects not only individuals but may also have social consequences. People may refrain from conversation in noisy environments because conversation is too uncomfortable or impossible, or, more subtle, they may limit their conversation to simpler messages. This happens although it is not known at what scale. Noise sufficiently loud to induce this effect is so common that it may be a large effect.

ATTENTION ROUTE: CONCENTRATION DISTURBANCE

Attention involves selection of elements, usually from visual or acoustical impressions or from mental representation, persistence of attention to selected elements, and ending and redirecting attention. Attention can be focussed, or it may be divided over more elements. By attracting the limited attention resources, noise can negatively affect processes requiring attention. Especially, difficult to resist is sound that contains information (for a review of the effect of irrelevant speech, see Jones, 1990). Jones and Macken (1998) discuss the (acoustical) characteristics of sound that determine its distractive power.

Table 1: Highest noise level (L_{Aeq}, cells) that with a certain speech effort ($L_{SA,1m}$, rows) gives a certain level of comprehension (related to $D = L_{SA,1m} - L_{Aeq}$, columns), with the listener at distances of 1, 2, or 4 m (based on ISO-9921, 2003). Not taken into account is the reverberation that may be important indoors. The noise is assumed to have a constant level with a standard spectrum for road traffic. For other road traffic noise and for other environmental sources such as aircraft and trains, the values may be different due to differences in time pattern and in spectra

Voice speaker	Difference (D) Speech Minus Background	1 – Excellent $D \geq 13$ dB(A)			2 – Good $D = 10$ dB(A)			3 – Reasonable $D = 4$ dB(A)			4 – Poor $D = -2$ dB(A)			5 – Bad $D < -5$ dB(A)		
	$L_{SA,1m}$	1 m	2 m	4 m	1 m	2 m	4 m	1 m	2 m	4 m	1 m	2 m	4 m	1 m	2 m	4 m
1 – Relaxed	54 dB(A)	41	35	29	44	38	32	50	44	38	56	50	44	59	53	47
2 – Normal	60 dB(A)	47	41	35	50	44	38	56	50	44	62	56	50	65	59	53
3 – Raised	66 dB(A)	53	47	41	56	50	44	62	56	50	68	62	56	71	65	59
4 – Loud	72 dB(A)	59	53	47	62	56	50	68	62	56	74	68	62	77	71	65

The shift of focus from the sound level or loudness of noise to specific characteristics as causes of distraction was a consequence of the high sound levels required for demonstrating effects of noise on attention in much of the experimental work.

Smith (1991) reports impaired sustained attention in noise. The distracting effect of noise probably is most harmful when it is essential for the task that information is kept in the working memory. Reduction of the reproduction of items from working memory has been found in many experiments to depend on the priority and difficulty of the memory task, and the type of sound (e.g., Boggs and Simon, 1968; Glass and Singer, 1972; Hamilton *et al.*, 1977; Finkelman *et al.*, 1979; Smith, 1982; Broadbent, 1983). An experiment by Millar (1979) indicates that it is the rehearsal of the items in the working memory that is negatively affected by noise (also see Wilding and Mohindra, 1983). Rehearsal involves attention to the items rehearsed.

Much human information processing requires that information is kept in the working memory and, hence, is vulnerable to noise. For example, when a person is adding two large numbers, in addition to these numbers he must keep intermediate results of the addition (e.g., the added ones, the added tens, the added hundreds, etc.) in the working memory. If the intermediate results get lost because the attention is distracted by sound, then the subject must start again from the beginning. The effect of noise on an arithmetic task has been demonstrated to depend on the working-memory demand imposed by the arithmetic task (Park and Carr Payne, 1963; Woodhead, 1964; Glass and Singer, 1972) as well as the priority of the task (Bell, 1978).

Another important example is reading. When a text is read, the new information being read is being combined with information read earlier that is retained in memory. This combination of information is needed to derive the implications, and to restructure information from the linear sequence in which it is read into meaningful chunks. If noise distracts the attention from this information in working memory, then implications will not be determined and restructuring will not take place. Effects of noise on the detection of semantic errors (Weinstein, 1974, 1977), text comprehension (Hockey, 1979) and cued recall of a text from episodic memory (Hygge *et al.*, 2003; Enmarker, 2004) have been experimentally demonstrated. In most circumstances this effect is not so evident as the effect on, for instance, the addition of numbers, because reading is not an all or none task with either success or failure as outcome. The quality of the information retained after reading may be affected by the noise, but there is little feedback that this happens, that is, there are no obvious indicators of the quality of the result of reading.

Distraction by noise and blocking out noise by focussing attention are common. We are aware of effects if they lead to failure, but most often it only affects the quality in a gradual way and therefore remains unnoticed. Most experimental demonstrations of the effects of non-speech sound involve high-noise levels. It appears that persons can maintain their performance at lower levels if they are motivated to do so (as they presumably are in most experiments) but maintaining performance has a physiological cost. For example, Carter and Beh (1989) found that intermittent noise during task performance affected cardiovascular parameters. In daily life, we are not always motivated to invest the required effort and are not willing to pay the cost in the form of fatigue. Therefore, in daily life the most important consequence of the effect of noise on attention may be that we resort to less demanding strategies or do not engage in cognitive demanding tasks in noisy environments. As with communication disturbance, it may be speculated that these changes occur frequently and affect performance at the societal level.

AROUSAL ROUTE: SLEEP DISTURBANCE

A certain level of arousal is needed to sustain attention but a too high level of arousal may hamper attention. Low arousal can be counteracted by noise, and in that way noise can prevent long-response times (Corcoran, 1962) and lapses in attention (Smith, 1998). The higher the arousal, the lower the probability of falling asleep and not being disturbed during sleep. Because of its arousing potential, sound can thus prevent a person from falling asleep, or affect the sleep quality of a sleeping person and possibly wake the subject. Here, we concentrate on these arousal-related noise effects on sleep.

For indications of noise-induced effects on sleep in the population, field studies are of particular importance since laboratory studies consistently find much stronger effects, also when first-night effects are accounted for (Pearsons *et al.*, 1995; Basner *et al.*, 2004). Field studies found that the noise of a single passage can cause the following instantaneous effects: extra motility (Horne *et al.*, 1994; Fidell *et al.*, 2000; Passchier-Vermeer *et al.*, 2002, 2005), change in sleep state and EEG arousals (Pearsons *et al.*, 1973; Vernet, 1979; Vallet *et al.*, 1983; Hume *et al.*, 2003; Basner *et al.*, 2004), momentary change in heartbeat parameters (Wilkinson and Campbell, 1984; Carter *et al.*, 1994; Hofman *et al.*, 1995), and conscious awakening (Fidell *et al.*, 1995, 2000; Passchier-Vermeer *et al.*, 2002). In addition to the instantaneous effects related to single events, large field studies found for aircraft (Passchier-Vermeer *et al.*, 2002) and road traffic (Griefahn *et al.*, 2000; Passchier-Vermeer *et al.*, 2005) that also sleep latency and average motility during the sleep period increased monotonically as a function of the average noise exposure level during (the beginning of) the sleep period. Such associations were not found for railway noise (Griefahn *et al.*, 2000; Passchier-Vermeer *et al.*, 2005). In addition to increased sleep latency, it appears that road traffic noise extends the period of light sleep, and shifts the barycentre of the advanced sleep stages towards the end of the night (Vallet *et al.*, 1983; Griefahn and Gros, 1986; Eberhart and Akselsson, 1987). The results from field studies are mixed with respect to overall changes in heart rate related to road traffic noise. Since noise-induced motility and EEG micro-arousals were already found for sound events with a maximum level in the bedroom of 32–35 dB and the probability of occurrence increased monotonically at higher levels, instantaneous effects of traffic noise on sleep are very common.

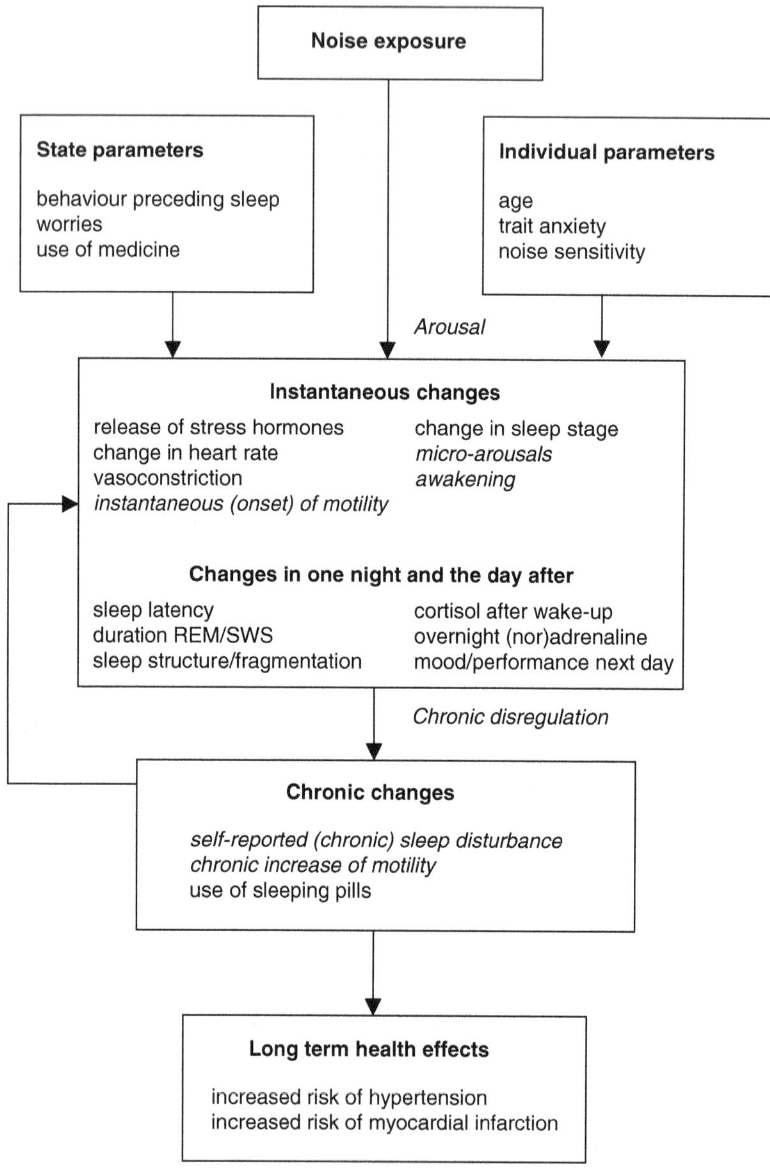

Figure 2: Framework for the Study of Noise-Induced Sleep Disturbance. The Effects Mentioned are Examples and Not Necessarily Proven Effects of Noise on Sleep. The Effects for Which There is a Sufficient Basis to Establish (Provisional) Exposure–Response Relationships, are in Italics

The consequences of instantaneous effects are not yet fully understood. A conceptual framework for noise-induced sleep disturbance is presented in Figure 2 (cf. Ising *et al.*, 1999). This framework gives a rough outline of steps in the development of effects of night-time noise. The framework does not imply that instantaneous effects necessarily contribute to chronic changes or long-term health effects, or that chronic changes necessarily contribute to long-term health effects. Recovery mechanisms can restore balances and prevent the occurrence of further effects. The effects in the figure for which there is a sufficient basis to establish (provisional) exposure–response relationships are in italics: the instantaneous effects (onset of) motility, micro-arousals and (conscious) awakening, and the chronic effects increase of mean motility during sleep and self-reported sleep disturbance. For example, an exposure–response relationship for (conscious) awakening has been assessed for civil aircraft (Passchier-Vermeer, 2003). The noise of a passage is described not by the maximum level of the passage, but by the total sound energy of the passage (SEL). In the range (indoor levels) $54 < \text{SEL} < 90$ dB(A), the probability of awakening induced by the noise of an overflight is: $p(\text{SEL}) = -0.00564 + 1.909 \times 10^{-6}\ \text{SEL}^2$.

In a number of cases, noise is likely to be a dominant factor interfering with sleep. More often it will cause a limited reduction of the quality of sleep that not always will be noticed as such. These noise-induced reductions of sleep quality may add to major causes of sleep problems that also appear to be mediated by increased arousal, such as social stress (e.g., divorce), medical stress (e.g., cardiovascular problems), circadian stress (e.g., shift work) and other environmental factors (e.g., high temperature). Altogether there are sufficient indications that exposure to traffic noise has a widespread influence on sleep, which by itself or in combination with other factors may reduce well-being and may have other specific consequences associated with poor sleep quality.

AFFECTIVE/EMOTIONAL ROUTE: FEAR AND ANGER

Many sounds are neutral, that is, lack an affective or emotional connotation. Such sounds may be simply perceived without causing any emotional reaction and associated behavioural response. Observations of animals exposing themselves to high levels of environmental noise without any overt aversive reaction suggest that the noise is a neutral sound to them. By two mechanisms, noise may cause (negative) affective/emotional responses.

Noise may frustrate communication, concentration, or sleep that lead to irritation or anger reactions. It appears that coping style, personal factors, and situational variables determine whether emotional responses will be elicited in a person experiencing interferences by noise, and how intense the responses are.

Fear reactions and worrying are elicited by noise if it is associated with danger that is threatening to the individual. Some of these reactions are primitive in the sense that they

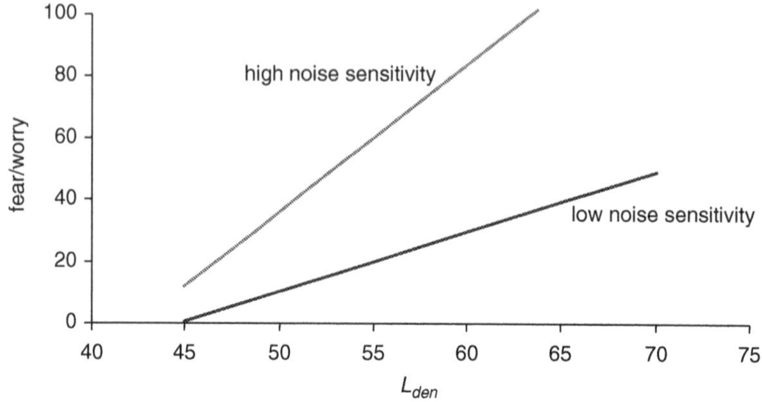

Source: Miedema and Vos (2003).

Figure 3: Relations between Fear/Worry and Noise Exposure (L_{den}) for Sensitive and Non-Sensitive Persons Found in a Large Aircraft Noise Effects Study

are innate or the result of early learning processes. Examples are the fear of the thunder or fear reactions to unidentified sounds in the dark. More important in the present context is that near airports fear is associated with aircraft noise in a substantial part of the population (Borsky, 1979; Moran *et al.*, 1981; Miedema and Vos, 1999). Also fear reactions depend on personal factors such as noise sensitivity. Figure 3 illustrates that fear/worry with respect to aircraft in the vicinity of an airport is related to the sound exposure level caused by the overflying aircraft and that this relationship is strongly modified by the noise sensitivity of the subject.

COGNITIVE IMPAIRMENT IN CHILDREN

Children appear to be especially susceptible to traffic noise exposure, presumably because of the immaturity of the auditory perceptual system and cognitive functions. In particular, their cognitive development may be adversely affected by exposure to noise. Noise effects through different routes (communication, concentration, sleep, and affective/emotional responses) may contribute in different ways to cognitive impairments in exposed children. Currently, it is not known to which extent and how exactly noise affects cognitive capacities in exposed children, but the evidence regarding effects on reading capacities are sufficiently strong to justify specific attention to children as a group vulnerable to noise.

Evidence of poorer auditory discrimination and speech perception in children exposed to apartment or traffic noise has been found by Cohen *et al.* (1973, 1986), Cohen (1980),

Moch-Sibony (1984), and Evans and Maxwell (1997), and evidence of poorer language skills, in particular reading ability, of children exposed to traffic noise has been presented by Cohen *et al.* (1973), Bronzaft and McCarthy (1975), Bronzaft (1981), Evans *et al.* (1995), Evans and Maxwell (1997), Haines *et al.* (1998, 2001, 2002), Hygge *et al.* (2002), and Stansfeld *et al.* (2005). The most convincing evidence comes from studies where children were exposed to aircraft noise.

Masking may play a role in negative effects of (aircraft) noise on reading by reducing the opportunities for developing the basic auditory perceptual capacities involved in speech perception and, indirectly, in reading. These effects may be partly caused by less frequent or simplified communication with the children as an adaptation to the noisy environment. Noise may be especially harmful for the communication in classrooms because the distances between speaker and listener are relatively large. Potentially extra vulnerable are children with temporary hearing loss due to otitis media, children with moderate permanent hearing loss, children with learning disabilities, and children for whom the language at school is their second language.

In addition to masking, effects of noise on attention probably play a role especially in the case of aircraft noise if it is associated with fear. When noise events chronically and frequently attract the attention and thereby interfere with ongoing tasks, this may reduce the acquisition of complex cognitive (reading) skills. This may occur without an influence of the noise on the attention skills themselves. Potentially extra vulnerable are children with attention problems, such as children with ADHD. One of the diagnostic criteria of ADHD is distractibility, and ADHD is known to be associated with reading problems.

In addition to masking and distraction effects of noise, also effects on sleep could play a role in cognitive effects. Through a negative effect on the quality of sleep of children, noise could deteriorate their general condition (fatigue) and thereby the capacity to acquire complex cognitive (reading) skills. More subtle would be an influence of noise on the consolidation during sleep of material learnt during the day.

CARDIOVASCULAR EFFECTS

Momentary stress responses may accompany interference by noise of behaviour or a desired state. Presumably as a result of the chronic occurrence of such noise-induced stress responses, high levels of environmental noise were found to increase the relative risk of ischemic heart disease.

Ising (1983) studied men who participated in a seminar for several days in which understanding of speech was essential. They were observed one day in a control condition and one day intelligibility of speech was reduced by exposure to road traffic noise.

It was found that the excretion of noradrenalin increased significantly by 10% in the noise exposure, while no effect was found on adrenaline. It is hypothesized that within limits persons can keep up their performance in noise by focussing attention. However, this leads to physiological costs. Tafalla and Evans (1997) and Miki *et al.* (1998) studied the effect of noise during arithmetic calculations. They found an increase in the excretion of cortisol in noise. Various reaction patterns of cortisol have been found in studies on the effect of night-time noise on hormone levels during sleep (Maschke *et al.*, 1995, 1997; Braun, 1999; Harder *et al.*, 1999). For example, Harder *et al.* (1999) found long-term habituation, hyperreaction, and hyporeaction of cortisol levels in an experimental study with night-time noise exposure for five weeks. The results of Braun (1999) indicate that long-term road traffic noise may induce a chronic increase in the cortisol and noradrenaline level, while there was no effect on adrenaline.

Early laboratory studies, reviewed by Passchier-Vermeer (1993), have shown momentary changes in the cardiovascular system induced by exposure to noise. Effects of long-term exposure to noise in the residential environment on cardiovascular health have been investigated in recent epidemiological studies (e.g., Rosenlund *et al.*, 2001; Babisch *et al.*, 2005). Maschke (2003) found a significantly higher prevalence of hypertension in individuals exposed to night-time road traffic noise levels (L_{night}) greater than 55 dB(A) compared to subjects exposed to noise levels (L_{night}) below 50 dB(A). Babisch *et al.* (2005) investigated the risk of road traffic noise for the incidence of myocardial infarction and showed an association between noise exposure and myocardial infarction in men. Noise-exposed women were not found to be at higher risk. The duration of exposure appeared to be important. In a subsample of men who lived for at least 10 years at their present address, the relative risk was higher. With respect to aircraft noise, Rosenlund *et al.* (2001) found an association between noise exposure levels (L_{Aeq}) greater than 55 dB(A) and the prevalence of hypertension.

In a meta-analysis, Babisch (2005) identified 60 epidemiological studies that had investigated the relationship between transportation noise and cardiovascular endpoints, assessed either objectively or by self-reports. Thirty-seven of those studies had assessed the prevalence or incidence of manifest diseases, including hypertension and ischemic heart diseases (angina pectoris, myocardial infarction, ECG abnormalities). On the basis of stringent criteria, five analytic studies were selected that could be used for deriving the relationship between road traffic noise and the risk of myocardial infarction. The selected studies were carried out in Britain (Caerphilly & Speedwell, pooled six years follow-up data) (Babisch *et al.*, 1999, 2003) and Berlin ("Berlin I", "Berlin II", "Berlin III") (Babisch *et al.*, 1994, 2005). In these studies the orientation of rooms (facing the street or not) was taken into account in the exposure assessment. The noise effect estimates obtained from a study were adjusted for the covariates considered in that study. It was found that for noise levels above 60 dB(A), the risk of myocardial infarction increases monotonously. Figure 4 shows the points that indicate the relative risk found

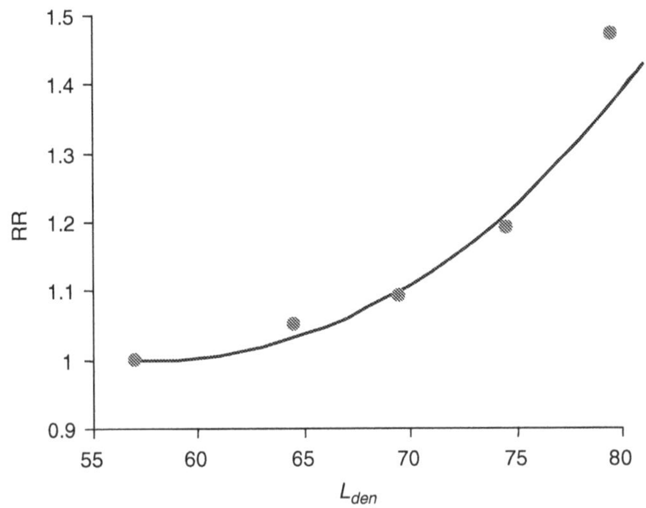

Figure 4: The Points That Indicate the Relative Risk (RR) of Myocardial Infarction for Different Noise Exposure Classes, and a Line Fitted through These Points (Based on Babisch, 2005). Assuming That in This Case Odds Ratios Can be Equated to Relative Risks and That for Road Traffic $L_{den} = L_{0622} + 2$)

for different noise exposure classes, and a line fitted through these points. It must be noted that the uncertainty regarding the exact values of the relative risks is large. The 95% confidence intervals of the individual data points encompassed 1.

EXPOSURE–ANNOYANCE RELATIONSHIP

Many noise annoyance surveys have been conducted, and attempts have been made to integrate the results through meta-analyses with the aim of establishing exposure-response curves for annoyance caused by traffic noise. The meta-analyses relate the percentage of persons annoyed by noise from transportation (aircraft, road traffic, railways) to the noise exposure level (see Schultz, 1978; Kryter, 1982, 1983; Fidell *et al.*, 1991; Miedema and Vos, 1998; Miedema and Oudshoorn, 2001). Miedema and Oudshoorn (2001) fitted a model of annoyance as a function of noise exposure to the data from a very large set of field studies in which noise exposure and noise annoyance were determined. The database used contains all studies examined earlier in the syntheses by Schultz (1978) and Fidell *et al.* (1991) provided that L_{den} and percentage highly annoyed could be assessed while meeting certain minimal requirements. In addition, many other studies were included. Consequently, the synthesis is more comprehensive than the earlier ones.

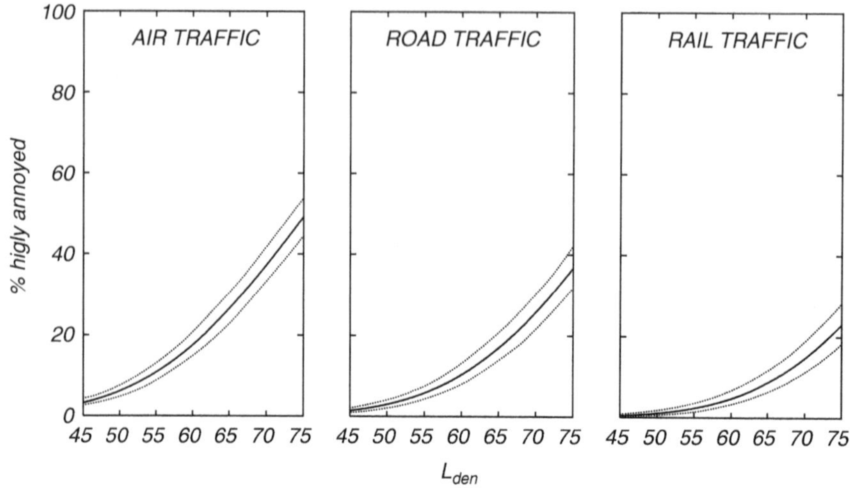

Figure 5: The Percentages Highly Annoyed (%HA) as a Function of L_{den}, for Air, Road, and Rail Traffic Noise. Broken Lines are the Boundaries of 95% Confidence Intervals

Moreover, the kind of errors and inaccuracies Fields (1994) found in the previous syntheses have been avoided. A model has been fitted to data from noise annoyance studies for aircraft, road traffic, and railways separately. Figure 5 shows relationships implied by the model for L_{den} using the percentage "highly annoyed" (%*HA*, cut-off at 72 on a scale from 0 to 100) as the annoyance measure. The percentage highly annoyed persons (%*HA*) is zero below 40–45 dB(A), and increases at higher levels monotonically as a function of L_{den}. Different functions were found for aircraft, road traffic, and railway noise. The rate of increase is higher for aircraft noise than for road traffic noise, which in turn has a higher rate of increase than railway noise. The 95% confidence intervals around the different functions do not overlap at higher exposure levels.

In many cases people are not exposed to either aircraft, road traffic or railway noise, but to a combination of these types of noises. The annoyance equivalent model (Miedema, 2004) describes how the annoyance caused by the total, combined exposure can be calculated. The above-mentioned publication also describes how the model is formally derived from a few basic assumptions, and discusses the plausibility of these assumptions. Briefly, the procedure based on the model is to translate the noise from individual sources into the equally annoying sound levels of a reference source (road traffic), and then to sum these levels. Figure 6 illustrates this for two different noise sources A and B. Thus, the model first translates the noise from the individual sources into the equally annoying sound levels of a reference source, road traffic, and then sums these levels giving total level L_{T}. The annoyance from the combined sources is found by substituting

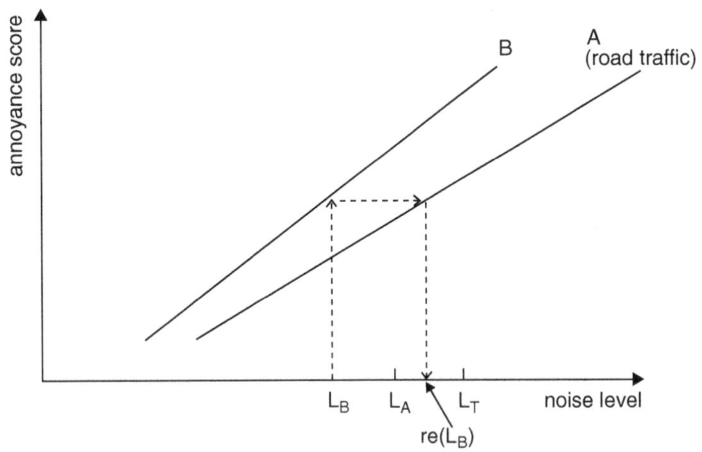

Figure 6: Illustration of the Annoyance Equivalents Model (See Text for Explanation)

exposure L_T in the road traffic exposure– annoyance relationship. A stepwise procedure for the assessment of the total noise level and the corresponding percentage annoyed can be found in the above-mentioned publication.

ADDITIONAL ACOUSTICAL AND NON-ACOUSTICAL FACTORS

In addition to the noise level at the most exposed facade (L_{den}), there are a number of factors that have an influence on noise effects in residential areas. A distinction can be made between acoustical factors and non-acoustical factors. Important acoustical factors are sound insulation and quiet side, and, to a lesser extent, ambient noise level in the vicinity of the dwelling. The exposure-effect curves for, for instance, annoyance, sleep disturbance, and cardiovascular risks are based on the average situation for each of the above factors (in the study samples on which the curves are based). If a dwelling has, for instance, a higher sound insulation than average, noise annoyance of its inhabitants will be less than expected on the basis of the exposure-effect curve. However, if a dwelling has a lower sound insulation than the average dwelling, noise annoyance of its inhabitants will be higher than expected on the basis of the exposure-effect curve.

Fields (1993) conducted a meta-analysis and Miedema and Vos (1999) used the original data from many studies in their meta-analysis to investigate the possible systematic effects of non-acoustical factors on noise annoyance. They found consistent results.

The latter publication presents quantitative estimates of the effects on annoyance. The results are based on analyses of the original data from various field surveys of responses to noise from transportation sources (number of cases depending on the variable varied between 15,000 and 42,000). It was found that fear has a very large impact on annoyance. Persons who experience fear related to the transportation that causes the noise report higher annoyance compared to persons who do not experience such fear. With three categories of fear, the annoyance difference between the lowest and the highest fear level is equivalent to a L_{den} difference of at most 19 dB(A). The effect of fear on annoyance is found for all three modes of transportation, but it appears that only few persons associate high fear with railway traffic. Another important factor is noise sensitivity. The effect of noise sensitivity on annoyance is reduced only very little if also age is taken into account. Demographic factors are much less important than fear and noise sensitivity. Noise annoyance is not related to gender, but age has an effect on noise annoyance. The largest difference in annoyance between (seven) age classes is equivalent to a L_{den} difference of approximately 5 dB(A). At the same noise exposure level, relatively young and relatively old persons are less annoyed than the ages in between. Higher annoyance is reported if the education is higher, the occupational status is higher, the dwelling is owned instead of rented, if a person does not depend on the noise source, and if the use of the transportation that causes the noise is low. In addition, a person in a household consisting of two persons reports more annoyance than a person in a household of another size, also if the age of the person is taken into account. However, the effects of these factors on noise annoyance are small, the equivalent L_{den} difference is equal to 1–2 dB(A), and, in the case of dependency, 3 dB(A).

There is at present no combined model of all the interrelations between noise exposure described by L_{den}, annoyance, and the other above-mentioned acoustical and non-acoustical variables that influence noise annoyance. Noise sensitivity is the single factor in addition to the noise exposure that has been most studied (e.g., see also McKennel, 1963; Langdon, 1976; and for overviews Job, 1988; Fields, 1993). The influence of noise sensitivity on noise annoyance is strong. It is discussed here on the basis of Miedema and Vos (2003). Noise sensitivity is a personal trait that is stable over time. Noise sensitivity has a weak relationship with noise exposure, which cannot explain the strong influence of noise sensitivity on various effects (communication disturbance, annoyance, and fear/worry related to the noise source), and it has been found not to change when the noise exposure changes. Noise sensitivity modifies the influence of noise exposure. For those who are less sensitive, the relation of annoyance with noise exposure is relatively flat while for those who are more sensitive the relation is steeper. A similar, but stronger modifying influence of noise sensitivity on the relation between L_{den} and fear/worry was illustrated in Figure 3. Noise sensitivity appears to reflect a general awareness of consequences of human activities for the environment and concern about harmful consequences (cf. Stansfeld, 1992) rather than differences in the basic perception of noise or the interference with information processing. Noise sensitive persons also react stronger to environmental factors other than noise (e.g., environmental odour).

The working of noise sensitivity is restricted to modifying reactions to environmental factors, so that it is not a general negative affectivity or neuroticism. Because environmental factors induce stronger aversive affective/emotional reactions in noise sensitive persons, these persons are likely to be vulnerable in the sense that they more easily develop stress reactions in response to them.

USAGE OF EXPOSURE–ANNOYANCE RELATIONSHIPS IN PRACTICE

The different degrees of noise impact in a neighbourhood after possible future reconstructions, alternative traffic policies in an urban area, or alternative realizations of an infrastructural project (extending an airport, building a new road or railway line) can be compared by calculating the noise exposures for the dwellings in the area concerned, and then using exposure–response relationships to quantify the noise effects. Figure 7 presents examples of noise maps showing the noise exposures in a city. The curves presented in Figure 5 can be used to estimate the expected number of highly annoyed persons in an area on the basis of such noise maps.

Noise maps combined with exposure–response relationships can be used in the context of the EU Environmental Noise Directive (EC, 2002), in target setting, in cost-benefit analysis, and in Environmental Health Impact Assessment (cf. Borst, 2001). When used in Environmental Health Impact Assessment, they give insight into the situation that is expected in the long term. They are not applicable to local, complaint-type situations were idiosyncratic features play an important role, or to the assessment of the short-term effects of a change of noise climate. The curves have been derived for adults. There are indications that annoyance from aircraft noise at a given exposure level has increased. In the future this may necessitate adoption of higher curves or curves that predict higher annoyance in particular circumstances, but sufficient evidence to do so is lacking at present.

DISCUSSION

Noise reduces speech comprehension. Noise affects attention and therefore mental processing of information involved in, for instance, reading. Noise increases arousal, thus disturbs sleep that may lead to fatigue, decreased performance, and depressed mood. Furthermore, noise may elicit emotional reactions when it interferes with behaviour or a desired state, or when it is associated with fear (e.g., aircraft noise), and thus acts as a stressor. These primary effects may in the long-term lead to annoyance, cognitive impairment, or cardiovascular effects. Chronic stress is likely to be important in some long-term effects, in particular the cardiovascular effects.

Figure 7: Four Noise Maps of the Same Part of a City (Leiden, The Netherlands) Showing the Levels of Road Traffic Noise (Upper Left), Railway Noise (Upper Right), Industrial Noise (Lower Left), and the Total Noise Levels Calculated with the Method Described in This Paper, Which Gives Railway Noise a Lower Weight Than Road Traffic Noise Because It Causes Less Annoyance at the Same Noise Level (See Figure 5). Note That the Total Noise Levels (Bottom Right) is Dominated by Road Traffic Noise, as is the Case in Many Cities

Noise annoyance is a sensitive indicator of adverse noise effects and by itself means that noise affects quality of life. For noise annoyance, extensive research has provided relationships that predict the level of noise annoyance that can be expected at a given level of noise exposure. This knowledge can be used in noise abatement policies to find the best practical ways of reducing noise through the reduction of the exposure. At present, it is possible to produce maps showing the exposures to noise from aircraft, road traffic, and railways, and with the relationships between exposure and annoyance, these can be translated in estimates of the number of people annoyed. These estimates can be made on the basis of the calculated existing noise exposures, but also, on the basis of scenarios for future noise levels so that the consequences of different scenarios can be explored.

Although there is some insight in other factors such as noise sensitivity that influence noise annoyance besides the noise exposure, currently this knowledge is not developed sufficiently to form the basis for policies targeted at such non-acoustical factors aimed at reducing noise annoyance through non-acoustical measures. In addition to more insight in the role of non-acoustical factors, it would be also very useful to improve the knowledge base for noise policies concerning the consequence of acoustical measures that do not change the L_{den} at the most exposed side of dwellings, but can be effective in reducing adverse noise effects in residential areas. Most important are the effect of insulation and the effect of a relatively quiet side, and in addition, also access to quieter areas in the vicinity of home may be beneficial.

The effects of traffic noise do not require extremely high or long noise exposures and can be assumed to occur, with different degrees of intensity, in a substantial part of the world population, especially in areas with a dense population and dense transportation networks. Owing to the widespread exposure, and the demonstrated and potential effects, noise is a global issue. It is also one of the global problems which continues to grow. It appears that it will be difficult to substantially reduce the noise exposition given the growth of transportation. Most effective are measures at the source on the basis of stringent noise emission criteria for aircraft, road vehicles, and trains. In addition, local improvements can be obtained by measures at the source (e.g., porous asphalt, speed limit, flight procedures), with respect to the transmission (e.g., noise barriers, office buildings shielding residential area), or at the receiver (e.g., insulation of the dwelling, layout of dwelling).

REFERENCES

Al-harthy, I. and A. Tamura (1999). Sound Environment Evaluation and Categorization of Audible Sounds – The First Survey of Human Response to Sound Environment in Muscat City (Oman), *Journal of the Acoustical Society of Japan*, **20**, 353–364.

Babisch, W. (2005). Updated Review of the Relationship between Transportation Noise and Cardiovascular Risk. In: *Proceedings of the 12th International Congress on Sound and Vibration* (J.L. Bento and D. Alarcao, eds), Lisbon.

Babisch, W., B. Beule, M. Schust, and N. Kersten (2005). Traffic Noise and Risk of Myocardial Infarction, *Epidemiology*, **16**, 33–40.

Babisch, W., H. Ising, and J.E.J. Gallacher (2003). Health Status as a Potential Effect Modifier of the Relation between Noise Annoyance and Incidence of Ischemic Heart Disease, *Occupational and Environmental Medicine*, **60**, 739–745.

Babisch, W., H. Ising, J.E.J. Gallacher, P.M. Sweetnam, and P.C. Elwood (1999). Traffic Noise and Cardiovascular Risk: the Caerphilly and Speedwell Studies, Third Phase – 10 Year Follow Up, *Archives of the Environmental Health*, **54**, 210–216.

Babisch, W., H. Ising, B. Kruppa, and D. Wiens (1994). The Incidence of Myocardial Infarction and Its Relation to Road Traffic Noise: The Berlin Case-Control Studies, *Environment International*, **20**, 469–474.

Basner, M., H. Buess, D. Elmenhorst, A. Gerlich, N. Luks, and H. Maass (2004). *Effect of Nocturnal Aircraft Noise, Volume 1: Executive Summary*, Cologne Institute of Aerospace Medicine of the German Aerospace Center (DLR), Cologne.

Bell, P.A. (1978). Effects of Noise and Heat Stress on Primary and Subsidiary Task Performance, *Human Factures*, **20** (6), 749–752.

Bell, P.A., T.C. Greene, J.D. Fisher, and A. Baum (2001). *Environmental Psychology*, Fifth Edition, Harcourt College Publishers, Fort Worth.

Boggs, D.H. and J.R. Simon (1968). Differential Effects of Noise on Tasks of Varying Complexity, *Journal of Applied Psychology*, **52**, 148–153.

Borsky, P.N. (1979). Sociopsychological Factors Affecting the Human Response to Noise Exposure, *The Otolaryngologic Clinics of North America*, **12**, 521–535.

Borst, H.C. (2001). The Role of URBIS Noise and Noise Effects Maps in Local Policy, *Proceedings of Internoise 2001*, The Hague.

Braun, D. (1999) *Nachlicher Strassenverkehrslärm und Stresshormonausscheidung beim Menschen*, Dissertation, Berlin.

Broadbent, D.E. (1983). Recent Advances in Understanding Performance in Noise. In: *Proceedings of the 4th International Congress on Noise as a Public Health Problem* (G. Rossi, ed.), Centro Ricerche e Studi Amplifon, Milan.

Bronzaft, A.L. (1981). The Effect of a Noise Abatement Program on Reading Ability, *Journal of Environmental Psychology*, **1**, 215–222.

Bronzaft, A.L. and D.P. McCarthy (1975). The Effect of Elevated Train Noise on Reading Ability, *Environment and Behaviour*, **7**, 517–527.

Carter, N.L. and H.C. Beh (1989). The Effect of Intermittent Noise on Cardiovascular Functioning During Vigilance Task Performance, *Psychophysiology*, **26**, 548–559.

Carter, N.L., P.N. Ingham, K. Tran, and S.N. Hunyor (1994). A Field Study of the Effects of Traffic Noise on Heart Rate and Cardiac Arrhythmia During Sleep, *Journal of Sound and Vibration*, **169**, 211–227.

Cohen, S. (1980). Aftereffects of Stress on Human Performance and Social Behavior: A Review of Research and Theory, *Psychological Bulletin*, **88**, 82–108.

Cohen, S., G.W. Evans, D. Stokols, and D.S. Krantz (1986). *Behavior, Health and Environmental Stress*, Plenium Press, New York.

Cohen, S., D.C. Glass, and J.E. Singer (1973). Apartment Noise, Auditory Discrimination and Reading Ability in Children, *Journal of Experimental Psychology*, **9**, 407–422.

Corcoran, D.W.J. (1962). Noise and Loss of Sleep, *Quarterly Journal of Experimental Psychology*, **14**, 178–182.

Eberhart, J.L. and K.R. Akselsson (1987). The Disturbance by Road Traffic Noise on the Sleep of Young Male Adults as Recorded in the Home, *Journal of Sound and Vibration*, **114**, 417–434.

EC (2002). Directive 2002/49/EC of the European Parliament and of the Council of 25 June 2002 relating to the assessment and management of environmental noise, *Official Journal of the European Communities 18.7.2002*, L189/12–25.

Enmarker, I. (2004). The Effects of Meaningful Irrelevant Speech and Road Traffic Noise on Teachers' Attention, Pisodic and Semantic Memory, *Scandinavian Journal of Psychology*, **45**, 393–405.

Evans, G.W., S. Hygge, and M. Bullinger (1995). Chronic Noise and Psychological Stress, *American Psychological Society*, **6**, 333–338.

Evans, G.W. and L. Maxwell (1997). Chronic Noise Exposure and Reading Deficits: The Mediating Effects of Language Acquisition. *Environment and Behaviour*, **29**, 638–656.

Fidell, S., D.S. Barber, and Th.J. Schultz (1991). Updating a Dosage-Effect Relationship for the Prevalence of Annoyance Due to General Transportation Noise, *Journal of the Acoustical Society of America*, **89**, 221–233.

Fidell, S., K. Pearsons, and B.G. Tabachnick (1995). Field Study of Noise-Induced Sleep Disturbance, *Journal of the Acoustical Society of America*, **98**, 1025–1033.

Fidell, S., K. Pearsons, B.G. Tabachnick, and R. Hower (2000). Effects of Sleep Disturbance of Changes in Aircraft Noise Near Three Airports, *Journal of the Acoustical Society of America*, **107**, 2535–2547.

Fields, J.M. (1993). Effect of Personal and Situational Variables on Noise Annoyance in Residential Areas, *Journal of the Acoustical Society of America*, **93**, 2753–2763.

Fields, J.M. (1994). *A Review of an Updated Synthesis of Noise Annoyance Relationships*, Georgia Institute of Technology, Atlanta.

Finkelman, J.M., L.R. Zeitlin, R.A. Romoff, M.A. Friend, and L.S. Brown (1979). Conjoint Effect of Physical Stress and Noise Stress on Information Processing Performance and Cardiac Response, *Human Factors*, **21**, 1–6.

Glass, D.C. and J.E. Singer (1972). *Urban Stress*, Academic Press, New York.

Green, K.B., B.S. Pasternack, and R.E. Shore (1982). Effects of Aircraft Noise on Hearing Ability of School-Age Children, *Arch Environ Health*, **37**, 284–289.

Griefahn, B. and E. Gros (1986). Noise and Sleep at Home, A Field Study on Primary and After-Effects, *Journal of Sound and Vibration*, **105**, 373–383.

Griefahn, B., A. Schuemer-Kohrs, R. Schuemer, U. Moehler, and P. Mehnert (2000). Physiological, Subjective, and Behavioural Responses During Sleep to Noise from Rail and Road Traffic, *Noise & Health*, **3**, 59–71.

Haines, M.M., S.A. Stansfeld, R.F.S. Job, B. Berglund, and J. Head (2001). Chronic Aircraft Noise Exposure, Stress Responses, Mental Health and Cognitive Performance in School Children, *Psychological Medicine*, **31**, 265–277.

Haines, M.M., S.A. Stansfeld, J. Head, and R.F.S. Job (2002). Multi-Level Modelling of Aircraft Noise on Performance Tests in Schools around Heathrow Airport London, *Journal of Epidemiology and Community Health*, **56**, 139–144.

Haines, M.M., S.A. Stansfeld, R.F.S. Job, and B. Berglund (1998). Chronic Aircraft Noise Exposure and Child Cognitive Performance and Stress. In: *Proceedings of the 7th International Congress on Noise as a Public Health Problem* (N. Carter and R.F.S. Job, eds), Noise Effects '98 PTY Ltd, Sydney, 329–335.

Hamilton, P., G.R.J. Hockey, and M. Rejman (1977). The Place of the Concept of Activation in Human Information Processing Theory: An Integrative Approach. In: *Attention and Performance* (S. Dornic, ed), Erlbaum, Hillsdale, 463–486.

Harder J., C. Maschke, and H. Ising (1999). *Längsschnittstudie zum Verlauf von Stressreaktionen unter Einfluss von nächtlichem Fluglärm*, Umweltbundesambt, Berlin.

Hockey, G.R.J. (1979). Stress and the Cognitive Components of Skilled Performance. In: *Human Stress and Cognition* (V. Hamilton and D.M. Warburton, eds), Chichester, Wiley, 141–178.

Hofman, W.F., A. Kumar, and J.H.M. Tulen (1995). Cardiac Reactivity to Traffic Noise During Sleep in Man, *Journal of Sound and Vibration*, **179**, 577–589.

Horne, J.A., F.L. Pankhurst, L.A. Reyner, K. Hume, and I.D. Diamond (1994). A Field Study of Sleep Disturbance: Effects of Aircraft Noise and Other Factors on 5,742 Nights of Actimetrically Monitored Sleep in a Large Subject Sample, *Sleep*, **17**, 146–159.

Hume, K.I., F. Van, and A. Watson (2003). *Effects of Aircraft Noise on Sleep: EEG-Based Measurements*, Manchester Metropolitan University, Manchester.

Hygge, S., E. Boman, and I. Enmarker (2003). The Effects of Road Traffic Noise and Meaningful Irrelevant Speech on Different Memory Systems, *Skandinavian Journal of Psychology*, **44**, 13–21.

Hygge, S., G.W. Evans, and M. Bullinger (2002). A Porospective Study of Some Effects of Aircraft Noise on Cognitive Performance in Schoolchildren, *Psychological Science*, **13**, 469–474.

Ising, H. (1983). *Stressreactionen und Gesundheitsrisiko bei Verkehrslärmbelastung*, Institut für Wasser-, Boden- und Lufthygiene des Bundesgesundheitsamtes, Berlin.

Ising, H., W. Babisch, and T. Günther (1999). Work Noise as a Risk Factor in Myocardial Infarction, *Journal of Clinical Basic Cardiol*, **2**, 64–68.

ISO-9921 (2003). *Ergonomics – Assessment of Speech Communication*, International Organization for Standardization, Geneva.

Job, R.F.S. (1988). Community Response to Noise: A Review of Factors Influencing the Relationship between Noise Exposure and Reaction, *Journal of the Acoustical Society of America*, **83**, 991–1001.

Jones, D. (1990). Progress and Prospects in the Study of Performance in Noise. In: *Proceedings of the 5th International Congress on Noise as a Public Health*

Problem (B. Berglund and T. Lindvall, eds), Swedish Council for Building Research, Sweden.

Jones, D.M. and W.J. Macken (1998). Acoustic Determinants of Auditory Distraction by Irrelevant Sound. In: *Proceedings of the 7th International Congress on Noise as a Public Health Problem* (N. Carter and R.F.S. Job, eds), Noise effects, Sydney.

Kryter, K.D. (1982). Community Annoyance from Aircraft and Ground Vehicle Noise, *Journal of the Acoustical Society of America*, **72**, 1222–1242.

Kryter, K.D. (1983). Response of K.D. Kryter to Modified Comments by Th.J. Schultz on K.D. Kryter's paper "Community Annoyance from Aircraft and Ground Noise," *Journal of the Acoustical Society of America*, **73** (3), 1066–1068.

Lambert, J. and M. Vallet (1994). Study Related to the Preparation of a Communication on a Future EC Noise Policy, Final Report, Inrets, Bron.

Langdon, F.J. (1976). Noise Nuisance Caused by Road Traffic in Residential Areas: Part III, *Journal of Sound and Vibration*, **49**, 241–256.

Maschke, C. (2003). Epidemiological Research on Stress Caused by Traffic Noise and its Effects on High Blood Pressure and Psychic Disturbances, In: *Proceedings of the 8th International Congress on Noise as a Public Health Problem* (R.D. Jong, T. Houtgast, E.A.M. Franssen, and W. Hofman, eds), Schiedam, the Netherlands: Foundation ICBEN, 96–101.

Maschke, C., D. Arndt, and H. Ising (1995) Nächtlicher Fluglärm und Gesundheit: Ergebnisse von Labor- und Feldstudien, *Bundesgesundhbl.*, **38** (4), 130–137.

Maschke, C., H. Ising, and K. Hecht (1997). Schlaf-nächlicher Verkehrslärm – Stress – Gesundheit: Grundlegen und aktuelle Forschungsergebnisse, *Bundesgesundhbl.*, **3**, 86–95.

McKennel, A.C. (1963). *Aircraft Noise Annoyance around London (Heathrow) Airport*, Central Office of Information, London.

Miedema, H.M.E. (2004). Relationship between Exposure to Multiple Noise Sources and Noise Annoyance, *Journal of the Acoustical Society of America*, **116**, 949–957.

Miedema, H.M.E. and C.G.M. Oudshoorn (2001). Annoyance from Transportation Noise: Relationships with Exposure Metrics DNL and DENL and Their Confidence Intervals, *Environmental Health Perspectives*, **109**, 409–416.

Miedema, H.M.E. and H. Vos (1998). Exposure Response Relationships for Transportation Noise, *Journal of the Acoustical Society of America*, **104**, 3432–3445.

Miedema, H.M.E. and H. Vos (1999). Demographic and Attitudinal Factors that Modify Annoyance from Transportation Noise, *Journal of the Acoustical Society of America*, **105**, 3336–3344.

Miedema, H.M.E. and H. Vos (2003). Noise Sensitivity and Reactions to Noise and Other Environmental Conditions, *Journal of the Acoustical Society of America*, **113**, 1492–1504.

Miki, K., K. Kawamorita, Y. Araga, T. Musha, and A. Sudo (1998). Urinary and Salivary Stress Hormone Levels While Performing Arithmetic Calculation in Noise Environment, *Industrial Health*, **36**, 66–69.

Millar, K. (1979). Noise and the 'Rehearsal-Masking Hypothesis,' *British Journal of Psychology*, 70, 565–577.

Moch-Sibony, A. (1984). Study of Effects of Noise on Personality and Certain Psychomotor and Intellectual Aspects of Children, After a Prolonged Exposure, *Travail Humane*, **47**, 155–165.

Moran, S.V., W.J. Gunn, and M. Loeb (1981). Annoyance by Aircraft Noise and Fear of Overflying Aircraft in Relation to Attitudes Toward the Environment and Community, *Journal of Audiological Research*, **21**, 217–225.

Park, J.F., jr. and M. Carr Payne jr. (1963). Effects of Noise Level and Difficulty of Task in Performing Division, *Journal of Applied Psychology*, **47**, 367–368.

Passchier-Vermeer, W. (1993). Noise and Health, Report number A93/02E, Den Haag, Gezondheidsraad.

Passchier-Vermeer, W. (2003). *Night-Time Noise Events and Awakening*, Netherlands Organisation for Applied Research TNO, Delft.

Passchier-Vermeer, W., B. Griefahn, and U. Moehler (2005). *Motility and Road and Railway Traffic Noise: Secondary Analysis of German Data*, Netherlands Organisation for Applied Research TNO, Delft.

Passchier-Vermeer, W., H. Vos, J.H.M. Steenbekkers, F.D. van de Ploeg, and K. Groothuis-Oudshoorn (2002). *Sleep Disturbance and Aircraft Noise Exposure. Exposure–Effect Relationships*, Netherlands Organisation for Applied Research TNO, Leiden.

Pearsons, K.S., D.S. Barber, B.G. Tabachnick, and S. Fidell (1995). Predicting Noise-Induced Sleep Disturbance, *Journal of the Acoustical Society of America*, **97**, 331–338.

Pearsons, K., R. Bennett, and S. Fidell (1973). Effects of Cessation of Late-Night Landings on Behavioural Awakening, Report no. 2439, BNN Systems and Technologies Corporation, Canoga Park.

Rosenlund, M., N. Berglind, G. Pershagen, L. Jarup, and G. Bluhm (2001). Increased Prevalence of Hypertension in a Population Exposed to Aircraft Noise, *Occupational and Environmental Medicine*, **58**, 769–773.

Rotton, J. (1990). Individuals under Stress. In: *Social Psychology: Living with People* (C.E. Kimble, ed.), W.C. Brown, New York, 505–543.

Schultz, Th.J. (1978). Synthesis of Social Surveys on Noise Annoyance, *Journal of the Acoustical Society of America*, **64**, 377–405.

Smith, A.P. (1982). The Effects of Noise and Task Priority on Recall of Order and Location, *Acta Psychologica*, **51**, 245–255.

Smith, A.P. (1991). A Review of the Non-Auditory Effects of Noise on Health, *Work & Stress*, **5**, 49–62.

Smith, A. (1998). Noise, Neurotransmitters and Performance. In: *Proceedings of the 7th International Congress on Noise as a Public Health Problem*, Noise effect, Sydney, 346–351.

Stansfeld, S.A., B. Berglund, C. Clark, I. Lopez-Barrio, P. Fischer, E. Öhrström, M.M. Haines, J. Head, S. Hygge, I. van Kamp, B.F. Berry (2005). Aircraft and Road

Traffic Noise and Children's Cognition and Health: A Cross-National Study, *The Lancet*, **365**, 1942–1949.

Stansfeld, S.A. (1992). Noise, Noise Sensitivity and Psychiatric Disorder: Epidemiological and Psychophysiological Studies, *Psychological Medicine Monograph Supplement*, **22**, 1–44.

Tafalla, R.J. and G.W. Evans (1997). Noise, Physiology, and Human Performance: The Potential Role of Effort, *Journal of Occupational Health Psychology*, **2**, 148–155.

Vallet, M., J.M. Gagneux, and V. Blanchet (1983). Long Term Sleep Disturbance Due to Traffic Noise, *Journal of Sound and Vibration*, **90**, 173–191.

Vernet, M. (1979). Effect of Train Noise on Sleep for People Living in Houses Bordering the Railway Line, *Journal of Sound and Vibration*, **66**, 483–492.

Weinstein, N.D. (1974). Effects of Noise on Intellectual Performance, *Journal of Applied Psychology*, **59**, 548–554.

Weinstein, N.D. (1977). Noise and Intellectual Performance: A Confirmation and Extension, *Journal of Applied Psychology*, 62, 104–107.

Wilkinson, R.T. and K.B. Campbell (1984). Effects of Traffic Noise on Quality of Sleep: Assessment by EEG, Subjective Report, or Performance the Next Day, *Journal of the Acoustical Society of America*, **75**, 468–475.

Wilding, J.M. and N.K. Mohindra (1983). Noise Slows Phonological Coding and Maintenance Rehearsal: An Explanation of Some Effects of Noise on Memory. In: *Proceedings of the 4th International Congress on Noise as a Public Health Problem* (G. Rossi, ed.), Centro Ricerche e Studi Amplifon, Milan, 759–771.

Woodhead, M.M. (1964). The Effect of Bursts of Noise on an Arithmetic Task, *American Journal of Psychology*, 77, 627–633.

Threats from Car Traffic to the Quality of Urban Life
Tommy Gärling and Linda Steg (Editors)

5

WAYFINDING IN URBAN ENVIRONMENTS

Gary L. Allen, Department of Psychology, University of South Carolina, Columbia, SC, USA
Reginald G. Golledge, Department of Geography, University of California, Santa Barbara, CA, USA

ABSTRACT

Human wayfinding in urban environments is a common yet important and relatively complex activity that occurs when urban structure meets individual cognitive and behavioral capabilities. Accordingly, this review begins with a quick overview of city structure, delves into the cognitive realm of spatial knowledge, and concludes with an examination of wayfinding in urban settings, including a look at situations in which orientation does not work well. The point is that urban traffic patterns, pedestrian habits, and, above all, efforts to change these must be anchored to some extent in an understanding of the cognitive and behavioral bases for our movement.

INTRODUCTION

This volume is concerned with car traffic and the problems, both the real and potential, associated with transportation of this type. Car traffic as a phenomenon is a function of many factors, some environmental, some psychological, and some sociological. We contend that an understanding of and an assessment of wayfinding performance in urban environments is essential to an evaluation of alternative means of transportation. We likewise contend that poor wayfinding among car drivers may result in unnecessary driving. Thus, this chapter addresses urban wayfinding. It begins with consideration of city structure, includes a review of spatial cognition relevant to wayfinding, and concludes with a discussion of wayfinding behavior.

CITY STRUCTURE

Evolution of Contemporary Structure

Today's urban places are structured around the concept of accessibility. In the late 19th century, steam driven locomotion, particularly by rail, began the process of suburbanization. People could now live well outside the city proper but still commute for work or other purposes in a reasonable time on a daily basis. The advent of the motorcar in the early 20th century accelerated this process remarkably. As a consequence, many of the cities that grew to substantial size in the 20th century have structures that were significantly different from those of the early periods of urban history where safety and government were dominant city-forming activities.

In the 20th century, city structure changed remarkably. By the mid-20th century, no longer were cities dominated by a monocentric structure in which business, commerce, government, education, religion, and other city-forming activities concentrated in the most accessible central area of the city. Perhaps, the first significant functions to decentralize were shopping and commerce. As suburbanization increased, the demand for suburban shopping escalated. At first, suburban centers were concentrated along railway lines and, later, along major arterials such as intercity highways. In the early stages, the monocentric structure of the city had produced what sociologists and geographers had termed the "concentric zonal" city (see sociologists Park *et al.*, 1925 and geographers Mayer and Kohn, 1960). The Industrial Revolution of the 19th century also had its impact as the large-scale movement of raw materials and finished goods focused much urban growth and activity along lines of transportation. This produced what land economist Hoyt (1939) termed a "wedge and sector" city (Figure 1).

In this city, a smaller central core remained as the hub of business, commerce and government, and segments of heavy and lighter industry, wholesaling, and transportation related activities drove wedges through the surrounding residential neighborhoods. However, as suburbanization increased, the demand for better access to facilities that provided for daily needs, such as food and other shopping, produced the first suburban shopping centers (Golledge and Stimson, 1997). Thereafter, a more complex pattern of decentralized planned shopping centers evolved. These varied from small centers with 5–20 functions (average area 50,000 square feet) built primarily to serve neighborhood customers, to larger centers (20–50 functions – community level centers, average 150,000 square feet), and finally a larger regional center (over 50 functions and averaging 500,000 square feet) that dominated a variety of communities and, in time, competed with the central business district for commercial dominance. In the latter part of the 20th century, other innovations, particularly in shopping (the freestanding one stop shopping center or big box store), developed, providing yet another significant focus of travel and interaction.

The result has been that many cities have evolved into what is termed a "multiple nuclei structure" or "polycentric" city. In these cases, many types of retailing – such as

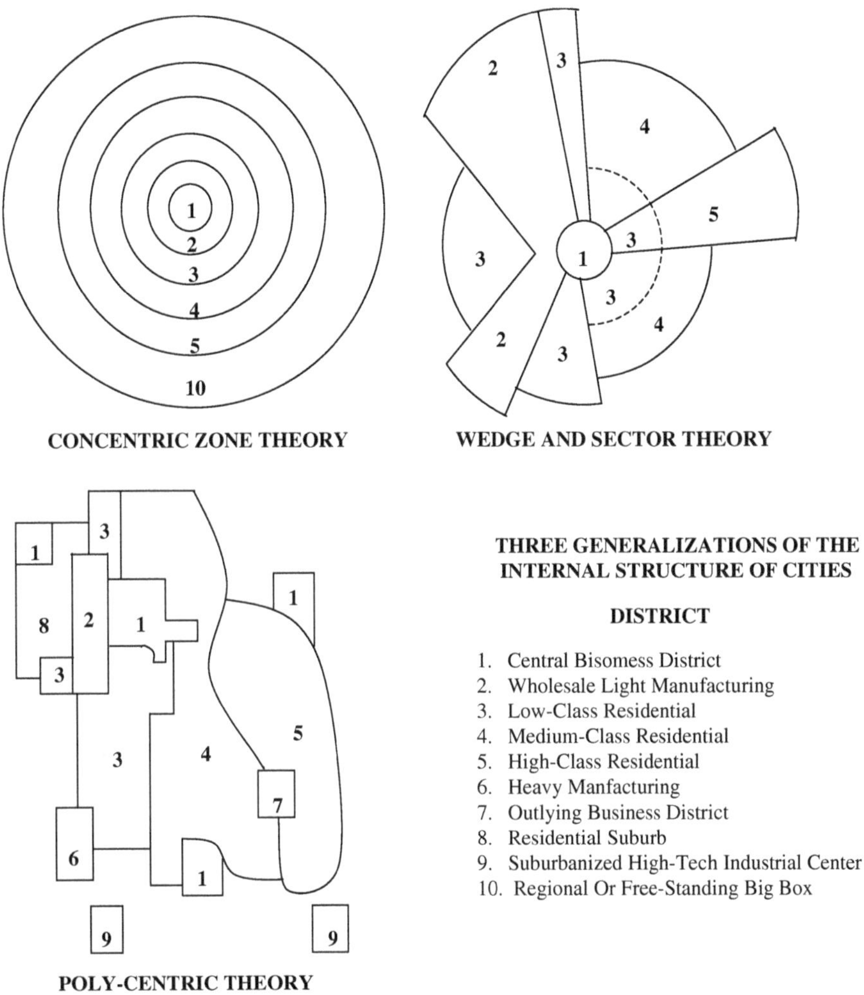

Figure 1: Generalizations of the Internal Structure of Cities

supermarkets, furniture, hardware, appliances, and vehicle sales – substantially departed from the central core area in response to consumer demands for more accessible places to shop. City cores remain the primary locations for government, business headquarters, law and order, religion, and those retailing activities that require extremely large-threshold populations to support them, such as high fashion clothing and jewelry shops and eating, drinking, and entertainment activities that serve the central city workforce (King and Golledge, 1978).

Models of Structure

Geographers have produced a series of descriptive models of city structure. These have been based on examination of the location and distribution of city elements including all types of land use, populations, and transportation systems (Berry, 1972).

A second line of research has related to individual behavior within urban environments, and includes a variety of purpose-driven activities such as journey to work, journey to shop, journey to school, search for new residence, recreation and leisure, search for health professional services, medical services, business services, and so on. An implicit assumption that lay behind the rationale for this approach was that, by identifying the arrangement of land uses in urban areas, general structural models can be developed, and these models can be used to help explain patterns of human spatial behavior. In other words, the assumption was that the behavior of urban residents would be constrained and partly determined by the nature of the city's physical structure.

By the late 1960s and early 1970s, planners and behavioral researchers in geography had begun to determine the geometry that lay behind people's perceptions of urban systems. The geometric properties of city structure initially concentrated on well-known points (e.g., landmarks). As popular city images were investigated and represented in greater detail by planners Lynch (1960) and Appleyard (1970), linear and polygonal facets of city structure were articulated. This was followed by an attempt to decompose cities into component parts (e.g., neighborhoods) with the major impetus coming from planners who desired to make planning decisions that were more person-responsible (Zannaras, 1968). In other words, cities were subdivided into arbitrarily defined regions (neighborhoods and communities), and planners and geographers began to investigate what people thought were the dominant features and boundaries of their own and other neighborhoods, communities, and regions (Lee, 1970).

Except for the pioneering work of Zannaras (1973), few attempts were made to look at cognitions of the entire spatial structure of cities. This is different from the work of Lynch and others who "assembled" city maps from the perceived point, line, and area components of perceived structures. Zannaras (1973; Zannaras and Golledge, 1974) concentrated on examining whether the physical structural models derived from objective examination of the geometry of urban land uses, population characteristics, and transport networks was structurally reflected in the perceptions of people, and if they could be related to specific types of behavior. Their assumptions were that individuals' conceptual structures of any specific urban area would conform to the physical models that had been derived from objectively analyzing locations of elements of the city.

In the United States, many of the "newer" cities (i.e., those that have developed largely in the 19th and 20th centuries) have adopted a single geometric structural form – based on a rectangular array of interconnecting streets, roads, and highways. The single most dominant area is still usually located at the core of the city but as growth continues in different directions, spider-like arms of development reach into the countryside while interstitial infilling proceeds at a much slower rate. Limited-access highways usually surround and bisect cities. Planners and designers have attempted to superimpose ideal structures on many of the original city forms (e.g., Appleyard, 1970), and unrestricted growth in any given direction is no longer an option.

Structure and Activity

By examining the relevant literature from planning, architecture, design, geography, sociology, and economics, a composite list of urban features that appear to influence the construction of cognitive images can be constructed (Table 1).

City dwellers are sometimes faced with a complex task in order to complete a schedule of daily activities. For those who work in intercity or international commerce, access to the center of business activity is still a dominant force in their daily planning. Consequently, knowledge of large cities is often constrained to corridors that can be visually experienced while traveling along major interstate or state-controlled-access highways between their residential neighborhood and their work environment (Appleyard *et al.*, 1964; Carr and Schissler, 1969).

Today's urban dweller, therefore, to satisfy a schedule of daily activities, must undertake a significant variety of trips to different parts of the city. Travel and the time it

Table 1: Universal Cues in Urban Environments

1. Shopping centers	17. Neon lights in business areas
2. Railroad crossings and freeway overpasses	18. Rivers, streams, or other water bodies
3. Direction signs	19. Hills
4. School buildings	20. Any freeway system
5. Banks	21. Number and spacing of freeway exits
6. Churches	22. Individual buildings (architectural style)
7. Movie theatres	23. Public buildings (functional characteristics)
8. Restaurants	24. Residential quality changes (housing submarket changes)
9. Open areas such as parks or green spaces	25. Residential density changes (housing spacing)
10. Speed limit signs	26. Smog
11. The city skyline	27. Increased density of multi-residential buildings
12. Traffic congestion	28. Major department stores
13. Traffic lights	29. Slums
14. Street width changes	30. Construction work
15. Billboards	31. Places of cultural significance
16. Statues	32. Transport terminals

takes to travel are important factors in choice of residence, choice of educational facilities (e.g., elementary, junior high, and high school location), and choice of that subset of the housing market that provides appropriate neighborhood quality for home life.

Summary

Over time, city structure has evolved from a simple monocentric concentric-zoned city to a much more complex polycentric structure, and, as a consequence, transportation systems and the relative importance of different types of landmarks and nodes have changed considerably. The impact of city structure on human behavior has become much more pronounced. However, despite the efforts of academics and planning professionals to clarify city structures, most people are unaware of the type of structure in which they live and operate. Rather, they develop cognitive maps that reflect their common needs to communicate and their idiosyncratic needs for daily activity completion.

SPATIAL COGNITION

Spatial Knowledge

Knowledge of urban structure, the outcome of the cognitive system in which most geographers, psychologists, and environmental planners are interested can take several forms. At a very basic level is the ability to recognize or recall specific places in the environment when they are encountered. Siegel and White (1975) referred to this as landmark representations. Acquiring and refining landmark representations is largely a matter of perceptual learning (Gibson, 1979), with the outcome being increasing differentiation with accumulated experience, provided there is sufficient motivation for such differentiation. In its crudest form, this knowledge involves only vague feelings of familiarity having encountered a place (or similar places) previously. In its most sophisticated form, landmark knowledge includes specific identification of a place.

Landmark knowledge is often embedded in another type of environmental representation, either route or configurational knowledge (Shemyakin, 1962; Siegel and White, 1975; O'Keefe and Nadel, 1978). Route knowledge is basically sequential information about behaviors to be performed in response to environmental cues or objects. Thus, whether or not the content of route knowledge is considered spatial per se, it results in the traveler being conveyed from one point in the environment to another. Learning a route is largely a matter of associative learning, that is, a matter of learning actions to be performed when certain conditions are experienced. As Hart and Berzok (1982) have pointed out, the first order of business in terms of route learning is ordinal mapping, or getting a sequence in the correct order. This is followed next by interval mapping, as relative distances between cues are established, and then, if necessary, by accurate mapping, which includes direction information (see Allen, 1982). In its most crude form, route knowledge consists of a vague serially bound set of actions to be performed, with the traveler confined to the immediate task of identifying the next landmark in the series and performing the correct response at that juncture. In its most sophisticated and well-practiced form, route knowledge involves an efficient package of condition–action pairs in which each pair primes the next in sequence automatically, the result being a flawless and low-effort trip from point of origin to destination.

Configurational knowledge (Siegel and White, 1975) is sometimes used synonymously with "cognitive map" (O'Keefe and Nadel, 1978), which consists of information specifying inter-location distance and direction relations for a set of places, as well as transformation rules that relate particular views of the environment to that overall relational structure. Configurational knowledge of this type is inherently spatial and has presumably evolved specifically to guide wayfinding efforts. Configurational learning is a type of pattern learning in which spatial relations among places are acquired (Tolman, 1948; O'Keefe and Nadel, 1978). In its least sophisticated form, a cognitive map consists of only a few prominent landmarks within a coordinated scheme, knowledge that can provide orientation in a general sense but leaves much of the environment undifferentiated. In it most well-developed form, the cognitive map provides an integrated, hierarchically organized network of place information. As Golledge (1993) indicated, the articulated cognitive map includes not only information about individual occurrences such as specific places and the temporal-spatial links between those, but also information about the distribution of occurrences, their connectivity, and their conceptual or semantic organization. This type of knowledge is much more sophisticated than landmark representations, involving a network of places and relations among them (Kitchin, 1996).

There has been theoretical debate concerning the relation between route knowledge and configurational knowledge. Siegel and White (1975), building on work by Shemyakin (1962) and Piaget and Inhelder (1956; Piaget *et al.*, 1960), maintained that configurational knowledge acquired through direct experience (rather than through viewing maps or models) was obtained only through a main sequence that began with landmark knowledge, continued through route knowledge, and finally resulted in an integrated cognitive map. In contrast, O'Keefe and Nadel (1978) posited route knowledge and configurational knowledge to be the result of different neurologically based learning mechanisms, route knowledge resulting from associative learning and cognitive maps resulting from place learning. Montello (1998) advanced an account that, as with O'Keefe and Nadel's (1978) theory, involves simultaneous nonmetric (landmark, route) and metric (configurational) knowledge acquisition, with no stage-like progression from one to the other. Research findings comparing these views have yielded somewhat equivocal results (e.g., Devlin, 1976; Holding and Holding, 1989; Magliano *et al.*, 1995). It is clear that, given sufficient peripheral or distal environmental cues, it is possible to learn a given place or even a set of places without first learning a particular pattern of movement to arrive at that place or to travel between the set of places. It is also clear, however, that travelers in a city often learn links between places and establish connectivity among them through movement (i.e., by learning routes) prior to getting "the big picture" (i.e., an integrated representation of how those places are related spatially).

Not all places within an area are equally well known, and this is particularly true of urban structure, where environmental legibility and frequency of travel commonly influence type and extent of spatial knowledge. Certainly, given the suburban structure referred to earlier, it is likely that a resident's spatial knowledge will include local configurations as well as route representations. Descriptively, knowledge of a city is probably something of a collage, as Tversky (1993) has pointed out, well defined and systematic in some places but sketchy and ad hoc in others.

Conceptual Knowledge

Spatial concepts play an important role in how spatial relations are represented in memory. Most frequently, these concepts are discussed as geometries. Topological space refers to relations based on proximity, enclosure, or dispersion. Projective space refers to straight-line relations, as in line of sight or pointing. Euclidean space basically refers to the concept of objects existing in coordinate planes consisting of points so that angles and distances may be determined. A very influential developmental point of view has traditionally held that the human infant is born into egocentrism and, as a function of neurological maturation and appropriate experience, first acquires the concept of topological space, and then the concepts of projective space and Euclidean space over the course of childhood (Piaget and Inhelder, 1956; Piaget *et al.*, 1960; Hart and Moore, 1973; Siegel and White, 1975). In contrast, classic cognitive mapping theory (O'Keefe and Nadel, 1978) posits a more nativist view, essentially that a Euclidean concept of space, essential to cognitive mapping, is part of human evolutionary heritage that emerges, similar to language, early in life. Recent empirical and conceptual work from developmental psychology has indicated that early emergence of spatial concepts appears to be a distinct possibility (Newcombe and Huttenlocher, 2000; Newcombe and Sluzenski, 2004).

Other concepts that are not inherently spatial play a significant role in the representation of spatial relations (Piaget and Inhelder, 1956; Piaget *et al.*, 1960; Liben, 1991). Class inclusion, basically the idea of hierarchical category organization, is fundamental to spatial thinking, as indicated by the next section dealing with hierarchical organization of spatial knowledge. The concepts of addition/subtraction and proportionality are also commonly applied to distances, as is the idea of transitive reasoning (i.e., if A is beyond B and B is beyond C, then A is beyond C). As with basic concepts of space, there is current debate as to whether these and other concepts are inborn at birth or whether they develop over childhood as the result of neural maturation and relevant experience (Newcombe and Huttenlocher, 2000). In adults, however, there is no doubt that these concepts are fully functional and exert clear influence on spatial reasoning when it comes to travel planning.

Hierarchical Organization

As information in knowledge structures increases, hierarchical organization tends to emerge spontaneously, and spatial knowledge structures are no exception (see McNamara, 2003). Increased comprehension of urban settings is reliably accompanied by an increase in hierarchically organized knowledge of city structure (Golledge and Spector, 1978). Evidence of such hierarchies is present in both route (Allen, 1981; Jansen-Osmann and Berendt, 2005) and configurational knowledge (Hirtle and Jonides, 1985; Stevens and Coupe, 1978). In route knowledge, hierarchical structure is clear in the representation of routes as consisting of distinct segments based on perceptual or conceptual similarity. It is also evident in the use of landmarks embedded in routes that guide wayfinding efforts. In configurational knowledge, hierarchical structure typically takes the form of regional organization, with smaller areas represented within larger areas (e.g., a home area, a neighborhood, a suburban area, a larger metropolitan region). Nonspatial hierarchical knowledge about geographic entities also plays an important role in reasoning about relations among places (McNamara, 1986). Reasoning on the basis of these nonspatial inclusion relations, we can reach certain conclusions about spatial relations (see Golledge and Stimson, 1997).

Anchorpoint theory (Golledge, 1978; Couclelis *et al.*, 1987) provides a powerful functional approach to the hierarchical organization of spatial knowledge and its concomitant effects. According to this view, certain locations within the environment are functionally significant to the individual – his or her home, place of work, shopping destinations, and other frequently visited sites. These places come to serve as primary nodes, or anchorpoints, and the links among them as primary pathways in residents' spatial representations. Sensibly, these primary anchorpoints come to dominate and define surrounding regions. Subordinate to these primary nodes and pathways are secondary and tertiary nodes and pathways, based on importance and frequency of travel. Together, the spatial network involving the nodes and pathways at various levels form a skeletal structure, which essentially is the basis of wayfinding efforts (see Golledge and Stimson, 1997).

Summary

Knowledge of actual settings is in the form of landmark representations, route representations, and configurational representations. In general, these representations can be strongly influenced by an observer's conceptual knowledge. Spatial knowledge, as with other forms of knowledge, tends to be organized hierarchically. Conceptual knowledge, especially hierarchical organization, can lead to errors or distortions during wayfinding, as will be addressed subsequently.

WAYFINDING

Wayfinding Tasks

Wayfinding is basically destination-guided locomotion in instances in which the locomotor affordances between point of origin and destination (i.e., basically, where one can go and where one cannot go) cannot be perceived all at once (Blades, 1991; Allen, 1999; Gärling, 1999). Functionally, most wayfinding efforts fall into three categories. *Commuting* has a familiar point of origin and a familiar destination, as with trips to work and back. A great deal of urban travel involves this type of behavior. Basically, wayfinding of this type is within the network specified by anchorpoint theory. *Exploring* involves "foraging" for spatial information, as the traveler begins in familiar territory and makes his or her way into unfamiliar surroundings, always with the intent of ending up back on familiar ground. Exploring is a common activity in new surroundings; it is one way in which the second and tertiary nodes and pathways in anchorpoint theory are expanded. *Questing*, perhaps the most interesting of the three cognitively speaking, involves travel to a novel destination, one that is known to the traveler only through symbolic means such as verbal directions or a map. Wayfinding of this type is quite common in urban settings; it also represents an expansion of the nodes and networks described in anchorpoint theory.

Wayfinding Means

Primary or Intrinsic Means

Primary or intrinsic means of wayfinding refer to those that are based on travel experience. The psychological tools available to travelers to accomplish wayfinding task are several and varied (Allen, 1999). Least sophisticated is oriented search (Heth and Cornell, 1985), a pattern of movement that is designed to bring the traveler into contact with the destination without explicit knowledge of where that destination is situated. Search of this type is most effective when the space is small or greatly constrained, as when a traveler knows that a store is in a particular neighborhood. It is pressed into service typically in exploring or questing.

More useful is "piloting," or route-based locomotion, a product of associative learning in which specific movements are associated with landmarks (Allen, 1982). "Piloting," which can include both distal and local landmarks, is an effective means of wayfinding over short and long distances, but by itself it is dependent upon recognition of familiar environmental cues. If a traveler cannot locate a landmark, such as a bank on the corner or a high building in the distance, then the wayfinding effort is in jeopardy. Wayfinding of this type is often the basis of commuting and questing.

Another means of wayfinding is path integration and the closely related activity of dead reckoning (Loomis *et al.*, 1999). The process basic to this form of wayfinding is keeping track of prior movements, either implicitly or explicitly. The spatial pattern of movement provides a way of computing direction and distance to previously visited areas. Path integration is effective, but error accumulates over travel. Thus, its accuracy is constrained over long journeys. It is quite useful during exploration and can be of use during quests to unfamiliar sites.

A final intrinsic means involves wayfinding by cognitive map. As mentioned previously, cognitive maps include direction and, to some extent, distance (time) information among locations. Information of this type can, of course, be pressed into service in the face of any type of wayfinding task. For an effective wayfinding system, the cognitive map is complemented by path integration for tracking movements and a global frame of reference for maintaining orientation within the environment (Jonsson, 2002; Cornell and Heth, 2004). With a system of this type, a traveler can navigate through a city, invent new routes, and overcome barriers to travel (i.e., "The regular route is blocked, so we are going north, which means we will have to turn east or to our right, to get there.") It is ideal for exploring and for questing.

Secondary or Symbolic Means

Secondary or symbolic means refers to those that involve spatial depictions or verbal descriptions of the environment. Wayfinding, particularly quests in which the destination has never been visited, frequently involves symbolic representations of space. Maps are excellent means of providing information regarding both routes and configurations (see Lloyd, 1993; MacEachren, 1995). The American Automobile Association, for example, provides motorists with both in the form of a TripTik©, consisting of a route-type road map with travelers always going forward (bottom to top) in a near linear fashion, which opens to reveal a conventional road map (north at top) on which travelers must locate their orientation and direction.

Wayfinding by map can be very effective, but potential shortcomings are frequently overlooked. Maps are means of communication, and although there are some fundamental aspects of mapping that may be inherent in all users (Blaut, 1991), some aspects of the communication process are fairly complex, and not everyone is equally facile in map use (Liben, 1991; MacEachren, 1995; Allen, 2000a). Maps used for pubic transportation such as buses, subways, and commuter trains; maps frequently present route information in largely linear formats, with little consideration to the geometric accuracies of such depictions. Such maps are easy to use, but Euclidean spatial relations between stops and between stops and final destinations cannot be determined. Maps inevitably must omit detail and concentrate on the communicator's purpose. In wayfinding, many details are omitted and scale violated in order to provide the traveler with basic information. With maps showing spatial configurations, there is also the matter of the alignment of map, map user, and the environment – the "alignment effect" (Levine *et al.*, 1984; Rossano and Warren, 1989). Many airports, hospitals, universities, and inner city areas post maps for visitors without regard for how those maps correspond to the environment in which they are embedded. In many cases, "you-are-here" indicators are circles without heading information. Misalignment between maps and the represented environment and insufficient orientation information results in misorientation for many travelers.

Verbal directions are even more common than maps as wayfinding aids, available in electronic format over the Internet, in electronic format in some transit systems, and in informal conversation. Although verbal information can take the form of configurational descriptions (Taylor and Tversky, 1992), it comes most frequently in the form of route directions (Couclelis, 1996; Denis, 1997; Daniel *et al.*, 2003). Route directions routinely consist of directives, statements that require movement of the listener (essentially reducible to "go to" and "turn"), and descriptives, statements that portray the appearance of the environment (Allen, 2000b). As mentioned previously, changes from directives ("Go straight for three stoplights") to descriptives ("At the third light, a bank is right across from the McDonalds on the corner") are common and accomplished without difficulty (Tversky *et al.*, 1999). Directives frequently consist of a series of coordinated movements, usually activity-landmark ("Drive until you see a stop sign") or landmark-activity-landmark ("From the front of the library, continue walking until you see the Town Theatre") statements (Denis, 1997; Allen, 2000 b). Interestingly, good directions, like nodes and links in anchorpoint theory, can be described as a skeletal structure (Denis, 1997; Denis *et al.*, in press).

Route directions tend to be more effective if certain basic principles are followed. For example, the order of statements should conform to the actual series of actions to be performed in the environment (Allen, 2000b). Route directions are basically verbal instantiations of route knowledge, and as such, they are very sensitive to violations of sequence; if statements are out of order or if they are remembered out of sequence, disorientation can result. In addition, descriptives are better placed at choice points along routes, and those giving directions should establish "common ground" for communication in terms of terminology and modes of expression (Allen, 2000b). In urban settings, it is common to communicate route information using street names. However, such names are relatively difficult to remember and should be complemented with ample references to environmental landmarks (Fontaine and Denis, 1999; Tom and Denis, 2004). Not surprisingly, landmarks play a crucial role in orienting travelers and providing course-maintaining information in urban settings (Denis *et al.*, in press).

Wayfinding Errors

Disorientation

For our purposes, disorientation refers to instances in which the traveler simply does not know where to go next. Regardless of maps and descriptions and regardless of environmental knowledge, he or she does not perceive a familiar landmark and has no sense of which way to proceed. In short, the traveler is, for the moment, lost. This is typically a terrifying state of affairs, and the traveler must cope with fear in addition to dealing with disorientation (Jonsson, 2002).

There are several ways for wayfinding, particularly during exploring or questing efforts, to result in disorientation. Searching can go awry when the area is too large or when the traveler has inadequate landmark knowledge, most likely due to perceptual similarity of specific building, street corners, and urban architecture. For those travelers finding their way by landmark-based "piloting," the problem is typically that the sequence of landmarks has either not been well established or has been disrupted (e.g., by construction). Before ordinal mapping is complete, the sequence in which landmarks appear along a route is not set (Allen, 1982; Hart and Berzok, 1982). Thus, the traveler may take the wrong action at the wrong time, for instance, take a turn too early when traveling by car or exit at the wrong bus stop. Commuting efforts that are based either entirely or substantially on route knowledge are particularly susceptible to disruptions of sequence. If a familiar suburban-to-urban highway or a routinely used subway line is temporarily closed, many users face the prospect of disorientation in addition to inconvenience. Typically, travelers rely on alternatives provided by media, which sometimes are insufficiently elaborated or incorrect in part.

Disorientation in the event of path integration is likely when the path segments taken are lengthy or circuitous. As mentioned previously, error accumulates with each segment traveled. To some extent, error can be overcome through the use of an internal cognitive map or external cartographic map to aid wayfinding, but in urban settings, principally rectangular city structure makes crow-fly travel impossible and the third dimension makes landmarks hard to detect. In addition, heading and turns are difficult to detect accurately on public transport.

Wayfinding efforts based on a cognitive map can also lead to disorientation, particularly when the configuration of landmarks is not well known. The process of learning about a city's structure is typically ongoing, and as knowledge is acquired, various locations become better and better specified (Golledge *et al.*, 1976). Incomplete or partial knowledge of a city's configuration, as with inadequately sequenced route knowledge, can readily lead to errors, in this case, misidentification of distal landmarks or the lack of local landmarks where they are expected.

Misorientation

Typically, wayfinding errors of this type occur during exploring or questing efforts in individuals with a good sense of direction (i.e., considerable experience in applying a global frame of reference and in using path integration within a cognitive map). Misorientation refers to constant error in one's cognitive map or cognitive mapping process. Errors of this type are difficult to study experimentally, but self-reports and case studies are not uncommon in the older scientific literature. An excellent case in point involving an urban setting was provided by Jonsson (2002) in his account of his being misoriented in Köln, Germany. Although he was an accomplished wayfinder and an experienced traveler, Jonsson found that in Köln the sun "rose" in the west and the Rhein River "flowed" south. He was misoriented by a constant 180°. Fortunately, the entire downtown configuration was part of the illusion.

Such is not always the case. Misorientation is also the consequence of map misalignment, as mentioned previously. Often when publicly displayed maps are misaligned with the environment they represent, observers are improperly oriented within that environment. For example, after consulting a 180° misaligned map in an airport, a traveler sought his exit improperly to the left instead of to the right. After consulting a 90° misaligned map of a university campus, visitors seeking the Visitor's Center went in the wrong direction.

Direction Bias

This type of wayfinding problem, which involves a consistent displacement of direction, is similar in some ways to misorientation. However, the traveler need not have much of a cognitive map or sophisticated cognitive mapping skills in order to experience this type of bias. Basically, systematic distortion of this type is the result of hierarchical knowledge. For example, the city of Reno, Nevada, was remembered as being east of the city of San Diego, California in the United States, when, in fact, the city is west-northwest of San Diego (Stevens and Coupe, 1978). This bias in direction is in part the result of observer's knowledge of the state of Nevada is regarded as being East of the state of California in terms of American geography. In urban settings, knowledge that a particular destination is in a zone or area of the city can lead to similar biases in direction.

Another way in which hierarchical organization leads to directional bias is incorporated into models of anchorpoint theory (Golledge, 1978; Couclelis *et al.*, 1987). According to a "tectonic plate" model, distortion is on a whole area basis. Dislocation of a major anchorpoint draws represented locations toward the dislocated position. Accordingly to a "magnifying glass" model, there is basic contrast within areas. Features near the anchorpoint are represented as being significantly farther away from the anchorpoint, with less distortion as distance increases. The final possibility is a "magnet model", where there is basic assimilation within the area. Nearby cues are represented strongly toward the anchorpoint, with somewhat less "pull" as distance increases. In all three models, there is bias of both direction and distance. Just which model applies to a given situation depends upon an individual's experience with the area, the structure of the area, and the task that is to be performed.

Distance Bias

Perhaps, the most pervasive distortions in wayfinding involve distance bias because it applies to all types of task and to all means, tactics, and strategies for finding one's way. As in the case of direction bias, hierarchical organization of spatial knowledge can lead to distance bias. In short, locations within the same region are generally judged as closer in distance than are locations that cross regional boundaries (Allen, 1981; Jansen-Osmann and Berendt, 2005).

In addition to this type of bias, there is a number of other effects known to influence judgments of distance. Gärling and Loukopoulos (in press) discussed how some of these might relate to choices between driving and walking. Perhaps the best-known bias of this type is the route-angularity effect, which states that routes with more turns tend to be judged as longer than routes of equal length that have fewer turns (Sadalla and Montello, 1989; Montello, 1997). This effect is related in some cases to the feature-accumulation effect, which states that distance judgments increase as a function of the number of turns, intersections, and namable features along the way (Montello, 1997). Clearly, this can account for route-angularity effect in cases of turns, but it is also the case that namable features along a route tend to increase estimated distance on their own (Luria *et al.*, 1967).

Direction of travel effects has also been documented in studies of distance judgment. Sadalla *et al.* (1980) found that estimates of distance from a substantial landmark (anchor-point) to another building were smaller than estimates of the distance from the other building to the landmark. This brings to mind the "magnifying glass" (increased distance) and "magnet" (decreased distance) models of anchorpoint theory introduced earlier.

Relevant for urban wayfinding is the effect of effort on judged distance. In general, a more effortful route is judged to be more distant than is a less effortful route (Proffitt *et al.*, 2003). In terms of increasing pedestrian traffic vis-à-vis car traffic, this is a finding with significant implications.

Strategies and Tactics in Urban Wayfinding

Cities are complex phenomena, covering large spatial domains, and, by virtue of their built nature (based on demand for space and accessibility), have a pronounced vertical dimension – even in suburban situations. A direct result is that view sheds are restricted, often not extending beyond immediate body-centered perceptual space. Distant features are obscured at ground level, and upper levels of urban structures can be observed only if and when they arise beyond the immediate skyline. When traveling, environmental information is restricted perceptually to a limited geometry – the road ahead (until perceptual distortion narrows the view along the road), views down intersecting streets, and views of distant high-rise features or buildings on natural slopes or humanmade gradients. How does one travel in such circumstances?

When the location of a destination is known (as in commuting or questing) and when secondary or symbolic sources provide information regarding the relevant transportation network (mass or individual), then a navigation process can be used to lay out a travel plan. Technically, navigation is a process of laying out a fixed route between an origin and destination, then strictly following that route in order to successfully complete the travel plan. For some travel modes such as airplane and ship, strict adherence to the predetermined route is essential for safety conditions. Obviously, this process involves selection of network segments, choice points, turn (directional) specifics, and intermittent location checking using on- and off-road landmarks. Following a specified route thus obviates the need for knowledge of what is obscured by the urban landscape – whether it be phenomena a block or so to either side of the route or features that can be perceived at greater distances.

But travelers in urban settings do not always have the opportunity to determine a well-planned route and follow it. Usually, available information is piecemeal or fragmented. This is usually the case when sketches or verbal descriptions supply the basic information on which to develop a travel plan. Under these circumstances, navigation per se is unlikely, and the traveler must rely on wayfinding, which does not restrict the traveler to specified route segments and specified choice points. Instead, it can be influenced by a variety of means and strategies, as indicated previously, and can produce multiple alternate solutions to the same travel problem, particularly when guided by a cognitive map. For example, in a city dominated by a rectangular transport network, a traveler could elect to take a route that has the shortest leg first, the longest leg first, or to travel stepwise through the network, trying to approximate a diagonal path.

Wayfinding decisions can be based on tactics such as using freeways as much as possible, avoiding freeways, minimizing stop lights and stop signs, minimizing left turns (in the USA), choosing an aesthetic route, choosing a route that allows you to use specific attributes of your vehicle (e.g., cornering ability), and so on. But even given the many strategies that can be implemented to find one's way, there are still constraints to be overcome, such as deciding which segment to take, which choice to make at a waypoint, and how best to recall the outward journey so as to make the return trip safely, easily, and timely. In urban settings, travelers are often reliant on signs. These include near and distant landmarks, on- and off-route city structural features, written or graphic road signs, street addresses, and so on. This allows "piloting," or but it can also support path integration. However, unlike "piloting" or formal navigation, this process does not require the traveler to store (or recall) the exact route that was to be followed.

In the complex, variegated, three-dimensional world of the city, navigation practices are used by public (mass) transit and by selected business traffic (e.g., postal delivery systems and waste disposal collection). When using public mass transit, the traveler leaves most of the navigation problem to the transit provider, effectively limiting one's travel decisions to how to access pick-up points and how to get from a drop-off point to the final destination. Often, the result is that, between pick-up and drop-off, the city becomes a "black hole" with only fragmentary information about the intervening area being processed and stored in memory.

Summary

Wayfinding as destination-guided locomotion is an everyday activity for most of us. It includes a variety of tasks, including commuting, exploring, and questing, and can be accomplished through a variety of means (e.g., search, landmark-based "piloting," path integration, and route planning based on a cognitive map). Because of incomplete knowledge and a number of other factors involving the representation of the environment, all travelers are susceptible to certain error in wayfinding, including disorientation, misorientation, direction bias, and distance bias. In urban settings, wayfinding involves attempts to reach desired destinations efficiently through reliance on private vehicles, through use of public transit, and by means of pedestrian behavior. It typically involves knowledge-based choice of route, but urban structure and incomplete knowledge of the environment frequently impacts our choices.

CONCLUSION

Urban wayfinding occurs as individuals make their way through city structure. Urban structure has historically dictated wayfinding efforts, but clearly advances in public and private transportation have influenced city structure as well. In the late 20th century, as urban areas grew in size and complexity, polycentric structure emerged, particularly around cities where outward growth was possible. Highways and other components of the road system became increasingly important, and individuals in late-growth cities and areas came to rely to a large extent on individually owned automobiles for transportation to necessary and desired locations. It remains to be seen what impact 21st century advances in communication technology will have on urban structure; the need for transportation to the workplace may be reduced somewhat, but travel is still involved in obtaining most goods and services.

Today, wayfinding in urban settings can be a cognitively, perceptually, and physically challenging activity. For all wayfinding tasks, the destination cannot be perceived from the point of origin, buildings typically restrict viewpoints, and idiosyncratic factors (one-way streets) add to the challenge. For commuting, the idea is to establish a routine, whether by public or private transportation. Route knowledge is basic for "piloting" of this sort, and with repeated travel, commuting becomes more automatic and less reliant on cognitive resources. In exploring, the traveler enters unfamiliar territory for the purpose of identifying potential destinations and wayfinding affordances. This is typically done when an individual is new to an area. So, the new urban resident takes a train or bus to the end of the line, or a commuter leaves a known path to explore an alternative. A variety of wayfinding means (search, "piloting," path integration, and wayfinding by cognitive map) is available to the urban explorer, but each has some degree of risk. Individuals vary greatly in their ability and desire to explore. However, most residents engage in questing, that is, traveling to new destinations based on symbolic communication by map or verbal directions. Typically, this type of activity is based either on "piloting" or on wayfinding by cognitive map. Again, each has its shortcoming. On the one hand, "piloting" depends largely upon an unbroken series of landmarks, and yet buildings and areas can have similar appearances, and proper names (of streets, buildings, and squares) are not inherently memorable. Cognitive maps are usually works in progress, incomplete in some aspects, and vague in others. As with exploring, individuals vary considerably in their willingness to engage in questing beyond certain limits.

At some point in wayfinding tasks, the traveler is typically left to his or her own devices. It is interesting to note that under such circumstances, individuals tend to rely on what they know about the environment in addition to what they see. In that regard, it is worthwhile to keep in mind that distortions and biases are common in representing spatial relations. Incomplete or vague knowledge can result in disorientation, and being "out of phase" with one's cognitive map or with a cartographic map can result in misorientation. Hierarchical organization can lead to both direction and distance bias, and a variety of distance biases have been documented. Given such shortcomings, some observers may think it amazing that travelers ever reach their destinations. However, for typical urban wayfinding, people are rather good, and it useful to keep in mind that biases are to a large extent error-limiting phenomena, not errors in orientation.

REFERENCES

Allen, G.L. (1981). A Developmental Perspective on the Effects of Subdividing Macrospatial Experience, *Journal of Experimental Psychology: Human Learning and Memory*, 7, 120–132.

Allen, G.L. (1982). The Organization of Route Knowledge. In: *Children's Conceptions of Spatial Relationships* (R. Cohen, ed.), Jossey-Bass, San Francisco, 31–39.

Allen, G.L. (1999). Spatial Abilities, Cognitive Maps, and Wayfinding: Bases for Individual Differences in Spatial Cognition and Behavior. In: *Wayfinding Behavior: Cognitive Mapping and Other Spatial Processes* (R. Golledge, ed.), The Johns Hopkins University Press, Baltimore, 46–80.

Allen, G.L. (2000a). Men and Women, Maps and Minds: Cognitive Bases of Sex-Related Differences in Reading and Interpreting Maps. In: *Spatial Cognition: Foundations and Applications* (S.O. Nuallain, ed.), John Benjamins, Amsterdam, 3–18.

Allen, G.L. (2000b). Principles and Practices in Communicating Route Knowledge, *Applied Cognitive Psychology*, **14**, 333–359.

Appleyard, D. (1970). Styles and Methods of Structuring a City, *Environment and Behavior*, **2**, 100–117.

Appleyard, D., K. Lynch, and J. Myer (1964). *The View from the Road*, MIT Press, Cambridge, MA.

Berry, B.J.L. (1972). *City Classification Handbook*, Wiley Interscience, New York.

Blades, M. (1991). Wayfinding Theory and Research: The Need for a New Approach. In: *Cognitive and Linguistic Aspects of Geographic Space* (D. Mark and A. Frank, eds), Kluwer Academic, Dordrecht, The Netherlands, 137–165.

Blaut, J.M. (1991). Natural Mapping, *Transactions of the Instute of British Geographers*, **16**, 55–74.

Carr, S. and D. Schissler (1969). The City as a Trip: Perceptual Selection and Memory in the View from the Road, *Environment and Behavior*, **1**, 7–35.

Cornell, E.H. and C.D. Heth (2004). Memories of Travel: Dead Reckoning within the Cognitive Map. In: *Human Spatial Memory: Remembering Where* (G. Allen, ed.), Lawrence Erlbaum Associates, Mahwah, NJ, 191–215.

Couclelis, H. (1996). Verbal Directions for Way-Finding: Space, Cognition, and Language. In: *The Construction of Cognitive Maps* (J. Portugali, ed.), Kluwer, Dordrecht, 133–153.

Couclelis, H., R.G. Golledge, N.D. Gale, and W.R. Tobler (1987). Exploring the Anchor-Point Hypothesis of Spatial Cognition, *Journal of Environmental Psychology*, **7**, 99–122.

Daniel, M.-P., A. Tom, E. Manghi, and M. Denis (2003). Testing the Value of Route Directions through Navigational Performance, *Spatial Cognition and Computation*, **3**, 269–289.

Denis, M., P.-E. Michon, and A. Tom, A. (in press). Assisting Pedestrian Wayfinding in Urban Settings: Why References to Landmarks are Crucial to Direction-Giving. In: *Applied Spatial Cognition: From Research to Cognitive Technology* (G. Allen, ed.), Lawrence Erlbaum Associates, Mahwah, NJ.

Devlin, A. (1976). The "Small Town" Cognitive Map: Adjusting to a New Environment. In: *Environmental Knowing* (G.T. Moore and R.G. Golledge, eds), Dowden, Hutchinson, & Ross, Stroudsberg, PA, 58–66.

Denis, M. (1997). The Description of Routes: A Cognitive Approach to the Production of Spatial Discourse, *Current Psychology of Cognition*, **16**, 409–458.

Fontaine, S. and M. Denis (1999). The Production of Route Instructions in Underground and Urban Environments. In: *Spatial Information Theory: Cognitive and Computational Foundations of Geographic Information Science* (C. Freksa and D.M. Mark, eds), Springer, Berlin, 83–94.

Gärling, T. (1999). Human Information Processing in Sequential Spatial Choice. In: *Wayfinding Behavior: Cognitive Mapping and Other Spatial Processes* (R.G. Golledge, ed.), The John Hopkins University Press, Baltimore, MD, 81–98.

Gärling, T. and P. Loukopoulos (2007). Preference for Driving Related to Cognitive Distance. In: *Applied spatial cognition: From research to technology* (G.A. Allen, ed.), Hillsdale, NJ, Erlbaum, 3–24.

Gibson, J.J. (1979). *The Ecological Approach to Visual Perception*, Houghton-Mifflin, Boston.

Golledge, R.G. (1978). Learning about Urban Environments. In: *Timing Space and Spacing Time I: Making Sense of Time* (T. Carlstein, D. Parkes and N. Thrift, eds), Edward Arnold, London, 76–98.

Golledge, R.G. (1993). Geographical Perspective on Spatial Cognition. In: *Behavior and Environment: Psychological and Geographical Approaches* (T. Gärling and R.G. Golledge, eds), Elsevier Publishers, Amsterdam, 16–46.

Golledge, R.G., V.L. Rivizzigno, and A. Spector (1976). Learning About a City: Analysis by Multidimensional Scaling. In: *Spatial Choice and Spatial Behavior* (R. Golledge and G. Rushton, eds), The Ohio State University Press, Columbus, OH, 95–116.

Golledge, R.G. and A. Spector (1978). Comprehending the Urban Environment: Theory and Practice, *Geographical Analysis*, **10**, 403–426.

Golledge, R.G. and R.J. Stimson (1997). *Spatial Behavior: A Geographic Perspective*, Guilford Press, New York.

Hart, R. and M.A. Berzok (1982). A Problem-Oriented Perspective on Children's Representations of the Environment. In: *Spatial Abilities: Development and Physiological Foundations* (M. Potegal, ed.), Academic Press, New York, 147–169.

Hart, R. and G. Moore (1973). The Development of Spatial Cognition: A Review. In: *Image and Environment* (R. Downs and D. Stea, eds), Chicago: Aldine, 246–288.

Heth, C.D. and E.H. Cornell (1985). A Comparative Description of Representation and Process During Search. In: *Children's Searching: The Development of Search Skill and Spatial Representation* (H. Wellman, ed.), Lawrence Erlbaum Associates, Hillsdale, NJ, 215–250.

Hirtle, S.C. and J. Jonides (1985). Evidence of Hierarchies in Cognitive Maps, *Memory & Cognition*, **13**, 208–217.

Holding, C.S. and D.H. Holding (1989). Acquisition of Route Network Knowledge by Males and Females, *Journal of General Psychology*, **116**, 29–41.

Hoyt, H. (1939). *The Structure and Growth of Residential Neighborhoods in American Cities*, Washington, DC, Federal Housing Administration.

Jansen-Osmann, P. and B. Berendt (2005). What Makes a Route Appear Longer? An Experimental Perspective on Features, Route Segmentation, and Distance Knowledge, *Quarterly Journal of Experimental Psychology*, **58A**, 1390–1414.

Jonsson, E. (2002). *Inner Navigation*, Scribner, New York.

King, L.J. and R.G. Golledge (1978). *Cities, Space and Behavior*, Prentice-Hall, Englewood Cliffs, NJ.

Kitchin, R.M. (1996). Methodological Convergence in Cognitive Mapping Research: Investigating Configurational Knowledge, *Journal of Environmental Psychology*, **16**, 163–185.

Lee, T.R. (1970). Perceived Distance as a Function of Direction in the City, *Environment and Behavior*, **2**, 40–51.

Levine, M., I. Marchon, and G. Hanley (1984). The Placement and Misplacement of You-Are-Here Maps, *Environment and Behavior*, **16**, 139–157.

Liben, L.S. (1991). Environmental Cognition though Direct and Representational Experiences: A Life-Span Perspective. In: *Environment, Cognition, and Action: An Integrated Approach* (T. Gärling and G. Evans, eds), Plenum Press, New York, 245–276.

Lloyd, R. (1993). Cognitive Process and Cartographic Maps. In: *Behavior and Environment: Psychology and Geographical Approaches* (T. Gärling and R.G. Golledge, eds), North-Holland, London, 141–169.

Loomis, J.M., R.L. Klatzky, R.G. Golledge, and J.W. Philbeck (1999). Human Navigation by Path Integration. In: *Wayfinding Behavior: Cognitive Mapping and Other Spatial Processes* (R.G. Golledge, ed.), The John Hopkins University Press, Baltimore, MD, 125–151.

Luria, S.M., J.S. Kinney, and S. Weissman (1967). Distance Estimates with "Filled" and "Unfilled" Space, *Perceptual and Motor Skills*, **24**, 1007–1010.

Lynch, K. (1960). *The Image of the City*, The MIT Press, Cambridge, MA.

MacEachren, A.M. (1995). *How Maps Work: Representation, Visualization, & Design*, Guilford Press, New York.

Magliano, J.P., R. Cohen, G.L. Allen, and J.R. Rodrigue (1995). The Impact of a Wayfinder's Goal of Learning a New Environment: Different Types of Spatial Knowledge as Goals, *Journal of Environmental Psychology*, **15**, 65–75.

Mayer, H.M. and C.F. Kohn (1960). *Readings in Urban Geography*, University of Chicago, Chicago, IL.

McNamara, T.P. (1986). Mental Representations of Spatial Relations, *Cognitive Psychology*, **18**, 87–121.

McNamara, T.P. (2003). How are the Locations of Objects in the Environment Represented in Memory? In: *Spatial Cognition III: Routes and Navigation, Human Memory and Learning, Spatial Representation and Spatial Reasoning* (C. Freksa, W. Brauer, C. Habel, and K. Wender, eds), Springer, Berlin, 174–191.

Montello, D.R. (1997). The Perception and Cognition of Environmental Distance: Direct Sources of Information. In: *Spatial Information Theory: A Theoretical Basis for GIS* (S.C. Hirtle and A.U. Frank, eds), Springer, Berlin, 297–311.

Montello, D.R. (1998). A New Framework for Understanding the Acquisition of Spatial Knowledge in Large-Scale Environments. In: *Spatial and Temporal Reasoning in*

Geographic Information Systems (R.G. Gollege and M.J. Egenhofer, eds), Oxford University Press, New York, 143–154.

Newcombe, N. and J. Huttenlocher (2000). *Making Space: The Development of Spatial Representation and Reasoning*, MIT Press, Cambridge, MA.

Newcombe, N. and J. Sluzenski (2004). Starting Points and Change in Early Spatial Development. In: *Human Spatial Memory: Remembering Where* (G. Allen, ed.), Lawrence Erlbaum Associates, Mahwah, NJ, 25–40.

O'Keefe, J. and L. Nadel (1978). *The Hippocampus as a Cognitive Map*, Oxford University Press, Oxford.

Park, R., E. Burgess, and R. McKenzie (1925). *The City*, University of Chicago, Chicago, IL.

Piaget, J. and B. Inhelder (1956). *The Child's Conception of Space*, Routledge and Kegan Paul, London.

Piaget, J., B. Inhelder, and A. Szeminska (1960). *The Child's Conception of Geometry*, Norton, New York.

Proffitt, D.R., J. Stefanucci, T. Banton, and W. Epstein (2003). The Role of Effort in Perceiving Distance, *Psychological Science*, **14**, 106–112.

Rossano, M.J. and D.H. Warren (1989). Misaligned Maps Lead to Predictable Errors, *Perception*, **18**, 805–816.

Sadalla, E.K., W.J. Burroughs, and L.J. Staplin (1980). Reference Points in Spatial Cognition, *Journal of Experimental Psychology: Human Learning and Memory*, **6**, 516–528.

Sadalla, E.K. and D.R. Montello (1989). Remembering Changes in Direction. *Environment and Behavior*, **21**, 47–69.

Shemyakin, F.N. (1962). General Problems of Orientation in Space. In: *Psychological Science in the U.S.S.R.* (B.G. Ananyev, ed.), Office of Technical Services, Washington, DC, 186–255.

Siegel, A.W. and S.H. White (1975). The Development of Spatial Representations of Laarge-Scale Environments. In: *Advance in Child Development and Behavior* (W.H. Reese, ed.), Vol. 10, Academic Press, New York, 9–55.

Stevens, A. and P. Coupe (1978). Distortions in Judged Spatial Relations, *Cognitive Psychology*, **10**, 422–437.

Taylor, H.A. and B. Tversky (1992). Spatial Mental Models Derived from Survey and Route Descriptions, *Journal of Memory and Language*, **31**, 261–282.

Tolman, E. (1948). Cognitive Maps in Rats and Men, *Psychological Review*, **55**, 189–208.

Tom, A. and M. Denis (2004). Language and Spatial Cognition: Comparing the Roles of Landmarks and Street Names in Route Instructions, *Applied Cognitive Psychology*, **18**, 1213–1230.

Tversky, B. (1993). Cognitive Maps, Cognitive Collages, and Spatial Mental Models. In: *Spatial Information Theory: A Theoretical Basis for GIS* (A.U. Frank and I. Compari, eds), Springer, Berlin, 14–24.

Tversky, B., P.U. Lee, and S. Mainwaring (1999). Why Speakers Mix Perspectives, *Spatial Cognition and Computation*, **1**, 399–412.

Zannaras, G. (1968). *An Empirical Analysis of Urban and Neighborhood Perception.* Unpublished M.A. Thesis, Ohio State University, Columbus, OH.

Zannaras, G. (1973). *An Analysis of Cognitive and Objective Characteristics of the City: Their Influence of Movements to the City Center.* Unpublished Ph.D. Dissertation, Ohio State University, Columbus, OH.

Zannaras, G. and Golledge, R. (1974). Methodological Approaches to the Cognition of Urban Structure. In: *On Determining Cognitive Configurations of a City: Volume I – Problem Statement, Experimental Design and Preliminary Findings* Final Report No. NSF Grant #GS-37969, Department of Geography, Columbus, OH.

Tommy Gärling and Linda Steg (Editors)

6

CONGRUENCE AND CONFLICT BETWEEN CAR TRANSPORTATION AND PSYCHOLOGICAL RESTORATION

Terry Hartig, Institute for Housing and Urban Research, Uppsala University, Gävle, Sweden

ABSTRACT

Car transportation can diminish the quality of urban life by constraining opportunities for psychological restoration in everyday settings. Yet many urban residents also count on their cars as aids to restoration, as with travel from the city to outlying natural areas or restorative interludes between work and family demands. In this chapter, I discuss instances of congruence and conflict between car transportation and psychological restoration for people in cities. More specifically, I refer to the ways in which car transportation enhances or diminishes the restorative quality of environments that people can access in and around cities during time available for restoration.

INTRODUCTION

Many people use a car to meet the demands of everyday life in urban areas. They may need one to drive children to school, themselves to work, and to bring the groceries home. They may also rely on a car during leisure time. It opens to them a broader range of recreational settings than they could reach from their urban residence with collective transportation, particularly places in a beautiful but sparsely populated countryside poorly served by trains and buses. During obligatory and recreational trips, the car may also afford them experiences they value in and of themselves. For example, many people savor the time on the commute, transitioning psychologically as well as spatially between their family and work roles (Mokhtarian and Salomon, 2001).

Just as many urban adults would agree that car transportation has a variety of instrumental and affective values, many would also agree that widespread use of cars has significant costs reckoned in terms of life quality (Steg and Gärling, this volume). As with the benefits of car use, the costs have to do with the ability to meet everyday demands, the quality of accessible residential and recreational settings, and the enjoyment of travel by car in and of itself, among other outcomes. The costs, like the benefits, can be reckoned on aggregate levels as well as the individual level, and they may relate to one another across levels of aggregation in ways that are complex and dynamic.

Although the benefits and costs of car transportation are many, diverse, and systemically interrelated across multiple levels of analysis, their complexity does not defy organization. In this chapter, I identify and organize some benefits and costs of car transportation with regard to a metric useful for describing and evaluating human environments. This metric, *restorative quality,* has fundamental relevance for life quality for all individuals. It concerns the ability to meet the demands of everyday life, and, in that people tend to like environments with high restorative quality, it reflects on possibilities for enjoying everyday life.

In the following, I identify a variety of instances in which car transportation relates to the restorative quality of environments available to people living in and around cities. I organize the various instances with regard to congruence and conflict. For some people, car use entails an improvement in the restorative quality of accessible environments; it is congruent with restoration. Conversely, for some people, car use reduces restorative quality; it conflicts with restoration. As with the costs and benefits of car use more generally, instances of conflict and congruence can be discerned looking across as well as within levels of aggregation, so that congruence at one level of aggregation may entail conflict at another. Before going into the various instances of congruence and conflict, however, I will first elaborate on the restorative quality metric and environmental conditions that affect restorative quality, for good or ill.

RESTORATIVE QUALITY

In meeting everyday demands, a person draws on a variety of psychological, physiological, and social resources. Some of these can become diminished with use, like the capacity to mount a fight-or-flight response or an ability to focus attention. Such adaptive resources must then be renewed if the person is to continue meeting his or her everyday demands. The processes through which individuals' renew their resources can be gathered under the rubric *restoration* (Hartig, 2004). The resources most commonly discussed in the restoration literature include psychophysiological response capabilities (Ulrich *et al.*, 1991) and an ability to direct attention (Kaplan, 1995). A person who needs to restore these resources is described as experiencing stress or directed attention fatigue (sometimes simply mental fatigue), and the restoration process is referred to as stress recovery or attention restoration.

Mental fatigue and stress can impair effectiveness and life quality in the short run and harm physical and mental health over the long run. Historically, to prevent such problems, unions, employers, legislative bodies, and religious authorities have sought to ensure that people routinely have enough time for renewing adaptive resources, both with sleep and during waking hours (Westman and Eden, 1997). However, having enough time for restoration does not guarantee that a person will adequately restore depleted resources. Restoration always occurs within the context of some activity, and activity always occurs within some environment. Environments vary in the degree to which they support restoration; that is, they vary in restorative quality. During time available for restoration, if the immediate environment offers little support for restoration, then restoration will not proceed as quickly, or, possibly, as completely, as it would in an environment with higher restorative quality.

Restorative quality is a general metric for describing human environments. It depends on the social and physical environmental characteristics that influence the potential for restoration that a person perceives and may realize there. Those characteristics can combine in numerous ways to determine a particular level of restorative quality. As a metric, then, restorative quality has its complexities. A thorough treatment of those complexities is beyond the scope of this chapter, but some account of relevant environmental characteristics is needed to support a discussion of congruence and conflict between car transportation and psychological restoration. As a starting point, I will distinguish environmental characteristics here according to whether they ordinarily promote or merely permit restoration.

The distinction between restoration permitting and restoration promoting characteristics of environments reflects three necessary features of a theoretical account of restorative quality in environments. Two of those features are negative, the third one positive. First, in its specification of a condition of resource depletion from which a person can restore, a theory about restorative environments must implicitly or explicitly characterize the circumstances that engendered the given need for restoration. Second, it stands to reason that restoration will not proceed under circumstances that engender a need for restoration, so a theory about restorative environments must also specify some form of separation or removal from the circumstances in which a person's need for restoration arose. Yet, that separation or removal, whether psychological or geographical or both, may only permit restoration; the environment that the person has gone on to may or may not have characteristics that promote restoration by helping it along more quickly to completion. Thus, a third and positive feature of theories about restorative environments is an account of environmental characteristics that promote restoration. That is to say, restorative quality cannot be defined only in negative terms, as separation or removal from conditions that would engender a need for restoration.

Table 1: Some Characteristics of Environments that Deny, Permit, and Promote Psychological Restoration

Deny	*Permit*	*Promote*
Noise	Quiet	Music, sounds of friendly voices, laughter, breaking waves, birds
Visual blight	Absence of visual blight	Beautiful scenery
Malodorous air	Absence of unpleasant smells	Pleasant smells
Physical discomfort	Absence of discomfort	Exceptional comfort
Presence of threat	No apparent threat	Safety assurances
Crowding	No crowding	Pleasant company
Pressing role demands	Support from others in meeting present role demands	Freedom to step out of the role for a time
Constraints on desired activity	No constraints on desired activity	Supports for desired activity
Time pressure	No immediate demands	Vacation time

In Table 1, I have listed a variety of environmental characteristics that can cause a need for and deny restoration, or permit restoration, or promote restoration. In constructing this list, I have not aimed to give an exhaustive account, but rather to convey that restorative quality can involve more than one sensory modality, that it has social and temporal aspects, and that it has some grounding in the person's ongoing activity.

What this list does not make apparent is just how the various environmental characteristics might come into play in restoration. Whether the restorative quality that a person perceives in an environment remains latent or actually manifests in restoration will depend in part on that person's restoration needs on the given occasion. Given a person who has a depleted capacity to direct attention, for example, environmental characteristics might come into play in the manner set out in attention restoration theory (ART) (Kaplan and Kaplan, 1989; Kaplan, 1995). According to ART, one requisite for directed attention restoration is being away, or getting some psychological distance from the work one usually does, from the pursuit of given goals and purposes, and from other circumstances in which directed attention is called into play. When away, a person can restore a diminished capacity for directing attention when he or she experiences fascination, a mode of attention assumed to be involuntary, effortless, and without capacity limitations. When able to rely on fascination in ongoing activity, the person can relax the inhibitory mechanism upon which directed attention depends, allowing for renewal of the capacity for directing attention. Fascination could be engaged by some of the positive environmental characteristics listed in Table 1, such as natural scenery, as well as by processes of exploring and making sense of the environment. Another factor specified by ART is extent, or the sense that the environment available for restoration is sufficiently large in scope to entertain continued exploration, and sufficiently coherent that one can make sense of what is going on around oneself and relate it to some larger frame of reference. A fourth factor, compatibility, encompasses a person's inclinations, the environmental supports for his or her activities, and environmental demands; it rests on the match between what a person wants to do, can do, and must do in the given environment.

Before going on to discuss instances of congruence and conflict between car transportation and restoration, I should deal with one more feature of the restorative quality metric here. It bears emphasizing that the restorative quality a person perceives in an environment has a relative character; it stands in relation to those settings in which that person's restoration needs arise and to the other environments that the person could want to access during time available for restoration. Researchers and design professionals might identify generally valid upper and lower bounds for restorative quality using some set of objective or broadly consensual criteria, but for a given person the practically meaningful range of variation in restorative quality holds with regard to the set of environments within which he or she moves. Some fortunate people have the means to concentrate their activities in settings of objectively or consensually high restorative quality. In those settings, they may quickly and readily satisfy the mundane restoration needs that arise from work, social demands, and so forth. Many people, however, spend their time within objectively or consensually stressful circumstances and with poor access to settings of high restorative quality, even during their leisure time. For those people, improved possibilities for fulfilling their restoration needs may require securing some means for increasing mobility. Substantially improving the restorative quality of the environments to which they already have access may cost far more than they can afford.

As a metric, then, restorative quality seems a useful tool for identifying costs and benefits of car transportation. For one, it encompasses and then extends beyond a general "stressfulness" metric, in that environmental stressors may counteract restoration but their absence can only permit restoration. Also, the restorative quality metric allows the car itself and the dynamic traffic environment to enter consideration; it does not only apply to the fixed environments joined with travel. Further, it supports discussion of non-obligatory and recreational travel as well as obligatory travel; restorative quality has relevance for incidental restoration (as on the commute) as well as intentional restoration (as with travel to recreation destinations and with driving as an activity in and of itself). Finally, and more subtly, the restorative quality metric implicitly factors in a consequence of a person's car use, namely, expansion of the set of environments that he or she can regularly and readily move among. Again, for a given person, the practically meaningful range of variation in restorative quality is established with regard to the environments moved among. If use of a car expands the practically meaningful range of restorative quality for a person, then this broader range must be considered when assessing the restorative benefits of car use for that person.

CONGRUENCE BETWEEN CAR TRANSPORTATION AND PSYCHOLOGICAL RESTORATION

In the following, I discuss several instances in which car use entails an improvement in the restorative quality of environments available to people. I concern myself primarily with the environments that people can or must enter during time available for restoration. People may intentionally dedicate some of that time to restorative activities in particular settings, as with evening hours for socializing outdoors in the neighborhood, nighttime hours for sleep at home, weekend days for outings in the countryside, or extended vacations in remote locations. However, some of the time available for restoration has an incidental character; people may find it along the road, in their cars.

In Table 2, I summarize the various instances of congruence discussed in the following. In some instances, the congruence holds at the level of the individual, such as a solitary driver. I discern other instances of congruence between car transportation and psychological restoration at an aggregate level. For all instances, I have distinguished restoration permitting and promoting aspects of car transportation and driving. For none of these instances do I provide an exhaustive treatment, but the discussion does suffice to illustrate the utility of my approach to identifying and organizing some of the benefits of car transportation.

Table 2: Instances of Congruence between Car Transportation and Psychological Restoration

Instance	*Level*	*Restoration Permitting*	*Restoration Promoting*
On the road	Individual	Safety, withdrawal, separation from demands	Enjoyment of driving, experience of variety in environments traversed
Residential relocation to the urban periphery	Individual	Lower exposure to urban stressors	Better access to large natural areas
Leisure mobility	Individual	Faster, more convenient departures; freedom from constraint by bus or train timetables	Access to a broader selection of environments of high restorative quality than available with other transportation modes
Extension and articulation of the road network	Aggregate	Safe roads open to travel by ordinary cars, even in difficult terrain; broader range of residential location options	Access to a broader selection of environments of high restorative quality
Technological refinement of cars	Aggregate	Technologies and design features that satisfy customer demands for safety, quiet, and comfort in the car	Entertainment, seating and communication options to fulfil customer demands for comfort and pleasant diversions in the car

On the Road

For a driver and passengers, the use of a car may improve the possibility for restoration in a most immediate way. They may find in the car a safe and isolated haven from pressing role demands or aversive surroundings, and so a place that permits restoration. Also, the pleasures of driving may promote restoration.

These two forms of improvement may occur with both obligatory and non-obligatory travel, but to different degrees. The permission of restoration seems more salient with obligatory travel, given that the locations joined are familiar settings for demands and travel between them is routine, to some degree compulsory, and often under time constraints. Getting away from the demands of the setting and into the car may be the most restorative part of the trip.

The commute can serve as the prime example of obligatory travel. I have already mentioned that many people value their commute as a time for transitioning between family and work roles. With that transition and in other ways, the commute opens possibilities for restoration that may counteract its more frequently discussed stressful aspects. For example, Kluger (1998) asked car commuters in New Jersey, New York, and Pennsylvania about the strain and enjoyment they experienced in relation to their commute. The survey measure of enjoyment included statements that reflected on restoration, such as "My commute gives me ... time to relax" and "My commute ... reduces my stress level" (p. 155). For enjoyment as for strain, the mean rating fell near the midpoint of the scale, revealing the ambivalent feelings that many respondents had about the commute. The two measures correlated negatively, as one would expect, though not particularly strongly ($r = -0.22$).

With non-obligatory travel, the restoration promoting aspects of driving presumably have greater salience than the restoration permitting aspects of car use. The idea of non-obligatory travel may strike some people as odd, but a substantial body of research affirms that many people value the activity of driving in and of itself (Mokhtarian, 2005). That is, car travel has much more than simply instrumental value. For example, in one survey, Steg (2005) found that Dutch adults distinguished symbolic and affective motives for driving from instrumental motives. Those people who drove the most kilometers per year gave higher attractiveness ratings to symbolic and affective aspects of car use than did those who drove less (cf. Kaiser *et al.*, 1994). Pleasures of driving for its own sake can include adding variety to one's daily experience, viewing beautiful scenery, experiencing speed, and feeling a sense of mastery in operating the car (Ory and Mokhtarian, 2005). These potential aspects of car use may promote restoration by engaging a driver's or passenger's attention, drawing thoughts away from stressful demands elsewhere and otherwise evoking positive emotions.

Of course, non-obligatory travel by car may be difficult to distinguish from obligatory travel by car (cf. Mokhtarian and Salomon, 2001). The fuzziness of the distinction may owe in part to the possibilities that a car affords for meeting restoration needs. For example, consider a person who wants to escape from the workplace or residence and into the haven of the car, to enjoy a period of solitary driving. This restoration strategy, whether conscious or not, may require that the person justify the travel, for him- or herself or for a "gatekeeper" in the to-be-departed setting. A reason then given may include some sanctioned purpose and destination ("We're out of milk, I should drive over to the grocery store and get some"). Nominally, the trip has an obligatory character, but its initial motivation stems from a need for restoration.

Residential Relocation to the Urban Periphery

Car transportation may enable a lasting improvement in the restorative quality of everyday environments by opening up additional alternatives for a place to live. To understand how the improvement is made, consider what the residential relocation process involves. When searching for a new residence, a household will evaluate available alternatives with regard to the various attributes of the housing and neighborhood (Lindberg *et al.*, 1992; Gärling and Friman, 2002). Some of those attributes have to do with exposure to stressful circumstances, such as crowding if the housing has too few rooms and the threat of victimization if the neighborhood has a high crime rate. Other attributes of the housing and neighborhood have to do with restoration possibilities, including views from the windows, space for a garden, and access to a park or green space. When searching for a new residence, a household will also consider the location of each available alternative relative to workplaces, schools, shopping, and other important settings. The relative location of the new residence will have implications for the frequency and characteristics of trips among settings, and so for the stressful demands confronted and restoration opportunities found in connection with mundane travel (Hartig *et al.*, 2003a).

Through change in the attributes of their residential environment and through change in the attributes of trips that start and end at the residence, households that relocate may reduce their exposure to environmental demands that cause a need for restoration to arise. They may also escape environmental conditions that do not permit restoration, and improve their access to environments that promote restoration. However, whether they can improve the restorative quality of their everyday environments in these ways depends on the residential alternatives available to them. The availability of alternatives depends in part on transportation options. The broader the reach a household has with available transportation modes, the greater the number of residential alternatives from which to choose, assuming no other constraints.

In some countries, a long-standing pattern of residential relocation involves moving out from an urban center to a peripheral location. It appears that households want to move from a place they perceive as visually blighted, noisy, crowded, polluted, and dangerous.

They want to move toward a place that they perceive as scenic, quiet, and uncrowded, with clean air and safety from crime. These perceptions, accurate or not (Lofland, 1998), presumably enter into evaluations of residential alternatives. Over time, households have located at an ever longer edge between an expanding city and a receding countryside, trying to maintain proximity to urban employment and services upon which they rely and yet maintain access to the amenities of the countryside, including those that promote restoration. The ability of households to make this tradeoff has depended substantially on car transportation. Critics of collective transportation have argued that the low density of residential development at the urban periphery is too expensive to serve adequately with buses and trains, and development of collective transportation has lagged behind that of car transportation.

Leisure Mobility

I have already indicated two ways in which leisure mobility exemplifies congruence between car transportation and psychological restoration. For one, because driving can afford escape and various pleasures, it can permit and promote restoration. Research has long recognized that many people view driving as a form of recreation under some circumstances, entirely aside from activities such as racing. For example, the first comprehensive assessment of the outdoor recreation needs and practices of people in the USA (Outdoor Recreation Resources Review Commission, 1962) found that driving or riding in the automobile for pleasure ranked first in participation, and it ranked first as a preferred activity when time available for leisure was limited to 2–3 hours.

Also, as with residential alternatives, car transportation expands the number and variety of potential leisure settings that a person can reach during time ordinarily available for leisure. The additional settings the person can reach may include some of higher restorative quality than those that he or she could reach by other transportation modes available at the point of departure, presumably the residence.

The pleasures of driving and the broader range of leisure settings accessible by car, stand as two ways in which car transportation promotes restoration through leisure mobility. Use of the car may also permit restoration to a greater degree than other transportation modes, in particular by allowing for easier get-aways. Whether departing from the workplace or some other setting, a person can initiate a trip with relative convenience, unconstrained by bus or train timetables. This convenience becomes more pronounced to the degree that the time of departure falls in a period with low frequency of service by bus or train. The convenience of departure is further enhanced by the ability that a person has to store recreational equipment in the car. Without this ability, a person traveling by some other mode might have to make additional trips to collect the requisite equipment from the residence or some other location.

Anable and Gatersleben (2005, Study 2) reported empirical results that well illustrate the points I have just made. They surveyed visitors to countryside leisure settings in the UK's National Trust system. Their respondents rated the importance of instrumental and affective attributes of transportation that might be used for a day trip for leisure outside of one's local area. They also rated the perceived performance of different transportation modes regarding those attributes. As instrumental attributes, flexibility, convenience, and cost all had mean ratings near the top of the scale ("very important"), as did relaxation, freedom, and absence of stress as affective attributes. Anable and Gatersleben found that they had few respondents who came by modes other than car or public transportation, a result in line with visitor data gathered by the National Trust. Looking at the ratings of performance for the transportation mode that their respondents used most regularly for day trips, they found that the car drew higher performance ratings from its users for flexibility, convenience, cost, and predictability than public transportation received from its users. It also had higher performance ratings for control and freedom, though not for relaxation or the absence of stress.

Extension and Articulation of the Road Network

I turn now to two instances in which car transportation has on the aggregate level improved the restorative quality of environments that people may enter during the time they have available for restoration. Let me make clear that I speak here of a general increase in the restorative quality of the environments broadly *available* to the members of a given population. I do not mean to imply that, because of car transportation, those people have actually come to spend more time in environments of high restorative quality.

Both instances of congruence on the aggregate level – extension and articulation of the road network and technological refinement of the car – have resulted from the steadily increasing demand for car transportation. Demand has increased for reasons already mentioned, such as the pleasures of driving; the desire for a larger set of residential alternatives; the desire to reach more, and more varied, leisure settings during one's limited free time; and perceptions regarding the relative convenience and flexibility of other transportation modes.

Involving as they do the two essential components of the car transportation system, the two instances of aggregate improvement emerged together through a co-evolutionary process. Certainly, extension and articulation of road networks in many regions predates transport by car. People have over millenia laced their landscapes with roads. Yet, even in regions with a long-established system of roads, the advent of car transportation has led to major changes, including straightening, widening, changes in grades, diversification in surface materials, improvements in road bed construction, sophisticated signage systems, and, of course, new road construction (e.g., Forman *et al.*, 2003).

Not least, peri-urban residential development predicated on car use has involved highly articulated extensions to the existing urban road systems.

The increasing availability of reliable, inexpensive, and fast personal vehicles also stimulated the extension of road systems into unpopulated regions that historically did not have roads. Construction of good roads in such places, some with difficult terrain, became justifiable for some authorities when large numbers of people had vehicles with which to travel into and then out of them. One motive for such travel and the roads to support it has been access to places with outstanding scenery that offer relative solitude. For example, until the early 1950s, the Sierra Club promoted the development of roadways to "render accessible" the Sierra Nevada, so that California's urban residents could more easily get away to enjoy beauty and tranquillity among its peaks and meadows (Sierra Club, 1989). Some roads, such as the Blue Ridge and other parkways in the eastern United States, were constructed to provide opportunities for enjoying natural scenery while driving (Forman *et al.*, 2003).

On the way to a distant destination, a traveler may need to refuel, use a bathroom, buy food, possibly even sleep. The infrastructure that supports travel by car helps travelers meet such needs. In addition to roads, the fixed components of the car transportation system include gas stations, rest stops, restaurants, and motels. The availability and quality of such infrastructure may, like the availability and quality of the roads, figure into the decisions that people make about where to travel during extended periods of leisure.

Technological Refinement of Cars

The earliest cars looked like the horse-drawn carriages that their descendants would quickly come to replace. After decades of refinement, cars resemble carriages in few respects other than that they have wheels and seats. The pace of refinement has at times been rapid; "plain vanilla" cars of today out-perform top-of-the-line vehicles of the not-so-distant past in some surprising ways, such as acceleration (Ealey and Troyano-Bermudez, 1996).

The industry's response to aggregate demand for transportation by car and its efforts to stoke that demand have involved enhancing the comfort and pleasure of car travel, both for the driver and for passengers. Beyond the intended outcomes for individuals, these efforts may have had aggregate effects that deserve consideration. For example, the improvements in the interior environment of the average car presumably have had real import for people who spend much time on the road. Given the number of car commuters in many urban areas, and their often long commutes, one could speculate that relatively good insulation from exterior noise, comfortable seating, climate control, and so forth yields a public health benefit, all other factors aside. Such a health benefit would to some extent be mediated by better stress mitigation and enhanced restoration relative to earlier vehicles.

CONFLICT BETWEEN CAR TRANSPORTATION AND PSYCHOLOGICAL RESTORATION

For many people, car transportation appears to enhance access to restorative environments during time available for restoration. This enhanced access would seem to recommend the car as a transportation mode. However, an honest account of the value of that access must acknowledge the costs of enabling it, which also can be reckoned in terms of restorative quality. Some of those costs fall only on those participating in the traffic environment at the moment, including car drivers, their passengers, pedestrians, and bicyclists. Other costs are paid by people who are not using the car transportation system at the time when the restorative quality of their immediate environment is diminished by the car use of others.

In the following, I treat some costs of car transportation as instances of conflict with psychological restoration. As with the instances of congruence, I consider each instance of conflict with regard to permission versus promotion of restoration. As with the instances of congruence, I do not give an exhaustive treatment of any of the conflicts, but the discussion should suffice to illustrate the utility of the present approach to identifying and organizing some costs of car transportation. I summarize each instance of conflict in Table 3.

Stressors in Transit

The traffic environment is not always pleasant or tranquil. Drivers and passengers may have restorative interludes during a trip, but they may also have periods of impatience, anger, fear, frustration, worry, and boredom. Many people do not travel even a large fraction of their total miles under conditions like those so often portrayed in the advertising for new cars. Instead of smoothly gliding along on an open country road through fields by the seashore, they move haltingly along in traffic on urban streets and highways, now and then startled by a honking horn or screeching tires, sometimes put off by the smell of exhaust, and sometimes passing through visually blighted areas. In contrast to the freedom promised in advertising, drivers and passengers may find themselves trapped, unable to do much but stick it out, and denied restoration.

Variability in the duration of trips between two locations may add to or perhaps exacerbate other stressful aspects of any one trip between them. For example, Kluger (1998) studied commute variability in relation to the strain and enjoyment reported by commuters. He defined commute variability for each person as the standard deviation of the estimates in minutes for travel *to* work on an average day, a bad day, and a good day, and *from* work on an average day, a bad day, and a good day. Commute variability contributed substantially to explanation of self-reported cognitive/affective strain, above and beyond the contributions of commute length and route choice (cf. Stokols *et al.*, 1978;

Table 3: Instances of Conflict between Car Transportation and Psychological Restoration

Instance	*People Affected*	*Manner in Which Restorative Quality Affected*	*Type of Conflict*
Stressors in transit	Car drivers and passengers	Experience of stress in connection with behavioral constraints and loss of control, noise and unpleasant emissions, threats from traffic or other drivers, long and variable commute, driver incompetence or rage	Denial of restoration
Stressors for other road users	Pedestrians and bicyclists	Experience of stress in connection with behavioral constraints and loss of control, noise and unpleasant emissions, threats from traffic or specific drivers, startling sounds	Denial of restoration
Surrender of outdoor recreational spaces for transportation infrastructure	Residents, recreationists and others	Recreational settings and street trees sacrificed or modified to meet demands for construction of roads, parking lots, gas stations and so forth; unwanted sounds of cars in places valued for quiet and solitude	Loss of access to restorative environments or compromised restorative quality
Effects on the residential context	Residents	Traffic noise and vibrations disrupt relaxation and sleep; traffic hinders neighboring across streets; traffic and construction of new transportation infrastructure disrupt or inhibit attachment to the residence	Compromised restorative quality and unrealized restorative potential
Impoverishment of the public realm	Citizens generally, urban residents in particular	Fewer people populating public spaces; proportionally more "aversive" people in public spaces; less commitment to making the public realm safe, fun, visually attractive, and so forth	Compromised restorative quality and unrealized restorative potential
Constraint of agendas in large parks	Car-bound park visitors	Extensive road system and placement of facilities encourage overreliance on the car for exploring a large park	Unrealized restorative potential

Novaco *et al.*, 1979). Kluger attributed the association to the unpredictability, cognitive load, and reduced control engendered by commute variability.

The traffic environment is a social environment for drivers, passengers, pedestrians, and cyclists. Some uncertain balance holds between acts of kindness and civility and acts of hostility and aggression. When one participant in the traffic environment attacks another one in an act of "road rage," it may seem a rare and extraordinary act, but it might as well be viewed as the push of harsh feelings beyond a threshold that many people have approached but never yet crossed. Hostility and fear also arise in the passenger compartment, as reactions to the behavior of the driver. Careless or aggressive drivers can set their passengers on edge. A passenger frightened by the driver's incompetence or anger toward other drivers and worried about his or her own safety can hardly experience the drive as a restorative interlude.

Stressors for Other Road Users

Car traffic affects the qualities of the urban street environment experienced by walkers, runners, and bicyclists. Some of its effect apparently involves restorative quality. A number of experimental studies performed in field settings have found that subjects who walk or run along urban streets show less restoration from immediately preceding demands than subjects who walk or run in a park. For example, subjects in one field experiment walked in an area of medium-density commercial and retail development or in a nature preserve, equipped with ambulatory blood pressure monitors that collected measures at 10-minute intervals (Hartig *et al.*, 2003b). In each setting the 50-minute walk followed an essentially flat route. On average, those who walked in the urban setting had their systolic and diastolic blood pressure increase over the first half of the walk, while those who walked in the nature preserve had their blood pressure decline during the same period. When they performed an attentional test at the midpoint of the walk, those in the urban setting showed a decline in performance relative to the pretest they had performed in a quiet laboratory. This performance decline may seem unsurprising, given the stream of traffic on the road beside the subjects as they performed the task. Those in the nature preserve slightly improved their performance on the task from the pretest to the midpoint of the walk. The between-groups performance gap that had opened up on the walk was still evident in the postwalk test scores obtained in a quiet field laboratory (cf. Hartig *et al.*, 1991; Bodin and Hartig, 2003). For people who need a break from work or who need to get out of the house for a little while, a heavily trafficked street may provide little relief from the demands they just left. It may even compound their experience of stress.

Surrender of Outdoor Recreational Spaces for Transportation Infrastructure

Green spaces in urban areas may provide a haven for those people denied restoration while walking on busy streets. Yet, growing numbers of drivers and cars and growing reliance on cars to reach residences at the urban periphery translate into increased demand for space in urban areas. Streets, the parking spaces along them, parking lots and structures, gas stations, and repair shops all require space. So do housing and other functions. The outdoor spaces available for recreation may receive little weight if decision makers see few other means for meeting the demand for space for more "essential" functions. Just as restoration needs may increase with worsening ambient conditions due to increasing car traffic, the possibilities for satisfying restoration needs in local greenspaces and other outdoor recreation sites may come under increasing threat.

Effects on the Residential Context

Car traffic can negatively affect the restorative quality of the urban residential context in several ways. For one, noise and vibrations can intrude on social exchanges and leisure activities, cause worry about traffic accidents, and in other ways disrupt relaxation and sleep at home. Öhrström (2004) studied the effects on sleep of a residential traffic intervention (construction of a tunnel) in Gothenburg, Sweden, that greatly reduced the amount of daytime and nighttime traffic. The traffic reduction translated into a substantial reduction in sound levels measured at house facades with different orientations to the road in question. Following the intervention, a sample of residents in the exposed area reported far greater improvement in sleep quality compared to a sample of residents from a control area, who showed little change in sleep quality. Before the intervention, some 26% of those from the exposed area reported deficient sleep quality.

Car traffic may also hinder neighboring and the development of an expansive territory, as illustrated in Appleyard's (1981) classic study on livable streets. Restriction of movement between homes on different sides of a street and restriction of the use of the street for play and other social exchange may mean less neighborhood socializing, less socially attractive outdoor spaces in the neighborhood, and so less restoration in a social context.

Diminished Enjoyment of the Public Realm

Lofland (1998) defines the public realm as "constituted of those areas of urban settlements in which individuals in copresence tend to be personally unknown or only categorically known to one another" (p. 9). People who travel by car in their daily rounds, she notes, are not sharing buses, trains, and public spaces with strangers while in transit. Without cars, people occupy the public realm in greater numbers and diversity. Lofland's discussion indicates some possible consequences of an urban transportation system not based on cars. People forced into contact with unknown others must learn to get along with them, even those they find aversive. They find ways to filter or avoid contacts, and they apply norms of tolerance toward others. Negative emotions of fear and loathing may occur more often in unwanted contacts with strangers, but people may habituate to them and so experience them less acutely. Moving by foot and more slowly, individuals may see more of what the city has to offer in the way of enjoyment. People-watching in the public realm, for example, may be a more engaging pastime, and urban spaces may be characterized to a greater degree by sociability. With more people on the streets, individual safety increases and fear of victimization declines. In a variety of ways, then, Lofland's portrayal suggests that reliance on car transportation may compromise the restorative quality of the public realm and hinder the realization of some greater restorative potential.

Constraint of Agendas in Natural Areas

Legal scholar Joseph Sax (1980) criticized national park planning in the USA because it supported an excessive degree of vehicular access to wild areas. Roads for cars, he said, set the agenda for people. Although many more people might come to a national park because the car transportation system made it easy for them to do so, most of those people, once in the park, would not leave the comfortable confines of their cars. So situated, they might not realize a more profound experience that would otherwise be available to them should they enter the backcountry with their provisions on their backs.

INTERRELATIONS AMONG INSTANCES OF CONGRUENCE AND CONFLICT

I have framed a variety of benefits and costs of car transportation in terms of restorative quality. With that metric in mind, I treated the various benefits and costs as instances of congruence or conflict with psychological restoration. To nuance the discussion, I described benefits of car transportation in terms of the permission versus promotion of restoration.

To this point, I have for the most part treated each instance of congruence and conflict in isolation. However, as I indicated at the outset, the costs and benefits of car transportation may interrelate systemically across multiple levels of analysis. In the following, I address several issues involving interrelations among instances of congruence and conflict, including tradeoffs, social equity, substitutability, and positive feedback.

Tradeoffs

Several forms of tradeoffs have apparent relevance here. Some concern the experience of the individual traveler. To begin with, for a given trip, those traveling by car are not traveling by another form, and the time they spend in the car is time they cannot spend doing other things. For each of these tradeoffs, the alternatives can be evaluated with regard to the restorative quality of the environment in which people travel. For example, based on his review of the then-extant literature, Kluger (1998) suggested that stress related to commute impedance may be more pronounced with travel by car versus train or subway. With regard to the commute, the mode and time tradeoffs may correlate, in that choice of one travel mode may also imply a relatively greater amount of time, as with a long commute by car from the suburbs versus a short commute by foot from an urban residential location close to one's workplace. That correlation itself depends on other tradeoffs that a person or household has made in selecting places to live and work, in which restorative quality matters may also have entered consideration (Hartig *et al.*, 2003a).

A superordinate tradeoff has implications for the population experience. Societal investments in car transportation may limit possibilities for development of other transportation modes. For example, urban space dedicated to streets for use by cars may no longer be available for rail lines. In favoring one versus another transportation system, a society can affect its members by affecting the restoration permitting and promoting qualities of the transportation environment itself, the environments adjacent to transportation infrastructure, and environments made accessible with the system. Given this, one could ask, for example, how the relative access to peri-urban environments afforded by car versus rail transportation weighs against the relative loss of restorative quality to car versus rail in the residential and public areas of city centers.

Social Equity

Questions about tradeoffs can in turn raise questions about social equity. Not least, those who benefit most from having a transportation system predicated on cars may bear relatively little of the costs for creating and maintaining it. For example, transportation planning may rob inner city areas of their scarce green space to improve parking possibilities or widen streets to meet demand created by suburban car commuters. Noise from the speeding cars of young late-night revelers may deprive older residents of their sleep at home.

As the examples I have given indicate, the distribution of benefits across people has spatial and temporal aspects. Most troubling, a concentration of restorative benefits with one group of people in a given period and locale may come at the cost of diminished restorative quality for a larger number of people, over a longer span of time and a broader area. Extension of the car transportation system into the urban periphery will open up residential alternatives for a relatively privileged group of people. Those alternatives will remain for them into the future – roads and houses tend to remain where people put them – but the commitment made to the particular set of transportation and residential options may translate into a lasting reduction in the amount of accessible green space that might otherwise have been used by a broader segment of the population.

Substitutability

Another issue that follows from the discussion of tradeoffs concerns the substitutability of one transport-based means to restoration for another one. For example, people may value their commute by car as a restorative interlude, but they might also realize the restorative benefit when commuting by another mode. For this reason, some results of commuting satisfaction studies in the USA (e.g., Mokhtarian and Salomon, 2001) must be regarded with caution; many of the respondents may have had little or no experience with commute modes other than the car. Such people may have difficulty saying whether they would have a better or worse commuting experience if they made their commute by means other than the car.

Positive Feedback

Finally, it appears that positive feedback characterizes not only the process through which the demand for car transportation has increased, but also the impact of car transportation on the restorative quality of urban settings. I have already noted that societal investment in car transportation implies less development for other transportation modes. Less development of other modes may in turn mean that the standard of service they can provide cannot satisfy the needs of some people, especially those wanting to reside at the receding urban periphery. When they in turn start relying on a car, then other transportation modes will have fewer people to support them, as riders and perhaps also as voters and taxpayers, unless new users are there to maintain demand. As societal commitment to car transportation accelerates, the costs of transportation in terms of restorative quality may increase. The increase in costs are insidious; just as restoration needs may increase with worsening ambient conditions due to increasing car traffic, the possibilities for satisfying restoration needs in local greenspaces and other outdoor recreation sites may come under increasing threat as demands for new transportation infrastructure become more pressing.

CONCLUDING REMARKS

Any transportation mode will yield benefits and impose costs that vary in kind. Here I have discussed benefits and costs of car transportation with regard to restorative quality. As an environmental metric, restorative quality has fundamental relevance to the quality of life of all people. It is however only one metric that might be applied in identifying and organizing benefits and costs of car transportation. That people consider it important for the quality of life in cities and elsewhere does not mean that they consider it of greatest importance. That people consider restorative quality important does not necessarily privilege it over other environmental metrics, in particular those metrics that help us to describe the integrity and functioning of the ecological systems on which humans and other living things rely.

A particular transportation mode may provide benefits that no other mode provides, just as it may have unique costs. Other benefits and costs may be imposed by transportation modes in general. Having here demonstrated an approach to identifying and organizing some costs and benefits of car transportation with reference to psychological restoration, it remains to apply the approach to multiple transportation modes in a comparative manner, and to do so with a view to benefits and costs in other domains, particularly the ecological domain.

REFERENCES

Anable, J. and B. Gatersleben (2005). All Work and No Play? The Role of Instrumental and Affective Factors in Work and Leisure Journeys by Different Travel Modes, *Transportation Research A*, **39**, 163–181.

Appleyard, D. (1981). *Livable Streets*, University of California Press, Berkeley.

Bodin, M. and T. Hartig. (2003). Does the Outdoor Environment Matter for Psychological Restoration Gained through Running? *Psychology of Sport and Exercise*, **4**, 141–153.

Ealey, L. and L. Troyano-Bermudez (1996). Are Automobiles the Next Commodity? *McKinsey Quarterly*, **4**, 62–75.

Forman, R.T.T., D. Sperling, J.A. Bissonette, A.P. Clevenger, C.D. Cutshall, V.H. Dale, L. Fahrig, R. France, C.R. Goldman, K. Heanue, J.A. Jones, F.J. Swanson, T. Turrentine, and T.C. Winter (2003). *Road Ecology: Science and Solutions*, Island Press, Washington, DC.

Gärling, T. and M. Friman (2002). A Psychological Conceptualization of Residential Choice and Satisfaction. In: *Residential Environments: Choice, Satisfaction, and Behavior* (J.I. Aragones, G. Francescato, and T. Gärling, eds), Bergin & Garvey, London, 55–80.

Hartig, T. (2004). Restorative Environments. In: *Encyclopedia of Applied Psychology*, Vol. 3 (C. Spielberger, ed.), Academic Press, San Diego, CA, 273–279.

Hartig, T., G.W. Evans, L.D. Jamner, D.S. Davis, and T. Gärling (2003a). Tracking Restoration in Natural and Urban Field Settings, *Journal of Environmental Psychology*, **23**, 109–123.

Hartig, T., G. Johansson, and C. Kylin (2003b). Residence in the Social Ecology of Stress and Restoration, *Journal of Social Issues*, **59**, 611–636.

Hartig, T., M. Mang, and G.W. Evans (1991). Restorative Effects of Natural Environment Experiences, *Environment and Behavior*, **23**, 3–26.

Kaiser, F.G., E. Schreiber, and U. Fuhrer (1994). Mobilität und emotionale Bedeutung des Autos: Ein Extremgruppenvergleich zwischen Viel- und Wenigfahrern [Mobility and Emotional Significance of Automobiles: A Known Groups Approach between Frequent and Occasional Drivers.] In: *Mobilitetsverhalten: Bedingungen und Veränderungsmöglichkeiten aus umweltspsychologischer Sicht* (A. Flade, ed.), Beltz, Weinheim, 113–130.

Kaplan, R. and S. Kaplan (1989). *The Experience of Nature: A Psychological Perspective*, Cambridge University Press, New York.

Kaplan, S. (1995). The Restorative Benefits of Nature: Toward an Integrative Framework, *Journal of Environmental Psychology*, **15**, 169–182.

Kluger, A.N. (1998). Commute Variability and Strain, *Journal of Organizational Behavior*, **19**, 147–165.

Lindberg, E., T. Hartig, J. Garvill, and T. Gärling (1992). Residential Location Preferences across the Lifespan, *Journal of Environmental Psychology*, **12**, 187–198.

Lofland, L.H. (1998). *The Public Realm: Exploring the City's Quintessential Social Territory*, Aldine de Gruyter, New York.

Mokhtarian, P.L. (2005). Guest Editorial: Travel as a Desired End, Not Just a Means, *Transportation Research A*, **39**, 93–96.

Mokhtarian, P.L. and I. Salomon (2001). How Derived is the Demand for Travel? Some Conceptual and Measurement Considerations, *Transportation Research A*, **35**, 695–719.

Novaco, R.W., D. Stokols, J. Campbell, and J. Stokols (1979). Transportation, Stress, and Community Psychology, *American Journal of Community Psychology*, 7, 361–380.

Öhrström, E. (2004). Longitudinal Surveys on Effects of Changes in Road Traffic Noise: Effects on Sleep Assessed by General Questionnaires and 3-Day Sleep Logs, *Journal of Sound and Vibration*, **276**, 713–727.

Ory, D.T. and P.L. Mokhtarian (2005). When is Getting There Half the Fun? Modeling the Liking for Travel, *Transportation Research A*, **39**, 97–123.

Outdoor Recreation Resources Review Commission (1962). *National Recreation Survey*, (ORRRC Study Report 19), United States Government Printing Office, Washington, DC.

Sax, J.L. (1980). *Mountains Without Handrails: Reflections on the National Parks*, University of Michigan Press, Ann Arbor, MI.

Sierra Club (1989). *The Sierra Club: A guide*, Sierra Club, San Francisco, CA.

Steg, L. (2005). Car Use: Lust and Must. Instrumental, Symbolic, and Affective Motives for Car Use, *Transportation Research A*, **39**, 147–162.

Steg, L. and T. Gärling (eds) (2007). Introduction. *Threats to the Quality of Urban Life from Car Traffic: Problems, Causes, and Solutions*, Elsevier, Amsterdam.

Stokols, D., R.W. Novaco, J. Stokols, and J. Campbell (1978). Traffic Congestion, Type A Behavior, and Stress, *Journal of Applied Psychology*, **63**, 467–480.

Ulrich, R.S., R. Simons, B.D. Losito, E. Fiorito, M.A. Miles, and M. Zelson (1991). Stress Recovery During Exposure to Natural and Urban Environments, *Journal of Environmental Psychology*, **11**, 201–230.

Westman, M. and D. Eden (1997). Effects of a Respite from Work on Burnout: Vacation Relief and Fadeout, *Journal of Applied Psychology*, **82**, 516–527.

Tommy Gärling and Linda Steg (Editors)

7

ASSESSMENT OF EXTERNAL COSTS: HOW AND WHY?

Philippe Domergue, Conseil Supérieur du Service Public Ferroviaire (CSSPF), Paris, France
Emile Quinet, Ecole Nationale des Ponts et Chaussées, Paris, France

ABSTRACT

This chapter first lists the various externalities generated by transport emphasising the importance of transport nuisances assessment and monetarisation. It further presents the general methods of valuing externalities, direct methods of costs of damages and indirect methods of willingness to pay or to receive, that is "revealed preference methods" based on surrogate markets and "stated preference methods" based on surveys. For each external effect, the valuation approaches used are described. The results of various studies made in Europe in this field are compared.

INTRODUCTION

When a car moves in the streets of one of our cities, it makes noise and generates vibrations, emits gases and particles and becomes a hazard for pedestrians, cyclists and other motor vehicles. The construction, maintenance and future end-of-life of the car produce many industrial wastes and the production of petrol or diesel is also a source of pollution. Cars slow one another down, thus each car causes other cars to incur extra travel time. The effects of traffic congestions in urban areas are experienced every day.

If environmental impacts were influenced by the rule of a competitive market, prices would ensure a correct allocation of resources and would correct the oversupply of these effects. The absence of markets for environmental damage is a clear case for public action; public authorities must devise mechanisms to respond to this absence of a competitive market. Assessment methods are needed to support the decisions they have to make. Going from the simpler to the more sophisticated, the assessment can provide general information for defining strategic orientations of transport policy, provide means to help decisions relating to the choice of infrastructure developments, inform questions of regulation and tariff systems, in order for instance to implement the "polluter pays" principle. In short, assessment of environmental impacts is of paramount importance for policy-making.

How to assess these environmental impacts is the topic of this chapter. The first section lists environmental impacts of transport. The second section describes various means and methods of assessment of these environmental impacts, among which monetarisation is the most elaborate. The third section analyses how these means are applied to assess different environmental impacts and reviews results of assessments for each listed impact and the extent to which monetarisation is possible. The fourth section presents results of the most important monetarisation studies conducted in Europe. The fifth section gives our conclusions.

OVERVIEW OF ENVIRONMENTAL IMPACTS

If we examine the system of transport from direct to indirect effects as would a good physician, the diagnosis gives a long list of serious forms of transport "illnesses":

- Power sources and their effects
  - Energy consumption;
  - Local and regional air pollution from vehicles exhausts;
  - Greenhouse effect caused by major emissions of greenhouse gases, especially CO_2;
  - Risks specific to electricity: electromagnetic fields, nuclear risks and radioactive wastes.
- Noise and vibrations
- Permanent effects of infrastructure
  - Land use and the effect of separation (community severance);
  - Obstacles to water flow;
  - Damage to the landscape and aesthetic effects;
  - Impact on fauna and flora.
- Congestion
- Prejudice to personal safety
- Discharge of effluents: industrial pollution
- Up- and downstream effects resulting from activities related to transport (manufacture of vehicles and infrastructure and of goods needed to operate them), impact of the life cycle of infrastructure, rolling stock, equipment and energy
- Water and soil pollution.

Assessment Methods

Several levels of ambition can be exercised in assessing the effects of transport on the environment (e.g., Quinet, 2003). At the very least, one can describe the phenomena. For certain impacts, for example effects on the landscape, it seems that it is not possible go beyond this (although, at least theoretically, it is possible to translate the corresponding impacts into monetary terms, for instance through stated preference questionnaires, see below).

Where it is possible to quantify, we obtain greater comparability and thus better guidance for decision-making. But quantification implies a choice of indicators, and hence a value judgment. In general, anything that affects human beings and their well-being would be favoured[1].

In its most complete form, the assessment attributes monetary values to external effects, which allows the corresponding impacts to be more easily incorporated in economic cost-benefit analyses. Monetarising non-market goods allows more consistent and equitable decision-making. For example, it could be inefficient and unfair to spend €10,000,000 per reduced human fatality in one situation (for instance, investment in a medical treatment) but not spend $100,000 to provide comparable human health benefits in another sector or location (for instance, by improving road safety).

A stepwise approach, often named the "Impact Pathway Approach", may be represented by a diagram such as that shown in Figure 1, taken from Friedrich and Bickel (2002). Two different aspects appear clearly in this diagram.

(i) A "physical" aspect, represented by the left side of the diagram, which expresses the relationships between transport activities and impacts on the environment. This is fed by the body of accumulated scientific knowledge that establishes the laws and parameters on which these relationships depend. The laws in question are often expressed by functional relationships between emissions and their effects, known as "dose-response relationships" or the "exposure-response relationships" highlighting how and in accordance with which parameters impacts are related to emissions.

(ii) An economic aspect is described on the right side of the diagram and introducing monetary valuation. This aspect takes quantity as its starting-point (number of deaths, quantities of crops destroyed and species disappeared) and applies unitary values to these quantities (value of a death, value of a crop).

[1] It should be noted that in doing this we make a choice concerning well-being and all that it depends on. In practice, we exclude the points of view of individuals whose preferences and attitudes are extreme. Thus, we make a sort of majority choice among tastes and attitudes, turning towards the tastes and attitudes of the *homo economicus* of our current societies – but we do not go as far as to substitute ourselves or to weight the impacts.

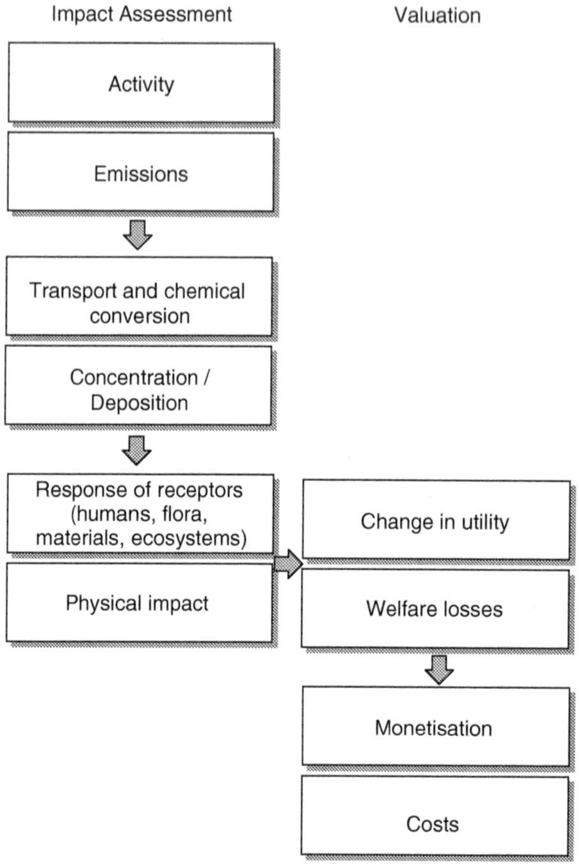

Source: Friedrich and Bickel (2002).

Figure 1: The "Impact Pathway Approach" for the Quantification of Marginal External Costs Caused by Air Pollution and Noise

The economic aspect of the Impact Pathway Approach referred to as monetary valuation implies several difficulties linked to the fact that there is no market for these impacts. In the case of commodities (i.e., goods commonly sold in a competitive market), the values are the market prices. For non-market goods, they are determined by methods that essentially aim to evaluate what individuals who suffer from the impact on the environment would be willing to accept (that is, the amount of financial compensation that victims would have demanded before they would "volunteer" to accept such damages). Several methods are used. One can distinguish between demand and supply methods.

Demand Methods Aim at Evaluating the Willingness to Pay of the Agents. Willingness to pay can be assessed either through agents' observed behaviour, known as "revealed preference" methods, or through asking agents questions about fictitious conditions, known as "stated preference" methods.

- Revealed preference methods analyse situations, where the individual has a choice between greater or lesser degrees of pollution or nuisance and more or less of a market commodity. These methods take advantage of the presence of surrogate markets; which can be dealt with in a number of different ways:
    - The cost of trips necessary to benefit from an amenity can be used to value that amenity; this method is used, for example, for the evaluation of leisure parks.
    - The use of hedonic prices: the price of some marketed goods depends on their exposure to characteristics such as air pollution or noise. Variations in the price of the marketed good with respect to variations in the environmental factor can thus be used to estimate the implicit value that individuals attach to the environmental quality.
    - Estimates of the cost of environmental protection. Observing the amount that individuals pay to reduce or to eliminate negative environmental effects (e.g., how much they pay to protect against noise through double-windows or other devices) provides an estimate of their implicit valuation.

These methods pose several difficulties (see Adamowicz, 2003). Each method implies that people are well informed and aware of the damage caused by the targeted nuisance. Hedonic relations are not true demand functions. Residents of the most noise-exposed houses may have chosen them partly because they are less sensitive to the environmental factor in which case the value of dwellings would decrease less than if the sensitivity was the same for everybody. Furthermore, there is the problem of distinguishing the individual impacts of various environmental variables, which are often closely linked to each other.

- Stated preference methods are based on survey questions about intended behaviour. Two forms can be used: either asking people what they are willing to pay to avoid the nuisance, or asking them what they are willing to receive in order to continue suffering from it. There are many difficulties of implementation:
    - To get reliable answers, it is necessary to devote a lot of care to devising the questionnaire, and to administer it through very sophisticated controlled procedures; nevertheless, it is difficult to avoid biases referenced in the literature as the strategic bias, the starting-point bias, the interviewer bias, the social norm bias, the payment vehicle or instrument bias, the hypothetical bias, the non-response bias and the whole-for-one bias.
    - Owing to psychological biases, willingness to pay to avoid a nuisance is consistently lower than the willingness to accept compensation to continue to suffer from it.

- As is the case for other methods, results can be biased by poor knowledge of the actual damage. This point is especially important in the case of value of human life estimates: who can conceive the meaning of a decrease of 10^{-6} in the risk of death?

It is generally assumed that revealed preferences methods are more reliable than stated preference methods as they reveal agents' "real" preferences. But their implementation must overcome several difficulties, such as statistical difficulties of assessing the impact of a factor that may be of minor importance.

The main *supply method* is known as the "cost of damage". It seeks to estimate the monetary cost of repairing the damage caused by the pollution or nuisance in question or of avoiding its effects. They are implemented in two stages. The first is technical aiming at estimates of the consequences of the nuisance in physical units, for instance, in the case of air pollution, the frequency and significance of any impacts on health or the damage to buildings. The second stage is the monetarisation of the damage, either through market prices for damaged goods, or through costs of repairing the damage, such as health care costs of injured or sick people, or through more subjective valuations (e.g., the value of human life). This method does not have the problem of lack of information, but monetary valuation of environmental damage is hazardous. For example, the cost of repair overestimates willingness to pay when the repair is not actually carried out and repair implies the definition of a zero level for the nuisance, and this definition is often arbitrary (in the case of noise for instance).

As already mentioned, the methods of revealed or stated preferences estimate the willingness to pay of those who bear the nuisance; they deal with demand. Avoidance or repair costs deal with supply. If the decisions were optimal, these methods would produce the same result and the double equality would hold:

$$\text{Price} = \text{marginal cost} = \text{marginal willingness to pay}$$

Figure 2 displays the total costs in the top part of the diagram and the corresponding marginal curves in the lower part. $C(N)$ represents the cost to the polluting producer, showing how marginal costs fall as the level of the polluting activity rises, whilst $D(N)$ is the external cost of the polluting activity (willingness to pay for pollution reduction). Without any intervention, the producer would seek to produce at N_m. The social optimum, where total costs, $T(N)$, are minimised is at N_0. If the polluting activity were reduced too far, say to N, then the gains from the loss of pollution would be outweighed by the loss of benefit to the consumers of the polluting activity.

However, in practice the demand and supply methods do not provide equal values. Thus, we have to take the decision as to which one to choose. The answer will typically depend on the nature of the problem that has to be solved (Infras-IWW 1995; Quinet,

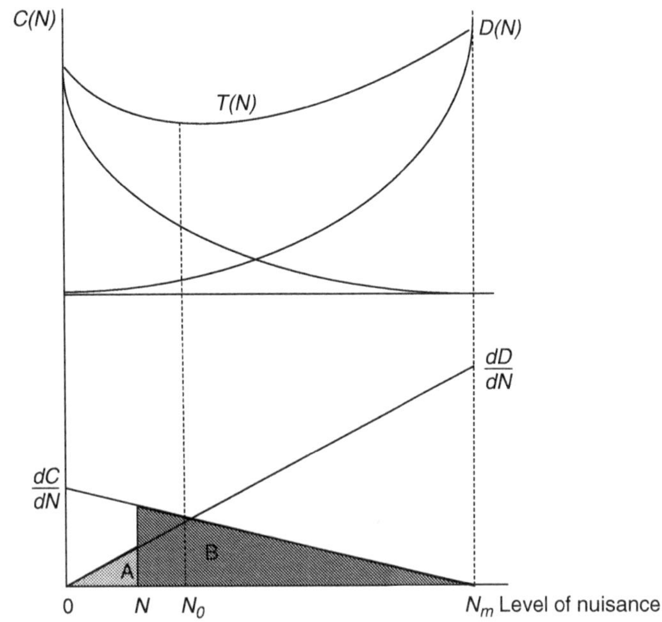

Source: Quinet and Vickerman (2004).

Figure 2: Environmental Optimum

1996). The majority of the studies choose estimates based on willingness to pay, that is, area A.

Moreover, both demand and supply methods imply several different actors. In the case of noise, the willingness to pay is based on the willingness to pay by agents who actually suffer from the noise. Similarly, the protection cost function includes both the cost of a noise barrier, paid for by the public authorities, and the costs of double-glazed windows paid for by the residents.

When the effect is permanent (in the case of global warming or effect on landscape, for instance), its valuation needs to include the effects on future generations. This raises two problems, how to estimate the value for these future generations and what discount rate to choose. It is often argued that three categories of values need to be taken into account, the sum of which is the total value of the good:

- The value in use, which corresponds to the actual revealed consumption of the good;

- The option value, which corresponds to the possibility of a future consumption of the good;
- The existence value, which corresponds to the value obtained by an individual who never consumes the good, but appreciates its existence so that other people can use it, or simply just as a resource.

These three values also exist for marketed goods, but it is often thought that option value and existence value are pre-eminent in the case of environmental goods, especially for the case of depletable goods. Cost of travel or hedonic prices only provide values in use; for stated preferences the outcome will depend on how the questions are asked.

THE USE OF ASSESSMENT METHODS

The Impact Pathway approach, which has been explained in the case of air pollution, constitutes an ideal that cannot be applied in its purest form for all impacts. Each type of impact requires different methods to quantify and monetarize, which face various barriers and obstacles.

- First, one may have to skip a stage, or the series may not be completed. It is, for example, obvious that the quantification of effects, such as aesthetic consequences, has no precise meaning. This does not mean that one cannot attribute a monetary value to a landscape, however, as we will see later.
- Monetarization is often difficult, at least in our present state of knowledge. For example, whilst we agree that biodiversity has significant value, it is difficult to place a value on a particular incremental change in the quantity or quality of wildlife habitat.
- Ideally, we would apply this approach to each unit of transport or for each isolated section of infrastructure, and then, to reply to questions on wider geographical areas, aggregate the elementary information gathered in this way (the "bottom-up" approach). But the implementation of this ideal often runs up against our inadequate knowledge and difficulty of obtaining the necessary data. So "top-down" approaches are also used. They apply overall results, for instance at the country level, then these results are disaggregated between several factors (such as rural and urban areas, or according to the type of vehicle). In this way, overall results are obtained, but we cannot distinguish with any degree of precision between local situations or various types of vehicles. The more dependent on local parameters the phenomena are, the more unsuitable "top-down" studies are, and as our knowledge develops, they are being used less and less.
- The implementation of the above-mentioned methods of evaluation runs up against another difficulty, that of synergies between impacts, which makes their separate analysis biased. For example, earth and air pollution combine to modify

the flora in a particular way. But our knowledge concerning these synergies is still very limited, except in certain particular cases.

We will now analyse these issues further for each of the impacts, distinguishing between physical and monetarised assessment

Physical Assessment

Each nuisance can be described as a complex system with many characteristics to be measured. *Accidents* can be represented by the numbers of fatalities, of severe injuries and of light injuries. National, regional and local statistics supply these indicators. Data are available for each category, that is, injuries, deaths, etc. Data show that road transport is by large the most unsafe mode, but there are large differences between the various types of roads (rural road or city street, motorways or usual highways).

Air pollution is defined as the emission of principal gases and fine particles harmful not only to human lungs but also for materials and buildings, agricultural crops and forests. The gases are CO (carbon monoxide gas), NO_x (different nitrogen oxides) and fine particles are noted PM10 (less than 10 microns). Based on a physical evaluation of nuisances, together with a comparison of air pollution (including the greenhouse effect) of heavy goods vehicles and combined transport, it may be concluded that heavy goods vehicles pollute 6–10 times more than combined transport, even including road emissions on terminal links as is shown in Table 1.

Climate change stems from the overall increase of temperature of the atmosphere; consequences are significantly complicated and various: increased desertification, raised sea levels, serious harm to agriculture and other destructive environmental and health-related side-effects. These effects result from very complex physical mechanisms that are the subject of many studies because of the very important impact for our world. The greenhouse gases (GHGs) that cause it – above all CO_2 – are well known. About one-third of greenhouse gases emissions come from transport.

Table 1: Comparison of the Pollution Produced by Heavy Goods Road Vehicles and Combined Rail Transport (Including Terminal Sections)

Emissions (in Grams per Tonne. Km)	*Heavy Goods Vehicles (15t. Payload Maxicode)*	*SNCF Combined Transport (15t. Payload Wagon)*
Carbonic gases (CO_2)	72.0	7.5
Carbon monoxide (CO)	0.32	0.05
Volatile organic compounds	0.18	0.02
Nitric oxides (NO_x)	1.04	0.11
Particles	0.1	0.1

Source: French Agency for the Environment and Energy Management (ADEME, 1994).

Consequences of the *noise* levels are undesired social disturbances as well as detrimental effects on individual well-being which can entail physical and psychological harm: consequences on hearing, stress reactions and increasing of cardiac infractions. Noise is a complex mechanism with a complex indicator – the A-weighted decibel (dB(A) – due to the fact that noise varies in time and space). There is consensus to differentiate significantly between the impact of a given noise level at night and in the daytime, southern European countries have higher levels than the northern because windows are often open. The European mean for day and night can be estimated at 50 dB(A). Serious consequences on health arise above 85 dB(A).

Congestion is the increase in travel time that each user causes to the others through slowing of the traffic flow. It can only be analysed and predicted using traffic models. The transition from free flow to congested traffic is very sudden. Time lost is a function of both traffic and the configuration of the available infrastructure.

Damages to nature and landscape can only be considered in the aggregate: waterproofing of lands is a typical indicator of the transformation of nature. *Additional effects of infrastructure in urban areas* such as separation (barrier effects), impacts for pedestrians and cyclists can be evaluated in terms of time lost due to lengthening pedestrian and cycle trips.

Impacts of up- and downstream processes include all negative effects attendant to the construction and maintenance of power plants, the production of wastes from energy generation or from vehicle or infrastructure construction and the use of those infrastructures. There are many indicators such as quantities of waste, for instance, oil refining wastes or wastes from nuclear generation of electricity.

Monetarization

Accidents. Methods for evaluating the cost of accidents generally involve multiplying the numbers of dead and injured people and other losses by the per unit cost of these deaths, injuries and losses. Material damages are usually assumed to be equivalent to the monetary costs of the damages. For deaths and injuries, the estimates typically cover the direct costs (such as medical care) and they involve evaluation of the value of human life. Two methods are used in this respect: evaluation based on willingness to pay through stated preferences or evaluations known as "human capital", based on production losses, either gross (without subtracting the avoided consumption) or net (from which the consumption has been subtracted). The principal parameter is the risk value by casualty. A common scale is 1.5 million euros per fatality, 200,000 euros per severe injury and 15,000 euros per light injury (as for instance in UNITE 2003). For unsafety, there is also a large uncertainty about the level of marginal cost, and more precisely about the relation between fatalities and traffic volume: in some studies the fatalities do not vary with traffic volume, while in others fatalities are proportional to traffic volume. Furthermore, studies differ according to whether personal unsafety risks are internalised or not.

Air pollution. The main external cost of air pollution is the impact on human health that can be evaluated by epidemiologic studies, such as WHO (1999). These studies estimate road transport-related health costs (mortality and morbidity) and are based on a link with a characteristic emission like PM10 particles. This relation allocates human impact to each category of traffic. Other studies assess the damages of emissions of NO_x, VOC and SO_2 on buildings, crops and forests.

Climate change. For climate change, the damage approach is not possible since long-term climate change risks and consequences are very difficult to estimate. Thus, estimates usually rely on the avoidance cost method based on specific reduction aims. However, several problems have to be faced. First of all, the reduction aim has to take into account both a scientific and a political dimension. How much to cut back within what timeframe? The present official commitments are those of the Kyoto Protocol, to reduce emissions by 8% in the European Union and by 5.2% in the industrialised countries of Annex 1 of the Protocol, between 1990 and the period 2008/2012. This is the base for the estimate of 20 euros per ton of CO_2 in the UNITE study (UNITE, 2003). This value represents a central estimate of the range of values to meet the Kyoto targets (5–38 euros in the European Union, based on an evaluation by Capros and Mantzos, 2000) with a modelling approach of emissions and economy (according to Capros and Mantzos, a full trade flexibility for GHGs involving all countries of the world leads to the low value while a no-trade scenario leads to the high value). But the scientists of the International Panel for Climate Change (IPCC) estimate that the minimal effort is to stabilise the stock of GHGs to a reasonable high level of 450 parts per million – thus explicitly more than the GHG stock of 1990 – and the corresponding value in their models is 140 euros per ton for a reduction of emissions of 70% in the 2050 time frame. This is also the aim in the Decision 1600/2002/EC (22 July 2002) of the European Parliament and of the Council for the Sixth Community Action Programme for the Environment. And now the first European discussions for the post-2012 period consider a next step of reduction of 15% to 30%. An INFRAS-IWW study (INFRAS-IWW, 2004) considers the two values of 20 and 140 euros per ton of CO_2 but recommends the higher level. These evolutions of the emission thresholds are summarized in Figure 3.

A second question is how to assess the level of reduction of emissions from the transport sector in the overall policy of emissions reduction. From an economic point of view, different studies show that efforts in the sector of transport are more difficult and more expensive than in other sectors like housing or industry. Yet transport is a big contributor to climate change and above all it has the highest growth rate.

Noise. The evaluation of noise is based either on stated preferences (estimation of persons disturbed by the noise of transport and their willingness to pay for reducing that noise to an acceptable level) or on revealed preferences (the loss of value of properties which are exposed to noise). Both types point to a correlation between the cost of a level of noise in dB(A) the per capita income category. To these, psychological effects are added health effects: the number of cardiac infarctions and number of persons suffering

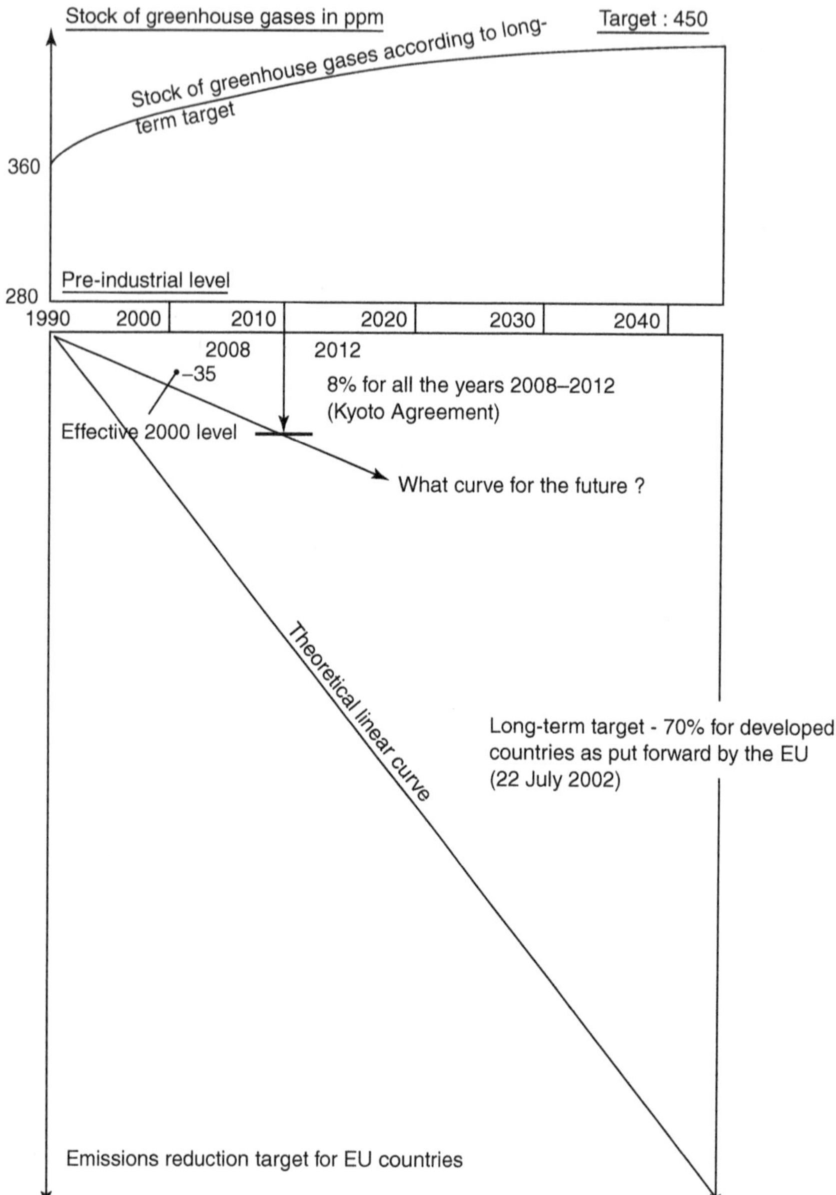

Source: Domergue (2003).

Figure 3: Green House Gas Emissions over the Years

physical and psychological harm can be evaluated using the same injury risk value of human life as for accidents.

Congestion. Congestion can only be estimated by traffic models designed for specific situations defined by the physical road network and the level of traffic on that network. Time lost on the road network is valued by taking an average value of time linked to the per capita income of each location. As far as public transport is concerned, the Mohring effect goes in the opposite direction and is a positive externality: when public transport patronage increases, the operating companies increase the frequency of services, then the headway and the total travel time decrease for the users; so a newcomer improves the situation of the ancient users, contrarily to what happens for road traffic.

Nature and landscape. Evaluation of external costs affecting nature and landscape is based on a "repair approach", as between the real situation and a reference situation such as the estimated state of the landscape in a prior period, for example before the construction of motorways after the Second World War. The "repair costs" are those associated with land "unsealing", restoration of biotopes, removal of soil or water pollution and other impacts. The repair approach method is used for all categories of areas and the costs are allocated for each transport infrastructure and level of traffic.

Urban effects. Separation effects in urban areas are estimated in terms of the delays incurred by pedestrians in moving across or around different types of roads or railway lines or other infrastructure. Nuisances for cyclists can be evaluated in terms of the costs of creating bicycle paths to avoid the barriers presented by the road infrastructure.

Up- and Downstream processes. The valuation approach considers additional air pollution and climate change effects due to energy generation processes, the construction and maintenance of vehicles, the construction and maintenance of infrastructure and also special risks associated with nuclear plant electricity generation. Each overall process is studied over the life cycle of the product.

EUROPEAN RESULTS

An Overview

In the long process that efforts to improve infrastructure pricing and to promote a sustainable transport policy constitutes, many estimates of external costs of transport have been made in Europe in recent years, supported by national governments, by international governmental organisations such as European Conference of Ministers of Transport (ECMT), or by the Commission of the European Union or by industry organisations such as the International Union of Railways (UIC). Among the most recent and most comprehensive studies are TRENEN II STRAN (1999), CAPRI (1999),

QUITS (1999), INFRAS-IWW (1995, 2000, 2004), RECORDIT (2001), PETS (2001), UNITE (2003), ECMT (2002), Samson *et al.* (1998).

Quinet (2004) analysed the sources of dispersion between the results of the main European external costs studies. These sources of dispersion are manifold:

- *The specificities of situations.* The situations differ according to the precise type of vehicle: we have seen previously that type of vehicle being evaluated may differ from a study to another (for instance a study concentrates on small cars, another on large cars). Furthermore, fuel consumption, and consequently air pollution, differs from one country to another. Similarly, the location and density of settlement have impacts on external costs.
- *The type of cost taken into consideration.* Some studies reckon average costs, other ones reckon marginal costs. Both concepts have an interest in economic analysis, the point is to know which one is calculated.
- *The types of external effect taken into consideration.* The studies do not take into account the same effects; most of them include noise, air pollution, global warming and unsafety; but not all of them include congestion, a very few of them include the Mohring effect, and a fewer number of them includes urban effects, up- and downstream effects, consequences on landscape, etc.
- *Physical relations.* For each of the effects, the calculation of costs is based on "physical" laws, which link the cause of damages to the effects: for instance, air pollution estimates use a chain of relations between gas exhausts, diffusion in the atmosphere, exposure of human beings, damages to health. Similarly, costs of unsafety use relations between the level of traffic and the number of fatalities. It happens that these relations include a large uncertainty, and that alternative relations exist for many of them. As an example, air pollution in Europe have been analysed through two main methodologies (stemming from the ExternE study and the WHO study) that give very different results.
- *The "secondary hypotheses" used by the modelling framework.* It is well known that huge models such as those which are used to estimate air pollution, congestion or global warming include, besides the general hypotheses which characterize them, a lot of secondary hypotheses which do not appear at first glance, but need to be made in the implementation of the model in a specific situation. These secondary hypotheses often concern to the data handling and adaptation to the needs of the theoretical framework of the model. Though difficult to assess without a deep insight in the model, these secondary hypotheses frequently have dramatic impacts on the numerical results.
- *Unit values.* Cost estimates use unit values such as value of time, value of statistical life. These values may differ from one study to another.

A meta-analysis of these various studies shows that the main causes of differences between assessments of external costs are the specificity of the situation (country,

Table 2: Share of GDP According to Two European Studies

Assessment Study	*INFRAS-IWW (October 2004)*	*UNITE (2001/2003)*
Studied countries	EU15 + Switzerland + Norway	EU15 + Switzerland + Hungary + Estonia
Year of the data and results	2000	1998
Externalities considered in % of GDP		
Accidents	1.8	0.5
Air pollution	2.0	0.6
Climate change	2.2	0.2
Noise	0.5	0.3
Subtotal	6.5	1.6
Congestion	3.0	1.0
Subtotal with congestion	9.5	2.6
Nature and landscape	0.2	No
Urban effects	0.1	No
Up- and Downstream processes	0.5	No
Grand total	10.3	2.6

Source: INFRAS-IWW (2004) and UNITE (2003).

location in an urban area or not, etc.) and the type of cost calculated (average or marginal). Compared to these causes, the scientific uncertainty, often advocated to disqualify the monetarisation, plays a small role in differences between cost estimates. It appears that estimates of external costs are reliable – provided that the location and type of cost of each estimate is clearly indicated and acknowledged – and can be used for political decision-making. As an illustration, let us compare the overall results of the two most comprehensive ones UNITE[2] and INFRAS-IWW[3] given in Table 2. They cover roughly the same geographical area: 15 countries for INFRAS-IWW, 18 countries for UNITE, the 3 extra countries being eastern European countries. Both studies use each of the three already-mentioned methods of monetarisation, and both use the same method for each item. Stated preference methods are used to estimate value of human life and noise. Revealed preference methods are used for noise and congestion. Cost of damage methods are used for air pollution and nature and landscape. The case of global warming is specific: the values are the prices deemed to equilibrate in the future the permit market which is set up in the Kyoto agreement framework.

[2] UNITE is a reference for the Commission for many assessments and estimates, for instance for the cost/benefit calculations of technical specifications for interoperability (TSIs) made for the Commission by the European Railways Interoperability Association (ERIA).

[3] INFRAS-IWW is a consortium composed of the Swiss INFRAS company and of the University of Karlsruhe which made three successive updating studies for the International Union of Railways in 1995, 2000 and 2004. These results are used worldwide and especially by the EU Commission, by the European Conference of Ministers of Transport (ECMT), by the European Environmental Agency (annual transport-environment TERM indicators) and by many trade associations.

A careful examination shows that the differences in the results stem mainly from differences in the definition of the costs; more precisely, UNITE reckons marginal costs while INFRAS-IWW reckons average costs. Furthermore, for safety, UNITE takes into account only costs that are not perceived by the user, while INFRAS-IWW includes also costs perceived by the user. UNITE has not taken into account several effects which are included in INFRAS-IWW. In UNITE, the shadow cost of CO_2 is low as it corresponds to the predictable value in the near future stemming from the Kyoto agreements, while in INFRAS-IWW it corresponds to the far future consequences of the Kyoto agreements. Of course, in these cases, no choice is intrinsically better than the other; it depends on the use of the estimate, more specifically on the kind of decision it is used for. For instance, marginal costs shall be usually preferred in questions of pricing, while average costs will be preferred in questions of strategic decisions or investment choice.

Evolution of External Costs with Time

The external values have risen considerably, as illustrated in the retrospective example presented by Hansson (1997) for the values for unsafety used to describe the roads in Sweden over a 24-year period (1971–1995), and summarised in Table 3. The values assigned to human life grow faster than the per capita GDP. This growth should be divided into distinct components:

- The physical evolution of each nuisance broken down into two factors, technological progress and penetration of that progress into the fleet of transport vehicles or the infrastructure.
- The evolution of the unit values, a factor that takes into account the effects of economic growth. Most studies assume that the unit values grow at an annual rate similar to the rate of per capita income growth; this assumption may be questioned because it overlooks the growing awareness of public opinion about

Table 3: Increase in Unit Values Used to Illustrate Unsafety in Sweden (in Millions of SEK of 1997)

	Fatalities	*Serious Injuries*	*Light Injuries*
1971	3.6		
1976	4.0	0.06	0.008
1981	4.6	0.095	0.011
1986	6.5	0.7	0.026
1989	8.4	0.91	0.039
1992	12.2	2.0	0.050
1995	13.0	2.0	0.090
Change	× 3.6 in 24 years	× 33 in 19 years	× 11 in 19 years
Average per year	+5.5%	+20.2%	+13.6%

Source: Hansson (1997).

environmental harm. Hansson (1997) shows, for instance, that the evolution in Sweden of the value of life over the previous 25 years was faster than the growth of per capita income, as shown in Table 3.

It is clear from Table 3 and from other evidence that sensitivity to environmental issues is growing at a much higher rate than other related economic indicators such as per capita GDP.

CONCLUSION

The transport system in the European Union is useful but unhealthy. According to the European Commission (2001) White Paper of September 2001, the transport sector accounts for 10% of GDP and its very low costs make it a key instrument for economic development to the extent that it is a variable for economic adjustment. This is to the detriment of individuals who take care of the day-to-day running of the system and of the environment that is being adversely affected by the nuisances caused by transport. In western Europe today, such nuisances account for around 10% of GNP according to the more complete studies made in western Europe.

Obviously, we cannot ignore the second figure though it represents external, "collateral" charges that are inevitable according to some. Unfortunately, they are the monetary translation of the upheavals described earlier. We all know people who have perished in road accidents. We are all subjected to the smell of exhaust fumes. Time and again, we witness natural disasters that experts now claim to be the direct effects of climate change. We do understand the dangers. Economists can now assess negative effects of transport system. They can monetarise most of them, and therefore assess the efficiency of various political instruments to remedy these dangers, from regulations to taxations, investment policies or new negotiable exhaust emissions contracts. This gives us some hope for the future: it makes sense to take decisions relying on these money valuations.

REFERENCES

Adamowicz, W. (2003). Valuation of Environmental Externalities. In: *Handbook of Transport and the Environment* (D. Hensher and K. Button, eds), Elsevier, Amsterdam, 375–389.

ADEME (1994). *Rapport Transport, Energie, Environment: Le Delfi,* Paris: Ademe.

CAPRI (1999). *Concerted Action on Transport Pricing Research Integration*, Project Funded by the European Commission under the Transport RTD Programme of the 4th Programme Framework.

Capros, P. and L. Mantzos (2000). Kyoto and Technology at the European Union: Costs of Emission Reduction under Flexibility Mechanism and Technology Progress, *International Journal of Global Energy Issues*, **14** , 169–183.

Domergue, P. (2003). Guérir notre système de transport – diagnostic, remèdes et ordonnances, *Revue Transports*, **417** (Janvier–Février), 21–30.

ECMT (2002). *Groupe on Fiscal and Financial Aspects of Transport, Optimal Transport Pricing*, ECMT, CEMT/CS/FIFI/2004/04, Paris.

European Commission (2001). *White Paper "European Transport Policy for 2010: Time to Decide,"* European Commission, Brussels.

Friedrich, R. and P. Bickel (eds) (2002). *Environmental External Costs of Transport*, Springer-Verlag, Berlin.

Hansson, L. (1997). *The Internalization of External Effects in Swedish Transport Policy*, Unpublished Paper, Lund University, Lund, Sweden.

INFRAS/IWW (1995). *External Effects of Transport*, Report for the UIC, Karlsruhe, Zürich.

INFRAS/IWW (2000). *External Costs of Transport: Accident, Environmental and Congestion Costs of Transport in Western Europe*, Report for the UIC Zürich/ Karlsruhe, Zürich.

INFRAS/IWW (2004). *External Costs of Transport Up Date Study*, Report for the UIC, Zürich/Karlsruhe, Zürich.

PETS (2001). *Pricing European Transport Systems*, Research Project Funded by the European Commission under the Transport RTD Programme of the 4th Programme Framework, Project Coordinator: C. Nash.

Quinet, E. (1996). Valuation of Environmental Externalities: Some Recent Results. In: *Infrastructure and the Complexity of Economic Development* (D.F. Batten and C. Karlsson, eds), Springer, Berlin.

Quinet, E. (2003). Evaluation of Environmental Impacts. In: *Handbook of Transport and the Environment* (D. Hensher and K. Button, eds), Elsevier, Amsterdam, 357–373.

Quinet, E. (2004). A Meta-Analysis of Western European External Costs Estimates, *Transportation Research D*, 9, 465–476.

Quinet, E. and R. Vickerman (2004). *Principles of Transport Economics*, Edward Elgar, London.

QUITS (1999). *Quality Indicators for Transport Systems*, Project Funded by the European Commission under the Transport RTD Programme of the 4th Framework Programme, Project coordinator: ISIS, Rome.

RECORDIT (2001). *Real Cost Reduction of Door-to-Door Intermodal Transport*, Project Funded by the European Commission under the Transport RTD Programme of the 4th Framework Programme, Project coordinator: ISIS, Roma.

Samson, T., C. Nash, P.J. Mackie, J. Shires, and P. Watkiss (1998). *Surface Transport Costs and Charges Great-Britain 1998*, ITS Leeds, Leeds.

TRENEN II STRAN (1999). Project Founded by the European Commission under the Transport RTD Programme of the 4th Framework Programme, Project

Coordinator: S. Proost, Kurt Van Dender, Centre for Economic Studies, Katholieke Universiteit, Leuven, Belgium.

UNITE (2003). *Unification of Accounts and Marginal Costs for Transport Efficiency*, Report Funded by the European Commission, 5th Framework – Transport RTD, Project coordinator: C. Nash, ITS Leeds.

WHO (1999). *Health Costs Due to Road Traffic-Related Air Pollution: An Impact Assessment Project in Austria, France and Switzerland*, Synthesis Report, WHO, London.

PART 2:
DETERMINANTS OF PRIVATE CAR USE

Threats from Car Traffic to the Quality of Urban Life
Tommy Gärling and Linda Steg (Editors)

8

THE USE OF THE CAR–MOBILITY DEPENDENCIES OF URBAN EVERYDAY LIFE

Bertil Vilhelmson, Department of Human and Economic Geography, School of Economics and Law, Göteborg University, Göteborg, Sweden

ABSTRACT

What adaptive potentials exist in the short and medium term if mobility has to be reduced? This chapter examines how tied people's activity patterns and the socio-spatial organization of modern society has become to high levels of mobility. It elaborates on four intertwined aspects of mobility dependence and adaptation: (i) time and time use, (ii) travel-related activities and goals, (iii) urban structure and land use, and (iv) virtual mobility alternatives. A main objective is to explore the structure of predominantly car-dependent urban activities according to their flexibility (or fixedness) in time and space.

INTRODUCTION

Extended Activity Space

The activity pattern of an individual essentially expresses his or her use of time for different purposes at various places. For many, this pattern is largely dependent on fast travel and is widely extended in urban space. This also implies that any change in the transportation options of an individual, for example, gaining or losing access to a car, has to be managed through changes in the distribution of time used for different activities and/or where these activities are performed (Gärling *et al.*, 2002). In modern societies, the spread of high-speed travel increased due to incomes and hence car availability among households has resulted in an extension of the activity space of individuals and has increased the accessibility of a range of life chances as well (Orfeuil and

Salomon, 1993; Pucher, 1999). Being able to travel fast has gained a stable position in the hierarchy of needs and priorities of individuals and households in most countries and cultures. Growing incomes have meant, and still mean, increasing demands for mobility, not least for car ownership and use.

The long-term development of mobility in Sweden depicts this process of extension. Going back to circa 1900, the average distance travelled was only around 1 km per person per day; in the early 1950s, the take-off stage of the diffusion of cars, mobility grew fast, passing an average of 10 km a day (Vilhelmson, 1999a). Today, average domestic travel totals 45 km per individual per day for the population as a whole[1]. So, over a century, the capacity of people to cover distances and interact with other places has increased by almost 50 times and the potential activity space covered by approximately 2,000 times. This development has affected the lifestyles and daily activity patterns of most people, in terms of where they live, work, shop, and spend leisure time, as well as affecting urban land use and city structure. However, these mobility capabilities are not evenly distributed throughout the population. For example, middle-aged male car users on average travel more than 60 km per day, greatly exceeding the average. This indicates a huge potential for mobility growth in years to come, if economic and political conditions allow it and the preferences of the population remain stable.

The process of extension is strongly associated with another main transformation of urban mobility, namely, shifts in ways of travelling and modal choice. In Sweden, since the early 1950s almost all growth in domestic daily mobility can be explained by increased car use. In most European countries, the car accounts for more than 80% of the total daily distance of passenger travel (Eurostat, 2005)[2]. Thus the relative importance of slower modes has continuously decreased, this decrease mainly involving urban/regional public transit by bus, train, and – not to be forgotten – on foot. Measures to counterbalance this trend are sometimes locally successful: for example, investments in rail-based rapid transit systems, urban bike routes, and bus-priority plans in certain larger cities (Turton and Knowles, 1998). Undoubtedly, a key element in the competition between different modes of urban transport is speed: all things being equal, if people have the opportunity, they generally choose the fastest mode of transportation.

Massive car use, however, does not only widen the range of opportunities. The social and spatial organization of a society also successively adapts to the increased mobility standards of a population. The form and functioning of a modern city – and citizens' lifestyles – may place profound demands on the mobility capabilities of individuals. The catchment areas of schools, shops, hospitals, and all kinds of leisure activities increase

[1] Details presented in this chapter concerning current Swedish travel are based on estimates derived from the national travel surveys.

[2] Note that car use, as a percent of total number of urban trips, is a more modest 45% in other European countries, while it is almost 90% in the USA (Pucher, 1999).

when the majority of people become more mobile. The need for mobility is thus built into the functioning and structure of the urban high-speed society. In the long run, an initially perceived freedom of action turns more or less into a pressing necessity, as indicated by traffic congestion and increased travel times during peak hours. Travelling extended distances is required in the daily life of many citizens, not least so that they can have appropriate work. This requirement is of course troublesome if travel demand has to be "managed". Still, it could be argued that there is freedom of choice from the individual point of view, not least regarding activities and activity locations during daily or weekly free time.

The Need for Decoupling and Adaptation

While physical transportation has continuously increased, problems have emerged that militate against maintained high levels and future mobility growth. These problems concern air pollution, energy use, spatial requirements, and risks associated with current transportation patterns (see Part 1 of this volume). Many studies show that necessary technical steps designed to reduce the negative impacts of cars (e.g., catalytic converters reducing exhaust emissions and more energy-efficient vehicles) risk being offset by a future total increase in traffic volume. This situation calls for a *decoupling* of material welfare (income) growth on the one hand, and distance travelled (energy use) and urban concentration of traffic (congestion, air pollution, and noise) on the other. Ongoing discussions in this regard concern various measures to reduce the need for travel, such as information and marketing, prices and taxes, telecommunications (e.g., telecommuting), and urban land-use planning (see Part 3 of this volume), and how they might influence the way people travel and carry out their everyday activities. So, while it is commonly agreed that overall the transportation system must be adapted to sound environmental standards, in the end such an adaptation may also restrict our daily mobility. Such mobility reductions inevitably affect the activity patterns, time use, and location of activities in the daily life of people, or perhaps enhance new ways of spatial interaction, for example, by replacing physical travel with the virtual mobility of telecommunication.

Questions of Dependence

A fundamental question then is how dependent the socio-spatial organization of modern society, and hence how *dependent* contemporary lifestyles, have become on high levels of mobility. What adaptive potentials exist in the short and medium terms if mobility has to be reduced? This question is extremely difficult to deal with, as the concept of mobility dependence denotes both inner psychological factors of the individual and external factors, for example, concerning the time–space environment of society. This chapter considers aspects of the latter, that is, mobility dependencies relating to external factors. The main objective is to explore the development, structure, and distribution of predominantly car-dependent urban activities, for example, according to their flexibility (or fixedness) in time and space. The chapter is organized around four intertwined aspects of mobility dependence and adaptation: (i) time and time use, (ii) travel-related activities and goals, (iii) urban structure and land use, and (iv) virtual mobility alternatives. First, we will briefly describe a joint theoretical approach in order to understand those issues better.

UNDERSTANDING MOBILITY DEPENDENCE – THE HUMAN ACTIVITY APPROACH

The human activity, or activity-based, approach to transportation research (Axhausen and Gärling, 1992; Fox, 1995; Jones, 1992, 1990) offers a useful conceptual framework for understanding mobility dependencies in everyday life. The human activity approach is rooted in time geography (Hägerstrand, 1974, 1984; Lenntorp, 1999, 2003), but also integrates aspects of environmental psychology and behavioural geography (Gärling and Golledge, 1993; Ettema and Timmermans, 1997) into the analysis of spatial mobility and human interaction. In brief, the activity approach regards everyday life as a sequence of activities performed by individuals at various places during the 24 hours of day and night. Almost all travel activities are derived from the need to carry out various stationary, or place-bound, activities. Travel is thus mainly a derived demand, not only in an economic sense (concerning the consumption of goods and services), but also in connection to the creation of everyday life and people's use of time and place. Activities performed to satisfy physiological needs (eating and sleeping), institutional demands (work and school), personal obligations (childcare, food, and shopping), and personal preferences (leisure activities) are more or less dependent on travel. Observed activity/mobility patterns thus are the outcome of needs and decisions at the individual level. Through scheduling, compulsory, and voluntary activities are linked into sequences of time use at various places, that is, situated activity programmes or routines. The choices are constrained, and there are different types and degrees of restrictions on when and where activities can be performed, such as the constraints of personal capacities and resources, coupling (the need to interact with other people and artefacts), and power (collective demands of society placed on individuals). The actual and desired activity of an individual is also dependent on the composition, roles, and tasks of the household to which he or she belongs and on the significance of the home as place of living, a place to which one always returns, and the prime location of people's time spending (Hanson and Hanson, 1993; Ellegård and Vilhelmson, 2004).

The activity-based approach has been challenged by Mokhtarian and Salomon (2000), who suggest that travel is not a by-product of an activity only, but can itself constitute an activity. At least under some circumstances it is desired for its own sake, in what could be labelled the phenomenon of undirected travel. A sense of speed, motion, control, and enjoyment – positive utilities – may motivate people to undertake "excess" travel, even in the context of mandatory or maintenance trips (Mokhtarian, 2005).

Lyons and Urry (2005) propose a related view, arguing that travel time is increasingly being used "productively" as activity time, a process enhanced by the spread and use of mobile information and communication technology (ICT).

Obviously, the combined needs and motives for travel comprise a complex phenomenon. From the perspective of multiple years or of the lifetime of an individual, daily activity patterns assume the shape of a more or less mobile lifestyle that integrates particular values, norms, behaviour, and consumption patterns. In categorical terms, and considering their spatial outcomes and extension, one might identify one lifestyle as characterized by geographical *stability*, another based on daily *commuting*, and a third featuring geographical *flexibility* (Vilhelmson, 1999a). These lifestyles occur simultaneously to various degrees in a population. Historically, the second has superseded the first, and the third the second as people have gained access to different means of transportation. A geographically stable lifestyle means that an individual adapts his or her life to suit the conditions in the immediate vicinity of his or her home. A lifestyle based on commuting means, in turn, that the individual expands the sphere of his or her activities in a way that is often repetitive and monotonous. Using fairly rapid means of transport, people can reach a single, more distant destination, such as their workplace, every day. Daily commuting to work became commonplace as railway and tram networks spread, and increased dramatically with the advent of the motor car. Even today, the journey to work is often considered a dominant feature of the extensive daily mobility of society, perhaps reflecting the appearance of the industrial society of yesterday. However, society of today has heightened the dependence of the individual on travel for all kinds of errands and activities. A geographically flexible lifestyle is where almost everything one does for work or recreation involves travelling long distances, usually by car. This lifestyle is now threatened by the need for decoupling and travel demand measures to obtain a sustainable transportation system.

MOBILITY AND TIME USE

One key issue – crucial from a decoupling perspective – is whether there are any limits to mobility other than those set by purely economic constraints, such as household income and the price of fossil fuel. One conceivable limit emphasized by the activity-based approach is *time*. Any future increase in mobility in society must be rooted in people's use of time and in their everyday activity patterns. Ultimately, there are great constraints: the day is 24 hours long, and much of that time is taken up by stationary activities that more or less must be done, and moreover, at fixed times and places. Generally people must balance the time allocated to travel in relation to time devoted to stationary activities during the day[3]. Consequently, the time available for travel is

[3] There are exceptions when travel is combined with a stationary activity, for example, working during a train journey or making telephone calls from the car (for possible implications of this, see Lyons and Urry, 2005).

severely limited – at least for large segments of the population – a fact that also limits the physical extension of people's activity spaces and the spatial expansion of urban regions.

Therefore, it comes as no surprise that the time factor is also critical in determining people's chosen mode of travel. The car – usually offering superior speed, door-to-door service, and relative freedom of action and choice as to when, where, how, and with whom a journey will be made – has the upper hand over other means of transport (see, e.g., Bruton, 1993; Naess, 2005). In Sweden, travelling by car is on average ten times faster than walking, four times faster than cycling, and twice as fast as using public transit. So the car is in essence a time-saving machine. Its speed should lead to increased efficiency and time-savings. It is thus worth noting how people's overall time use is affected by car ownership and use. An individual acquiring a car basically has two options concerning when to use it in relation to his or her everyday activity pattern. One is to fit the car into his or her existing activity pattern, a tactic that does not affect the total volume of travel. The speed of the car then releases time, making it available for stationary activities, for example, more time for socialising or using Internet at home. The other extreme is when all the time-saving generated by the increased travel speed is used for more and/or longer journeys, a pattern that increases the daily distance travelled. Travel time is not reduced in this scenario.

Research into travel habits indicates that the amount of time-spent travelling remains fairly stable over time, at least at the most aggregate (population) level (Schafer, 1998; Schafer and Victor, 2000). For example, the adult population of Sweden used an average of 80 minutes a day per person for journeys and travel during the 1978–2001 period (Vilhelmson, 1999b); meanwhile, the individual capacity to travel fast, that is, the number of cars, increased by almost 50%.

Average figures are deceptive, and of course the actual amount of travel time varies between different segments of the population (Höjer and Mattsson, 2000), women generally spending somewhat less time on travel than men do, and the elderly less than the middle aged do (Krantz, 1999). The empirical observation of stable total travel time at the aggregate level is poorly understood and often misinterpreted, especially when turning to the discussion of behavioural constants, invariants, and laws at the individual level (for an exhaustive review of empirical studies and a critical evaluation of this, see Mokhtarian and Chen, 2004). What is important here, however, is the dynamic rather than the cross-sectional aspect of this process, the *tendency* for aggregated travel time by and large to remain "unaffected" when a population gains access to faster means of transport. This resembles a kind of "rebound" effect with respect to time also found in other areas of technology usage, when people make use of time-saving, and often more energy-consuming, technologies, such as washing machines, microwave ovens, mobile phones, and computers (Binswanger, 2001). All in all this means that if mobility – overcoming geographical distance – can be produced using less time, households will demand more of such a time-saving service. This is further indicated by the fact that the aggregated daily travel time of car users in various age groups is similar to that of those travelling by other, much slower means of transport (Vilhelmson, 1999a). But how is this rebound process of speed, travel time stability, and escalating activity space to be understood?

MOBILITY, EVERYDAY ACTIVITIES, AND DEPENDENCE

This brings us to the matter of what people actually use travelling time for, and how dependent households have become on high mobility in their day-to-day affairs. This is of course difficult to establish objectively and incontrovertibly. Common sense and experience tells us that there exists a continuum of trips, from strictly mandatory to entirely discretionary. Arguably, journeys relating to the spheres of production and work are perceived as necessary, normally fixed in time and space and hence very difficult to influence and adapt – at least in the short and medium terms. There is probably somewhat greater freedom of choice and scope for influence in relation to maintenance trips, that is, those made for the purposes of housework, shopping, and other social services. Options and flexibility are almost certainly greatest when it comes to activities taking place in the remaining time, that is, leisure activities, visiting friends, and so forth.

An important observation should be made here. Regardless of how travel is measured, whether in terms of total time, distance or frequency, about half of all domestic travel in Sweden is associated with free-time activities: predominantly, visiting friends and family, and leisure activities such as sports, socialising, entertainment, and hobbies. At the same time, while trips relating to production and work account for only one-third of the total distance travelled, these are the trips that often dominate urban research, planning, and debate. Yet, arguing from an aggregated position, a large proportion of mobility activities appear to be more optional and susceptible to influence.

This highlights the importance of sorting out the more detailed composition of everyday activities, and free-time activities in particular, in order to elaborate on how dependent on mobility and tied down in time and space they really are[4]. If more exhaustive information on the spectrum and sequences of people's daily activities is needed, then time-use surveys offer interesting possibilities.

Data from the Swedish national time-use survey (covers a population of 20–64 years old, and concerns activities year round, summer period June–August excluded) make possible to link trips to a very large number of stationary activities, and to use very detailed information regarding trip purposes and mobility requirements. A straightforward method of classifying activities according to their flexibility and/or fixedness in time and space has been used, in order to address the complex question of mobility dependence (see Figure 1) (Vilhelmson, 1999b). More than 135 types of stationary activities

[4] Conventional travel surveys seldom give a precise picture of free-time activities.

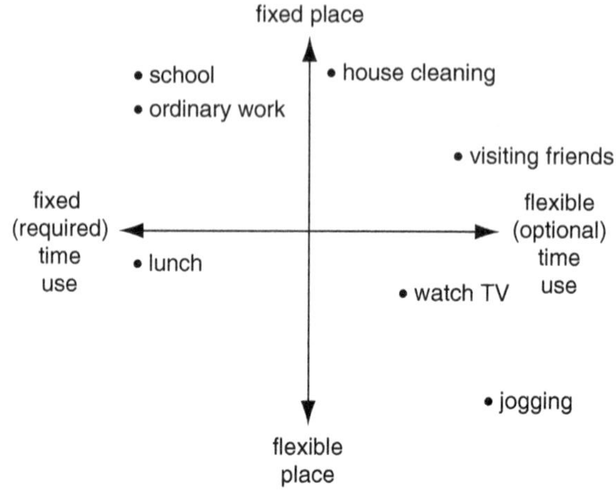

Source: Vihelmson (1999b).

Figure 1: Classification Scheme for Activities According to Their Time and Place Requirements; Examples of Stationary Activities

contained in the time-use survey (in this context also regarded as trip destinations) are classified into four basic categories according to their flexibility in time and space.

Basic to the classification scheme is a distinction between the place and the time for performing a certain activity. The timing of many activities is either fixed or dictated by the roles and obligations a person has to fulfil. Timing of other activities are optional or may be done voluntarily. The same applies to where a specific activity may be performed. Some activities are absolutely bound to a certain place, while others may take place in a range of alternative locations. The location of an activity could thus be either fixed or optional from the individual's point of view. In its basic form, the scheme defines four classes of activities along a continuum extending from those fixed in time and space, at least in the short term, to those characterized by total flexibility. When a longer-term perspective is taken into account, of course, the scheme breaks down. The inflexibility of a journey to work, for example, could be altered by a change of employment, residential area, or transportation mode. Another approach is to use an ICT-based alternative, to start teleworking and become virtually mobile. Clearly this method of classification focuses on "external" constraints and demands. It pays attention only to the social and spatial organization of the environment, and ignores the "inner", psychological dependencies that drive a person to perform certain activities demanding travel (Steg *et al.*, 2001; Jakobsson, 2004; see also Chapter 12 by Gatersleben in this volume). However, the objective here is merely to estimate the potential levels of flexibility or adaptation, rather than to predict behaviour.

All stationary activities examined in the time-use survey have thus been classified *a priori*, as illustrated in Figure 1, making it possible to estimate the associated number of trips and amount of travel time. On weekdays, 60% of all trips and of the total time spent on mobility are fixed, both in time and space. On weekends this proportion is lower, 30% of all trips and 25% of all travel time being fixed. Trips related to activities characterized by great flexibility in time and space amount to approximately 20% of the total on weekdays and 40% on weekends. If activities that are not so firmly fixed in time, though fixed in space, are included, the level rises to 40% on weekdays and 70% on weekends.

Additional results from this analysis show that the proportion of car use increases with the flexibility of the performed activities, while public transit is more closely connected with trips that are more fixed in time and place. This reflects the general character (and drawbacks) of public transit, which is limited to particular routes, stations, and timetables. Furthermore, household-related factors and factors related to time use and work significantly affect flexibility in travel. Having several children in a family means less flexible travel time. One-person households are more flexible in their time use than married couples are. People who wish to cut back on their working hours devote more time to flexible travel activities than other people do. Those who experience time pressure on weekdays, and therefore may have to cancel particular activities on a given day, also spend more time on flexible trips than others do. Several factors – primarily type of living region, car access, and gender – do not covary with the time allocated for travel for flexible activities. As regards the distance travelled for flexible activities during weekdays, the resulting significant variables are almost the same as regards travel time. Also, as expected, car access appears to be a factor of significant importance. Gender also plays a role here, women travelling less distance than men do for flexible activities.

To sum up, we estimate that roughly half of all trips made on weekends are of a flexible character with respect to time and space requirements, while on weekdays this proportion declines to approximately one-fifth. These estimates provide a capsule indication of the structural dependencies on mobility in modern society, of how speed is used to support more flexible lifestyles, and of how adaptive travel may be in the short term according to the spatial and temporal organization of society.

MOBILITY, URBAN STRUCTURE, AND LAND USE

Is it the particularly urban location patterns and land use – the spatial distribution of people, housing, workplaces, shopping centres, services, and so forth – which necessitate so much travel? It is commonly held that, with the help of the car, societies have maintained, developed, and then cemented a geographical structure which is difficult to change and escape from in the short and medium terms without serious losses of welfare. Thus, societies would appear to be very much locked into our day-to-day patterns of travel.

Urban land use in this context refers mainly to elements such as urban form and size, urban spread, population density, and mix of activities (e.g., housing/workplace-balances). From the perspective of a comprehensive policy of sustainable mobility, a crucial question is whether increased accessibility (or proximity) in urban areas could put a stop to the seemingly ever-increasing volume of travel (Mogridge, 1986; Giuliano and Small, 1993; Breheny, 1995; Wegener, 1996). Intuitively, it would seem reasonable to assume that if the physical distances between activities could be reduced, then the need for travel in everyday life could also be reduced. In recent years, research has intensified concerning matters of transportation, accessibility, and the sustainability of urban environments (Breheny, 1995; van der Waals, 2000). Concepts such as "compact cities" and "containment strategies" have been launched, emphasizing the spatial aspects of sustainable development. The "compact cities" concept denotes planning goals and measures that promote more densely populated cities, seen as a necessary condition for sustainability. "Containment strategies" imply that the future expansion of urban populations, activities, and interactions should be kept within the boundaries of existing built-up areas so as to limit further urban sprawl. The theory underlying this approach assumes close and direct links between the physical structure of the city (the land use) and people's activities in time and space (Newman and Kenworthy, 1991, 1999; Naess, 2005). Such ideas have influenced many planning documents at different levels in European countries. In Sweden, this planning aim has been further emphasized by a general opinion that the built-up areas continue to spread and that the entire urban system gradually develops into a North American pattern, where extensive urban areas, sparsely populated, generate immense transportation needs, absolute car dependency, and high energy consumption. Such "urban sprawl" is perceived as a major driving force behind an ever-increasing mobility.

At root, the matter is one of people's need for travel in different spatial structures. Do, for example, people living in large and densely populated cities travel less, covering shorter distances, in their everyday lives than do people living in smaller cities and more sparsely populated built-up areas? Perhaps the situation is more complex, as indicated above: general access to cars and high-speed infrastructures combined with fairly stable time-use patterns may yield similar amounts of travel, despite different regional characteristics – the rebound effect. The problem addressed in this section is whether certain urban sizes and densities are more mobility dependent than others are. This is done by examining some long-term changes in travel patterns generated in urbanized and semi-urbanized areas of Sweden over the twenty years between 1978 and 1997 (Vilhelmson, 2005).

The first notable observation is that the spatial expansion of urban areas and the thinning out of population densities had in fact already culminated in Sweden in the 1970s. Since then the spatial growth of built-up areas (villages, towns, and cities) has stagnated, and only minor spatial expansion has occurred. Likewise, the period up to the 1970s was one of quick change in urban population densities and degree of suburbanization: the thinning out of population densities happened quickly, but ebbed in the 1980s. In recent years, densities have even increased in medium- and large-sized cities. This development casts some doubt on the role of urban size and density as major driving forces of the increase in mobility experienced in Sweden over the past 25 years.

The second observation concerns whether people living within the boundaries of large, densely populated ("compact") urban areas travel less than do people living within the boundaries of more sparsely populated areas. There is indeed an observed relationship between population density and distance travelled; this relationship is, however, not linear but U shaped, representing the relationship between extent of built-up area and density, on the one hand, and distance travelled on the other (this is consistent with findings from other European countries reported in Orfeuil and Salomon, 1993). In general, less distance is travelled in more densely populated areas, but only up to a certain level. In 1997, people living in medium-sized cities (50,000–200,000 inhabitants) in Sweden travelled the least distance, on average. The medium-sized cities of Sweden – such as Karlstad, Linköping, Umeå, and Uppsala – are those associated with the lowest level of observed activity space; among the largest cities – Stockholm, Göteborg, and Malmö – this level is higher. Twenty years before, in 1978, the pattern was similar with one important exception: people living in medium-sized cities then travelled farther, on average, than they do now. A potential for reduced travel has thus been realized, though socio-demographic shifts in the population under study, related to an altered age composition, may explain part of the observed change.

Access to a car is another obvious and important factor in regional variations in activity space and travel. The extent of built-up area and population density are closely associated with car availability. In the small villages of Sweden, almost 9 of 10 people (20–65 years old) have access to a car, in their households, and hold driving licenses, a proportion that decreases in larger and denser population centres. In the built-up areas of the three largest cities in Sweden, only 6 of 10 people have access to a car in their households. The general increase in car access during the 1978–1997 period is concentrated, by and large, to the smaller population centres and villages, as well as in rural areas, and hardly changed at all among people living in medium- or large-sized cities. The gap in car dependency between different types of built-up areas thus widened during the period.

It is often argued that denser, more compact urban structures would promote the use of public transit, walking, and cycling (Fulford, 1996). The general impression is that between 1978 and 1997 there was no radical change in the use of different means of transportation among people living in different types of urban areas. The rates of change were rather slow compared to the rapid expansion of car use that occurred in the 1950–1975 period. The car is by far the dominant mode of travel, and car driving is still increasing in importance among people living in the urban areas under study, but with one important exception. Among people living in medium-sized Swedish cities there was a decrease in car use, accompanied by a marked increase in the use of bicycles in daily travel activities (from 10% to being used for nearly 25% of all trips), which helps to explain the simultaneous decrease in total travelled distance. Travel on foot and as car passenger decreased slightly in all types of regions, while the role of public transit has remained fairly stable.

All in all, our observations question the common assumption that increases in urban population density – as promoted by the compact cities and urban containment strategies – automatically reduce the amount of travel, at least in the context of Swedish urban structures and activities. Instead, one could claim that increased density harbours a *potential* – hitherto hardly realized – for travel reduction that could conceivably occur under specific conditions and in certain types of urban areas characterized by certain land-use patterns. Such conditions might include, for example, rising costs of energy/fossil fuels, changing values and preferences, and the development of attractive alternative modes of transportation. The observed increase in the use of bicycles in medium-sized Swedish cities over the past two decades is a good example of this. Thus, it can still be maintained that denser urban areas may be more robust in the face of changes in the transportation system, caused, for example, by such factors as energy shortages or air pollution.

VIRTUAL MOBILITY

A radical way to reduce dependency on physical travel is instead to use the space-transcending abilities – the virtual mobility – of ICT. People connected to computer networks and the Internet, using mobile phones and other ICTs, have more freedom to live and work wherever they like. In Sweden, more than 80% of the population aged 15–65 years have one or more computers at home, more than 70% are connected to the Internet, and 77% have a personal mobile phone, indeed signifying the reduced friction of distance in the information society. Concepts such as teleworking, telecommuting, tele- and videoconferencing, teleservices, teleshopping, and telemedicine suggest that the spatial consequences of the diffusion of virtual mobility may be far reaching and may also affect the need for car travel. On the one hand, people using ICT may seize the opportunity to become more stationary, reducing a large number of tiring and tedious journeys for work, business, and shopping purposes by teleworking, teleconferencing, and teleshopping, and thus become more stationary. On the other hand, they may also use ICT to become even more physically mobile than before – during work as well as leisure time – and step up the long-term process of activity space extension.

In terms of the resulting travel patterns, the first stationary lifestyle scenario is essentially based on the idea of the smooth substitution of space-transcending technologies (i.e., the interchangeability of physical travel and virtual communication). Human contacts realized by car travel are assumed to be easily replaced by virtual communication, thus reducing energy consumption, emissions, urban congestion, noise, traffic accidents, and so forth. In the second mobile lifestyle scenario, ICT is viewed mainly as complementary and does not fundamentally affect established travel patterns; instead, new contacts are added and old ones intensified through access to ICT. Thus, in the long-run virtual mobility might even intensify the need for travel and face-to-face meetings. Combinations of these two contrasting scenarios are possible, as are several indirect and long-term adaptive effects regarding the interaction between physical travel and virtual communication (Salomon, 1986; Mokhtarian, 1990, 1997).

The wide range of potential travel patterns arising from the acceptance of new information technologies raises the question of to what extent people's decisions regarding where to live and work are undergoing a profound change in the emerging informational society. Telework is here discussed as an important case, because ICT-based alternatives to commuting have for long been available, in fact since the end of the 1970s. Hopes have also for long been expressed that the increased use of ICT will, more or less spontaneously, reduce the actual volume of commuting and other work-related travel.

Commuting accounts for 20% of the total number of trips in Sweden. Commuting trips are usually of a highly routine character, and the concentration of large flows of commuters in time and space causes congestion and unhealthy air conditions in urban areas. Now, ICT makes it possible to organize work in new and more flexible ways in space and time, and several modes of ICT-based work can be distinguished (Lindström *et al.*, 1997). Teleworking – or telecommuting – from home or other sites outside the regular place of work, such as telecenters, telecottages, satellite offices, remote offices, and virtual organizations, belong to this category. Also, entirely mobile work is another IT-enabled mode of work, describing the situation in which an individual's normal work can be performed when travelling or from a combination of locations by using portable ICT equipment. In short, if telework represents the transportation substitution side of ICT, mobile work stands for the transportation generation side.

The actual number of teleworkers is partly a matter of definition (for a discussion, see Mokhtarian, 1998). In 2001, according to Swedish communication-use surveys, approximately 5% of all gainfully employed persons did telework regularly (Vilhelmson and Thulin, 2001). In these surveys, telework was given a broad, spatial-organizational definition: regular work done at a location other than an ordinary, fixed place of work, but during the scheduled work time of the day (people doing mobile or home-based work are not regarded as teleworkers). This means that approximately 200,000 of a total of 4 million employed people in Sweden did telework on a regular basis. During the 1997–2001 period, this number (and proportion) seemed to fluctuate slightly, though the differences are not statistically significant and no increasing trend was observable. Most telework is performed in the home, while telecenters, telecottages, or other types of ICT-based remote work sites are very rare. There are some regional variations, as telework is rather more prevalent in the large urban areas of Sweden than in the more rural parts of the country. This is noteworthy since one might expect telework to be more common in rural areas, where the distances between work and home are frequently very long.

If using ICT to keep in contact with the regular place of work is considered a necessary condition for referring to an activity as "telework", an even smaller fraction – about 4% – of all employed telework according to that definition. And if regularity over the week is taken into consideration, it is estimated that only 3% of the total workforce telework at least one day a week, while only 1% telework during a regular day of the week. These numbers constitute the basis for any evaluation of how teleworking might affect daily traffic and congestion levels.

Even mobile workers in Sweden have not increased in number in recent years, though analysts expect mobile work to expand, driven by new generations of mobile ICT innovations (3G phones, mobile Internet, wireless communication, and so forth) that will stimulate further the development of a less place-determined work community (Gareis, 2003). Results also show that the current level of mobile telecom equipment ownership among mobile workers does not significantly differ from that of more traditional commuting-based workers.

A main observation is thus that, despite a high level of access to ICT equipment in workplaces and households, telework in Sweden is not yet very common and most people are still attached to traditional modes of commuting-based work, a conclusion also reached in reviews based on small-scale surveys and experiments (Mokhtarian, 1998; Salomon, 1998; Sturesson, 2000; Skåmedal, 2004). This points out that a straightforward substitution potential in the interaction between transportation technology and ICT is seldom applicable. The decision makers view the situation differently and take many factors into consideration when evaluating the trade-offs between travel- and ICT-based activities. At the same time, it is not only paid work that motivates work travel as regards scheduling, routing, and modal choice; many other activities and secondary trips are linked to trips to and from work, for example, taking children to school, shopping, or accessing various services. In line with the activity-based approach, this implies that if the conditions for travel to work are altered, the conditions for other activities will also be affected. So far the potential for travel substitution or reduction (to reduce the time and monetary costs of travel) have not been so powerful that it alone has promoted the rapid spread of, for example, telework in society. And when telework is done, a number of studies show that its effects on travel are not substantial, even though some short-term substitution effects are generally observed (Mokhtarian, 1997; Skåmedal, 2004).

IMPLICATIONS AND CONCLUDING REMARKS

The goals of decoupling and sustainable mobility include both a technological side, that is, of promoting environmental friendly technologies, and a behavioural one, that is, of implementing measures curbing travel needs. The latter implies individual adaptations and responses, such as reduced demand for travel, shorter trips and smaller activity spaces, less time spent on travel or use of slower modes of travel, or substitution of virtual communication for physical transportation. All in all, these adaptations and responses should lead to a lifestyle emphasizing greater presence in both the local community and in cyberspace. The prospect of such behavioural responses focuses on the dependency on and susceptibility of physical transportation in modern societies. This dependence is affected by many factors working on various levels and scales, for example, the supply and cost of transportation alternatives, incomes and values of the population, and urban land use and accessibility (Vlek and Michon, 1992; Wegener, 1996; Gärling *et al.*, 2002). The relative freedom of choice and flexibility offered by access to rapid means of transport, such as the car, is *per se* a prime driving force in the processes of time–space convergence, compression, and distanciation of urban activity patterns.

Yet, the issues of mobility and car dependency are not just functional or practical ones; they are also matters of attachments, meanings, values, and symbols (Chapter 12 by Gatersleben in this volume). They concern how people value the relationship between individual welfare and ever-increasing physical mobility, and the extent to which this relationship can continue to be assigned a positive value in the longer term. This has specific implications, particularly for the relationships between local community and the extension of urban regions and conurbanization. Is there a conflict between them, in that mobility tends to sever local ties and community-based welfare, as explicitly argued by Putnam (2000)? Or is the weakening of local ties and neighbourhoods well compensated for by the increased range and distant contacts of the individual – the "community liberated" – as argued by, for example, Wellman (1999)? In any case, and from a more pragmatic policy perspective, influencing the extent of mobility by first achieving a radical shift in perceptions of welfare would appear to be a daunting task. There may thus be justification for considering options closer to hand. Some tentative inferences as to the scope that exists for influencing mobility may be drawn from the analysis here of the relationships between physical travel, time, activity patterns, distance, urban structures, and other, virtual means of communication.

An initial conclusion is that the total volume of mobility, in terms of distance travelled, is mainly influenced by the available speed – in reality, access to a car (and associated infrastructure). Access to cars and their use are both in turn heavily influenced by economic circumstances, such as incomes and fuel prices. The potentially effective instruments for influencing *overall* volumes of mobility are thus primarily have an economic character (e.g., environmental surcharges on fuel). It is likely to be difficult to use measures such as the geographical/spatial relocation of activities alone as a means of appreciably influencing future mobility trends. This does not contradict the fact that certain spatially extended urban structures and social situations/organizations may be more vulnerable, whereas others, characterized by propinquity, may be more robust if mobility levels have to be reduced for whatever reason.

A second implication of the examination presented in this chapter concerns the effects of rationalization *within* parts of the transport system. It may thus be expected that individuals will use the time saved by *faster* journeys to make more and/or longer journeys, which will increase the total distance they travel. This is a more likely outcome than the time-saved being used for stationary activities. Hence, measures aimed at reducing the time taken for certain journeys, such as journeys to work, tend to be nullified by increased travelling time relating to other activities or by a subsequent increase in the length of journeys to work. Investments in time-saving infrastructure in the form of new roads, bridges, new public transit facilities could, in principle, produce the same effect.

A third implication is that any structural changes, physical or organizational, made *outside* the transport system, but intended to reduce the amount of travel, may not influence overall mobility in the desired direction, and may not divert time-spent travelling to other, stationary activities. This applies to measures such as developing more compact, denser urban structures or changing the timing and location of work, for example, by implementing teleworking or shorter working hours. This and the preceding conclusion apply assuming that the marginal cost of travel does not simultaneously rise. Nor must this conclusion obscure the fact that changes in infrastructure, social planning, and time organization will nevertheless be necessary in order to affect the amount of travelling within a society; such changes will probably not suffice, however.

A fourth suggestion is that the large number of car journeys associated with free-time activities and errands will be influenced mainly if, regardless of reason, mobility is forced to stagnate or decline. There seems to be some freedom of choice here. The scope for adaptation will more likely to be determined by peoples' evaluation of the relationship between individual welfare and mobility than by the spatial and temporal dependencies of fixed activities.

Finally, it is a risk that the spread and effects of transportation and communication technologies in society being regarded as a predestined process, difficult to influence or control in any fundamental sense. Initial hopes for a rapid increase in telework, closely following the development of ICT, belong to such thinking. It is also clear that circumstances external to the system under study may also influence the development process and could realize about potential effects, embedded in technology, some of which may be more desirable than others. In coming years, environmental and other risks will reinforce the need to realize the substitution potential of ICT, hitherto difficult to realize, in order to reduce resource-consuming and environmentally detrimental physical travel.

Further research is undoubtedly needed in order to strengthen the theoretical and empirical foundations of the study of mobility dependence and adaptability. This should not obscure the fact that policy measures ought to be undertaken at present, with reference to the present knowledge base.

ACKNOWLEDGMENTS

The authors' research was financially supported by grants from the Swedish Agency for Innovation Systems and the Swedish Research Council.

REFERENCES

Axhausen, K. and T. Gärling (1992). Activity-Based Approaches to Travel Analysis: Conceptual Frameworks, Model, and Research Problems, *Transport Reviews*, **12**, 323–341.

Binswanger, M. (2001). Technological Progress and Sustainable Development: What About the Rebound Effect? *Ecological Economics*, **36**, 119–132.

Breheny, M. (1995). The Compact City and Transport Energy Consumption, *Transactions of the Institute of British Geographers*, **20**, 81–101.

Bruton, M.J. (1993). *Introduction to Transportation Planning*, UCL Press, London.

Ellegård, K. and B. Vilhelmson (2004). Home as a Pocket of Local Order: Everyday Activities and the Friction of Distance, *Geografiska Annaler B*, **86**, 223–238.

Ettema, D. and H. Timmermans (1997). Theories and Models of Activity Patterns. In: *Activity-Based Approaches to Travel Analysis* (D. Ettema and H. Timmermans, eds), Pergamon, Oxford, 1–36.

Eurostat (2005). *Themes/Transport*. Retrieved September 1, 2005, from http://epp.eurostat.cec.eu.

Fox, M. (1995). Transport Planning and the Human Activity Approach, *Journal of Transport Geography*, **3**, 105–116.

Fulford, C. (1996). The Compact City and the Market. In: *The Compact City: A Sustainable Urban Form?* (M. Jenks, E. Burton, and K. Williams, eds), E&FN Spon Publishers, London, 122–133.

Gareis, K. (2003). Home-Based vs. Mobile Telework: The Interrelationship between Different Types of Telework. In: *Organisation and Work Beyond 2000* (B. Rapp and P. Jackson, eds), Physica-Verlag, Heidelberg, 171–185.

Gärling, T., D. Eek, P. Loukopoulos, S. Fujii, O. Johansson-Stenman, R. Kitamura, R. Pendyala, and B. Vilhelmson (2002). A Conceptual Analysis of the Impact of Travel Demand Management on Private Car Use, *Transport Policy*, **9**, 59–70.

Gärling, T. and R.G. Golledge (eds) (1993). *Behavior and Environment: Psychological and Geographical Approaches*, North-Holland, Amsterdam.

Giuliano, G. and K. Small (1993). Is the Journey to Work Explained by Urban Structure? *Urban Studies*, **30** (9), 1485–1500.

Hägerstrand, T. (1974). *The Impact of Transport on the Quality of Life*, Technical Report No. 13, Department of Geography, Lund University, Lund, Sweden.

Hägerstrand, T. (1984). Human Interaction and Spatial Mobility: Retrospect and Prospect. In: *Transportation Planning in a Changing World* (P. Nijkamp and S. Reichman, eds), Gower, Aldershot, 11–28.

Hanson, S. and P. Hanson (1993). The Geography of Everyday Life. In: *Behavior and Environment. Psychological and Geographical Approaches* (T. Gärling and R.G. Golledge, eds), North-Holland, Amsterdam, 249–269.

Höjer, M. and L.G. Mattsson (2000). Determination and Backcasting in Future Studies. *Futures*, **32**, 613–634.

Jakobsson, C. (2004). *Motivational and Volitional Control of Automobile Use*, PhD Dissertation, Department of Psychology, Göteborg University, Göteborg, Sweden.

Jones, P.M. (1992). Some Recent Methodological Developments in Our Understanding of Travel Behaviour, *IATTS Research*, **16** (2), 18–29.

Jones, P.M. (ed.) (1990). *Developments in Dynamic and Activity-Based Approaches to Travel Analysis*, Gower, Aldershot.

Krantz, L.-G. (1999). *Rörlighetens mångfald och förändring. Befolkningens dagliga resande i Sverige 1978 och 1996* (Meddelanden från Göteborgs universitets Geografiska institutioner, Serie B, nr 95), Department of Geography, Göteborg University, Göteborg, Sweden.

Lenntorp, B. (1999). Time-Geography – At the End of Its Beginning, *GeoJournal*, **48**, 155–158.

Lenntorp, B. (2003). *The Drama of Real-Life in a Time-Geographic Disguise*, Paper presented at the Sixth Theo Quant Meeting, Besancon, France.

Lindström, J., A. Moberg, and B. Rapp (1997). On the Classification of Telework, *European Journal of Information Systems*, **6**, 243–255.

Lyons, G. and J. Urry (2005). Travel Time Use in the Information Age, *Transportation Research A*, **39**, 257–276.

Mogridge, M.H.J. (1986). If London is More Spread Out Than Paris, Why Don't Londoners Travel More Than Parisians? *Transportation*, **13**, 85–104.

Mokhtarian, P.L. (1990). A Typology of Relationships between Telecommunications and Transportation, *Transportation Research A*, **24**, 231–242.

Mokhtarian, P.L. (1997). The Transportation Impact of Telecommuting: Recent Empirical Findings. In: *Understanding Travel Behavior in an Era of Change* (P. Stopher and M. Lee-Gosselin, eds), Pergamon, Oxford, pp. 91–106.

Mokhtarian, P.L. (1998). A Synthetic Approach to Estimating the Impacts of Telecommuting on Travel, *Urban Studies*, **35**, 215–241.

Mokhtarian, P.L. (2005). Travel as a Desired End, Not Just a Means, *Transportation Research A*, **39**, 93–96.

Mokhtarian, P.L. and C. Chen (2004). TTB or Not TTB, that is the Question: A Review and Analysis of the Empirical Literature on Travel Time (and Money) Budgets, *Transportation Research A*, **35**, 695–719.

Mokhtarian, P.L. and I. Salomon (2000). How Derived is the Demand for Travel? Some Conceptual and Measurement Considerations, *Transportation Research A*, **35**, 695–719.

Naess, P. (2005). Residental Location Affects Travel Behavior – But How and Why? The Case of Copenhagen Metropolitan Area, *Progress in Planning*, **63**, 167–257.

Newman, P. and J. Kenworthy (1991). Transport and Urban Form in Thirty-Two of the World's Principal Cities, *Transport Reviews*, **11**, 249–272.

Newman, P. and J. Kenworthy (1999). *Sustainability and Cities: Overcoming Automobile Dependence*, Island Press, Washington, DC.

Orfeuil, J.-P. and I. Salomon (1993). Travel Patterns of the Europeans in Everyday Life. In: *A Billion Trips a Day: Tradition and Transition in European Travel Patterns* (I. Salomon, P. Bovy, and J.-P. Orfeuil, eds), Kluwer Academic Publishers, Dordrecht, 33–50.

Pucher, J. (1999). Transportation Trends, Problems, and Policies: An International Perspective, *Transportation Research A*, **33**, 493–503.

Putnam, R.D. (2000). *Bowling Alone: The Collapse and Revival of American Community*, Simon & Schuster, New York.

Salomon, I. (1986). Telecommunications and Travel Relationships: A Review, *Transportation Research A*, **20**, 223–238.

Salomon, I. (1998). Technological Change and Social Forecasting: The Case of Telecommuting as a Travel Substitute, *Transportation Research C*, **6**, 17–45.

Schafer, A. (1998). The Global Demand for Motorized Mobility, *Transportation Research A*, **32**, 455–477.

Schafer, A. and D.G. Victor (2000). The Future Mobility of the World Population, *Transportation Research A*, **34**, 171–205.

Skåmedal, J. (2004). *Telecommuting's Implications on Travel and Travel Patterns*, Doctoral Dissertation, Studies in Science and Technology, Linköping University, Linköping, Sweden.

Steg, L., C. Vlek, and G. Slotegraaf (2001). Instrumental-Reasoned and Symbolic-Affective Motives for Using a Motor Car, *Transportation Research F*, **4**, 151–169.

Sturesson, L. (2000). *Distansarbete: teknik, retorik och praktik*, Carlsson, Stockholm.

Turton, B. and R. Knowles (1998). Urban Transport Problems and Solutions. In: *Modern Transport Geography* (B. Hoyle and R. Knowles, eds), 2nd ed., Wiley, Chichester, 135–157.

van der Waals, F.M.J. (2000). The Compact City and the Environment: A Review, *Tijdschrift voor Economische en Sociale Geografie*, **91**, 111–121.

Vilhelmson, B. (1999a). Daily Travel: Trends, Limits, Susceptibility to Influence. In: *Changing Environmental Behaviour* (L.J. Lundgren, ed.), Swedish Council for Planning and Coordination of Research, Stockholm, 91–104.

Vilhelmson, B. (1999b). Daily Mobility and the Use of Time for Different Activities. The Case of Sweden, *Geojournal*, **48**, 177–185.

Vilhelmson, B. (2005). Urbanisation and Everyday Mobility. Long-Term Changes of Travel in Urban Areas of Sweden, *Cybergeo: Revue européenne de géographic,* No. 302, 17/02/2005, 13 p. (http://www.cybergeo.presse.fr)

Vilhelmson, B. and E. Thulin (2001). Is Regular Work at Fixed Places Fading Away? The Development of ICT-Based and Travel-Based Modes of Work in Sweden, *Environment and Planning A*, **33**, 1015–1029.

Vlek, C. and J.A. Michon (1992). Why We Should and How We Could Decrease the Use of Motor Vehicles in the Near Future, *IATTS Research*, **15** (2), 82–93.

Wegener, M. (1996). Reductions of CO_2 Emissions of Transport by Reorganisation of Urban Activities. In: *Transport, Land-Use and the Environment* (Y. Hayashi and J. Roy, eds), Kluwer, Dordrecht, 103–124.

Wellman, B. (1999). The Network Community: An Introduction. In: *Networks in the Global Village: Life in Contemporary Communities* (B. Wellman, ed.), Westview, Boulder, CO, 1–48.

Threats from Car Traffic to the Quality of Urban Life
Tommy Gärling and Linda Steg (Editors)

9

Concepts of Travel Behaviour Research

Kay W. Axhausen, Institute of Transport Planning, Swiss Federal Institute of Technology, Zürich, Switzerland

ABSTRACT

In the first part this chapter proposes a conceptual framework for travel behaviour research through a definition of the scope of the research topic, essentially human activity schedules, and a conceptualisation of the traveller as a network actor negotiating infrastructure and human networks and dealing with the social content of the activities undertaken. In the second part of the chapter an operationalisation of this framework through the dynamic microsimulation of daily life nested within the microsimulation of longer-term projects and choices.

INTRODUCTION

Travel behaviour research, now about sixty years old, draws for its concepts on a wide range of disciplines and on its own understandings, which are not necessarily consistent with each other, but often either overlay the same term with multiple, divergent meanings, or provide different terms for the same object or process. This is especially true, when one is talking about processes of change at the personal or system level. In addition, the field is tied to its professional forecasting and therefore its modelling remit, which generally leads to confusion between the conceptual understandings in its research domain and the impoverished, truncated, simplified forms these find in the numerical models due to measurement problems, resource availabilities or computational limitations. This conflict will be visible in this discussion as well. The purpose of the chapter is to propose a general conceptual framework for travel behaviour research. It is a starting point for discussion, but hopefully also a kernel around which researchers can start building a common understanding of the dynamics of travel and communication patterns today.

The discussion will start by defining the scope and the topics of travel behaviour research, which will be followed by an attempt to provide a framework of how travellers decide about the schedule of their day and by implication about their patterns of contacts. Two digressions look at two detailed issues: How to measure activity spaces? How do persons construct their choice sets? The final section turns to the elements of a modelling system translating these ideas into an operational system. An outlook highlights the open research questions.

SCOPE

Travel behaviour research studies the physical movement of persons outside their *reference locations* for any *purpose*. The movement of freight is only addressed in as far as people transport freight for work, shopping or other private purposes. Logistic as the study of the production and distribution systems of goods and services will be excluded here. Urry's scheme for the categorisation of "mobilities" (Urry, 2000, cited in Larsen *et al.*, 2006) is (1) physical travel of people for work, leisure, family life, and migration; (2) physical movement of objects; (3) imaginative travel elsewhere through images and memories; (4) virtual travel on the Internet, telephones, emails, etc. The categorisation highlights that a single-minded focus on the physical movement of persons might not serve the discipline well for much longer. The increasing quality of virtual environments, such as chat rooms, shared white-board systems, conference calls or the video conferences for substituted face-to-face meetings, and of (interactive) web sites for service delivery, such as shopping, banking, inspecting goods, for instance, homes, holiday flats, etc., raises the question whether the scope should not in a first step be extended to the physical and virtual movement outside the reference location. In principle, one can and should see physical movement as one of many modes allowing communication and exchange between persons and between persons and systems. Still, given that this would extend the scope vastly beyond the current range of the travel behaviour research literature, this step will not be taken here. Nevertheless, even with a more narrow extension to virtual movement, the problem of an appropriate generic term for such interactions arises. For the moment, I would suggest the term *contact* for any interaction between persons or between persons and a system or location, independent of the physical co-presence of the persons or person and the system or location. (The attributes of contacts will be discussed jointly with the attributes of activities below.)

In this chapter, the main focus will be on physical movement outside the reference location of a person. The *reference location* is the place to which the person returns at the end of the day. This is generally the home, but might be the room in a student dormitory, the hotel room or some other short-term base for the person involved. The first topic of study for travel behaviour research is therefore the share of persons leaving the reference location on any one day, or over any period of time (see Madre *et al.*, 2006, for a review of the evidence from travel diary surveys). The second object is then the total time spent out-of-home[1] on any one day (or period). Surprisingly, there is no travel behaviour literature to speak of about the time spent out-of-home.

The allocation of the time between the first departure of the day from home until the final return between movement and activities defines the outline of a person's daily *schedule*. The other basic elements, which need to be fixed or chosen by the traveller, are:

- *Purpose* of the activity;
- *Timing* of the activity and by implication of the associated movements;
- *Duration* of the activity;
- *Location* of the activity, which is normally requested at a level of detail convenient for its recall by the respondents[2];
- *Participants* of the activity, if any. Note that one should account for any animals, which influence behaviour, such as a dog, or more rarely a horse;
- *Expenditure* for the activity, that is costs incurred at the point of use;
- *Structure* of the associated *movement*, if any, is required between two subsequent activities.

These elements assume a definition of *activity* as a sequence of purposeful actions within the same spatial and social context, which is given by the set of persons interacting with each other. Where movement is the purpose of the activity, for instance, walking the dog, jogging, cycling, a drive with the car, it should be coded as an activity. The concept of the contact inherits these characteristics from the notion of activity. The two basic properties of the associated movement duration and distance covered and speed derived from them (see below for a more detailed discussion) apply therefore to the concept of contact via virtual movement as well. The speed of synchronous communication, via the phone or in chat rooms, is for all intents and purposes infinity, although small delays are noticeable for certain technical modes of communication, such as voice-over-IP telephony, certain ship-to-shore services, or calls over very long distances. For asynchronous modes, such as letters or emails, the speed can be calculated with the duration of the transmission starting with the posting of the message and ending with its reading by the addressee.

The elements of movement, defining its structure are (see also Axhausen, 2000, 2003):

- *Stage* (unlinked trip, segment) is the movement with one mode or means of transport, including waiting times during or after the stage. Walking is understood as a mode of transport. Means are the particular vehicles used. (Although strictly speaking an activity, it is useful to treat the time needed to get the vehicle ready

---

[1] Home is used here and later as a synonym for reference location since the private primary residence is the reference location in which most days end.

[2] In surveys of daily travel, the respondents are normally asked to recall street addresses or places; in long-distance surveys or tourism surveys, the locations are aggregated to the level of municipality, region or even country.

to leave [pack it, seat all passengers, check the vehicle], or to obtain a ticket as a special stage.)

- (Linked) *trip* (leg) is a sequence of stages between two activities. A walk-only trip has one stage, while all trips involving a vehicle generally have at least three (walking to it, using it, walking away from it to the activity).
- *Customer movement* is any aggregation of stages or trips, which supports operational, pricing or revenue allocation processes of public transport operators. For a trip, for example, consisting of two bus stages and a rail stage, the two operators involved would count two customer movements.
- *Tour* is a sequence of trips away from and then back to the same location.
- *Journey* is a tour away from and then back to the reference location.
- *Move* is special journey from one reference location to the next, normally reserved for the movement between two main homes.

Each of these elements is defined by:

- *Timing* (of its begin).
- *Duration.*
- The technical *means of transport* (stage only), which itself can be defined at various levels of detail depending on which characteristics of the vehicle or the service, in the case of *public transport*[3], are of interest to the analyst.
- Main *mode* of transport is identical to the mean of transport in the case of the stage, but otherwise needs to be chosen from among those of the stages involved by an appropriate rule. The convention is to select the means which binds the traveller most in terms of timing or covers the largest share of the total distance travelled. This means, that regular timetabled public transport means have priority in this ranking over private means, such as walking, cycling or the private car.
- The *start location* or origin.
- The *end location* or destination. Note that for car stages this is the parking space and not the activity location, a difference that is normally neglected, but leads to an underestimate of walking as parking space and activity location are not necessarily close to one another.
- *Route*, that is, the exact sequence of links and nodes used, or more general the *connection*, that is, the physical route, the *access points* (parking place, stops, stations, port, airports) plus the (timetabled) vehicle (run).
- *Accompanying persons* during the movement. Note that one should account for any animals here as well.
- *Expenditure* for the movement and any associated parking, that is the out-of-pocket costs.

---

[3] *Public transport* is any service, where a third party transports a person against payment. The different forms of public transport are defined by the type of vehicle employed and the detailed characteristics of the service (frequency, sequence of access points, service period, rules of carriage, quality of accommodation within the vehicle).

The main scope of travel behaviour research is the measurement, analysis, modelling and forecasting of the travellers' schedules, that is, the volume, structure and characteristics of the activities and movements involved.

The reader will have noted that the word *mobility* has been avoided so far. The term has acquired so many meanings and connotation, that it is difficult to use without a definition every time it occurs. This is obviously cumbersome and invites misunderstanding. The term is used to mean travel undertaken by travellers. It refers to the capabilities of a person with regards to movement and travel independent of his or her actual travel. The term is also applied to movement in the social space, for example between classes, milieux, groups or roles, some but not all of which involve substantial amounts of travel. The extension of the term through subdivision, as in the example of Urry above, or through extension, such as for example Kaufmann's (2002) *motility*, endangers its usefulness further. It will generally be avoided here.

EXPLAINING TRAVEL BEHAVIOUR

The travel behaviour literature generally documents the increasing refinement of the conceptual frameworks and models for the behaviours and choices of travellers on individual days[4]. Taste differences, lifestyles, attitudes and socio-demographics have been added to the description of choice situations, which are driven by the relative *generalised costs* of the activities and their associated travel. What is mostly missing is integration between the short- and long-term dynamics of travellers' behaviour. While the industrialised world will never again see a similar dramatic shrinking of its time-space system as it did during the last 50 years (see Figure 1 for the Swiss example), other major changes should force travel behaviour analysis to adopt fully dynamic frameworks, of which charging at the point of use is the most likely candidate next to improved real-time information. The recent difficulties and costs involved in expanding infrastructure capacity have led to increasing reliance on demand management approaches to balance the network loads. Transport telematics, on-line traffic control, road pricing, revenue management of the airlines, but also employer-based *green travel* come to mind. All of these work because travellers have some flexibility with regard to their timings, above and beyond other possible changes to their schedule such as route, mode or destination. These changed timings are likely to affect more days than just the current one and therefore suggest models that are able to address sequences of days.

Additionally, the limitations of the current set of variables to explain behaviour have become clear. While the models often have good fits, there remains an unease, especially

[4] See the proceedings of the international conferences on travel behaviour research: TRB (1974, 1983), Stopher and Meyburg (1976), Hensher and Stopher (1979), Stopher *et al* . (1981), Rijkswaterstaat (1986), International Association for Travel Behaviour (1989), Stopher and Lee-Gosselin (1997), Ortuzar *et al* . (1998), Mahmassani (2002), Hensher (2001) and Axhausen (2006b).

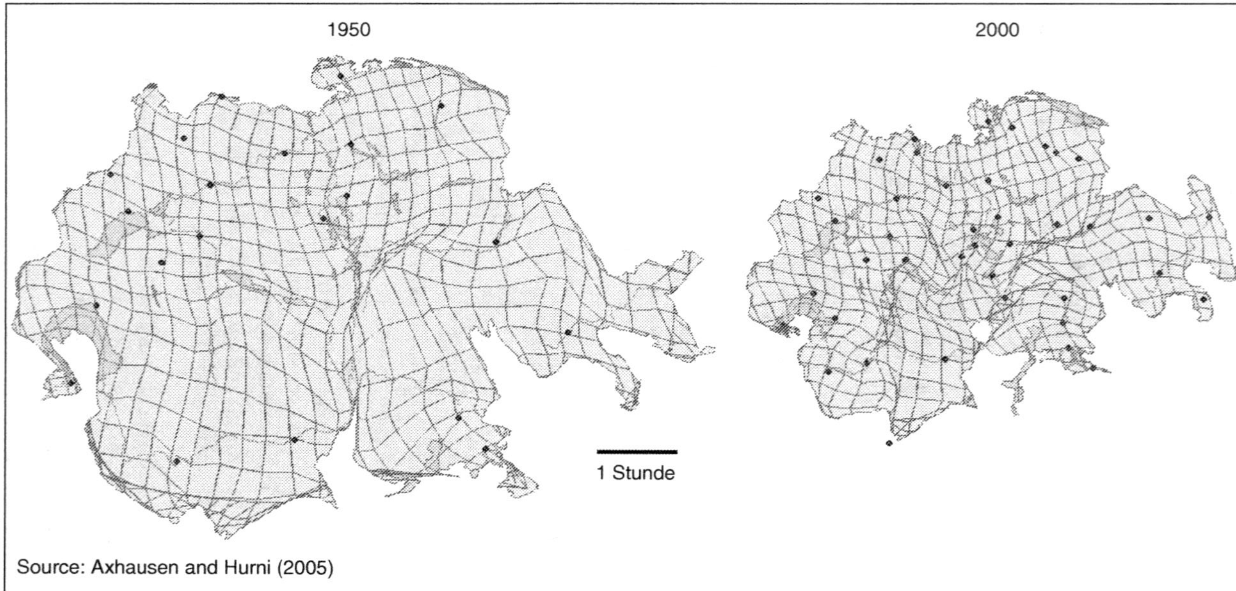

Source: Axhausen and Hurni (2005).

Figure 1: Road Travel Time-Scaled Mapping of Switzerland (Same Scale)

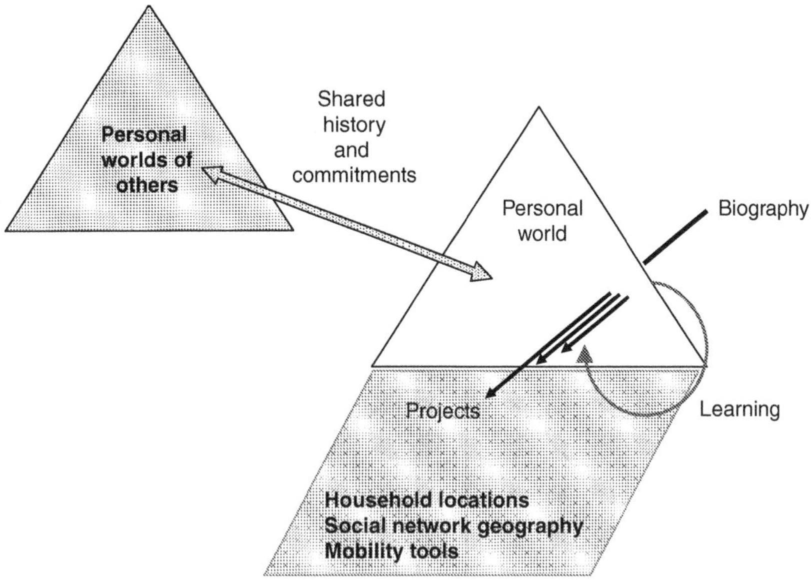

Figure 2: The Individual in a Dynamic Social Context

about *trip generation*[5], the stability of behaviour over time, the joint choices of groups and households, the formation of *choice sets* and finally the selection and formation of choice rules by travellers. Against a dynamic perspective, and also again visible in an analysis of omitted variables, two issues (and variable groups) seem obvious: the information assembled by travellers via their *biographies* and the information and abilities inherent in their *social networks*.

Figure 2 summarises a view of the traveller in a dynamic social context. The *personal world* of the traveller is perhaps better known as his or her *mental map*. However, this term emphasises the geography of the activity space at the expense of other important elements of knowledge: types of activities known, when and with whom to undertake them, and the opening hours of facilities. An alternative term for this knowledge would be *activity repertoire*, which in turn is partial because it does not refer to the geography of activity participation. Drawing on the knowledge accumulated over the course of a lifetime, the traveller selects *projects* against the background of his or her current *commitments*. Longer-term choices, such as household locations (including workplaces, regularly used

[5] In the most common approach to the schedule generation (see, e.g., Ortuzar and Willumsen, 2003), it is partially reconstructed through a sequence of partial analyses addressing the number and type of activities (trip generation), origins and destinations (trip distribution), modes (mode choice) and equilibrium allocation of routes and connections (assignment).

shops and other facilities), *social networks* as well as available *mobility tools* (such as licences, motorised vehicles, public transport season tickets and bicycles, which reduce the variable costs of travel in exchange for some upfront – capital – payment) are included in Figure 2. The members of the social network and their personal worlds are both resources and constraints. Thus, their knowledge, their abilities and their material resources can be drawn on within the limits of convention and proportional to the strength of the personal relationship, which gives the individual extra leverage in his or her daily life. Their locations, abilities and resources are also limiting, as they have to be considered in decision-making. Consider the trivial case of the joint choice of a restaurant, if one of the members of the group has particularly strong dislikes, such as an aversion to a cuisine, or is allergic to, say, seafood.

Still, the *social capital* of a person is embedded in their social network. At the microscopic level of human interaction in daily life, it needs to be given a specific definition to enable its measurement and operational use in conceptual, econometric and simulation models. Axhausen (2005, 2006) proposes that it is the joint skillful ability of the members of a network to perform, act and enjoy each other as the result of their joint history, commitments, references and understandings. In daily life, it encompasses both productive and hedonic aspects of joint human action. This capital is built up through joint activity and therefore travel, but also through technology mediated contacts such as letter writing, texting via SMS, emailing, chat room talk, instant messaging, phoning employing any number of technologies (land line, mobile, voice-over-IP, etc.). These tools, and the ability to use them, form together with the mobility tools, mentioned above, the *networking tools* of a traveller.

A person belongs to multiple social networks. A social network is defined as a set of persons who are linked pairwise, so that each person can reach any other through an active tie. The nature of the tie varies by the type of network, but always entails a certain minimum flow of resources and commitments between the two persons: being related to one another, working for the same group or firm, having studied with the same teacher, belonging to the same sport clubs, attending the same games, being a regular at the same bar, driving the same antique car model, having grown up in the same place (at the same time), etc. As the possible number of overlays of different strength is nearly endless, it seems pointless to speak just of strong and weak ties. Each tie entails certain rights and duties. The differences in these rights and duties also segment the principle universal net encompassing all human beings into a set of many smaller networks, which have different levels of awareness of their respective existence[6]. Nevertheless, their presence is felt through the general level of trusting and trust in a particular environment (see Seligman, 2000, or Offer, 2006).

[6] Some networks and their members are public knowledge, such as parliaments, the boards of listed firms and professors of a university. Other networks are known to exist, but their full membership is not generally known: social clubs, such as the Masons, Lions or Rotary, sports clubs, political parties, churches, families, etc.; finally, some networks can be assumed to exist, but their memberships are unknown: friendship groups, alumni groups, etc.

Given their limited understanding of the world and the time pressure of daily life, travellers have to schedule their day. Pre-commitments to work or regular activities, such as a gym, joint sport, family dinner, reduce the complexity of the choice problem by providing constraints, as well as tested routines (see Schlich, 2004, for a detailed analysis). Still, there are degrees of freedom even with these pre-commitments and there is free time. It seems reasonable to assume that the travellers construct alternative solutions from which they select that advances their interest best (see Offer, 2006, on the problem of how well persons can and do choose between longer-term commitments and pleasure now). For the outside observer and modeller, the time horizon of this choice, the set of alternatives constructed and considered and the complete frame of reference remain unknowable.

The utility of the choice is the weighted sum of the hedonic pleasure derived from the schedule (activities, contacts), any money earned, any contribution to a larger project, the activity expenditures and the generalised costs of the necessary movements. It is important to note that projects in the sense of a coherent set of activities undertaken to attain a goal provide additional structure over longer-time periods. They range from the trivial, such as preparing a dinner, to the profound, such as working towards a professional degree. The boundary to firm *commitments*, which one could see as open-ended or projects without a specific goal, is fluid.

The concept of *generalised cost* was advanced in the thinking about travel behaviour from the start, as all analysts recognised that travel costs are not reducible to just one variable. Travellers trade off different attributes of a movement, foremost, expenditure against travel time. It is also obvious that the acceptable trade-off will vary depending on the exact choice situation of the traveller defined by the time pressure of the choice, the time pressure of the movement, the physical strength of the traveller, the mobility tools at hand, the weather, and the purpose and social content of the activity at the destination of the movement. The generalised cost can be thought of as the taste-weighted sum of the perceived risk-adjusted expenditure and the comfort-adjusted and risk-adjusted travel time of the movement (see Xiang and Polak, 2005, for a similar reasoning). The risk-adjustment is necessary, as the traveller cannot know with certainty how large the expenditure will be, or how long the movement will last. The route taken and the means of transport offer different levels of comfort and enjoyment. The elements of quality include further safety from accident and violence, quality of information provided, smoothness of the ride, interiors of the vehicle, etc.

The perceived expenditures or costs will vary by person and the time horizon of the choices at hand. Generally, fixed costs, such as purchase prices of cars or season tickets will be ignored in day-to-day decision-making; even fuel bought irregularly might be ignored by car drivers. This discounting explains the disproportionate weights given to single fares, tolls and parking fees, which are a regular feature of any analysis of the impacts of these very visible out-of-pocket expenditures.

Given the field's focus on movement, travel behaviour research has learned relatively little about the utility derived from the activities for which the movement is undertaken. In principle, Becker's (1976) leisure time allocation model and its derivatives dominate the field (see, e.g., DeSepra, 1971; Jara-Díaz and Guevara, 2003). There are however few attempts to estimate the underlying utility functions and even fewer that experiment with alternative functional forms beyond the logarithm of duration. A decreasing marginal utility of any activity with regards to its duration stands to reason, or the common experience of satiation or boredom. Still, it is unclear, whether the marginal utility becomes negative, and if so, when and why. On the other hand, the argument made above about social capital as the product of joint learning would suggest that any social activity should produce some positive utility as long as it increases the joint skill level.

While studies about the activity performance (main and any interwoven secondary activities) are rare in travel behaviour (see Hensher, 1977, for example of research on value of travel time savings of business travellers, or Lyons *et al.*, 2006 for all journeys), work on the social content is essentially non-existent. Social content is the social implication of any activity above and beyond any actions performed. Axhausen (2005) proposes to capture the social content of an activity through the following items:

- A more detailed coding of the actions involved, perhaps at the level of detail typical for time-use studies;
- A description of the social purpose of the activity and of the obligations fulfilled with it;
- The beneficiaries of the activity;
- Composition of the party travelling together to the activity;
- Composition of the party participating in the event and having meaningful interactions with the traveller;
- The locations of the fellow travellers and participants prior to the trip or activity, to assess their costs to be involved in the activity or the movement;
- Distribution of the travel and activity costs among the participants and beneficiaries;
- The planning horizon of the activity;
- Number of previous visits to that location, in particular, if it was the first visit ever;
- The secondary activities undertaken during the trip and the activity, if any.

These items would give an indication of both the how and why of the activity, but also of the value attached to it. They would position it in the larger context of the person's daily life through the planning horizon and the information about the beneficiary.

Individual items of this list have been tested in recent diary surveys (see Axhausen *et al.*, 2002, 2006; Schlich *et al.*, 2003; Löchl *et al.*, 2005), but their joint potential to illuminate the social content of activities and travel is still unexplored.

Just as the social content has not been given much attention in travel behaviour research, the search for *behavioural innovation*, or its cousin *variety seeking*, which does not, strictly speaking, require new experiences, has been equally neglected. The term "innovation" or an invention transformed into an object or service in daily use is actually difficult to apply to travel behaviour research, which is not primarily concerned with new objects and services, but with the patterns of behaviour resulting from the interaction of these with the traveller. We could define behavioural innovation as any of its (detailed) elements never before observed pattern of activity and movement. The analyst has to limit the number of the elements and of their categorisations to be able to operationalise novelty (Schlich, 2004; Schlich *et al.*, 2004). The recent availability of long-duration diaries and of even longer GPS-based observational data of travel behaviour makes it is now possible to analyse the level of innovation in daily behaviour (see Figure 3 for an example).

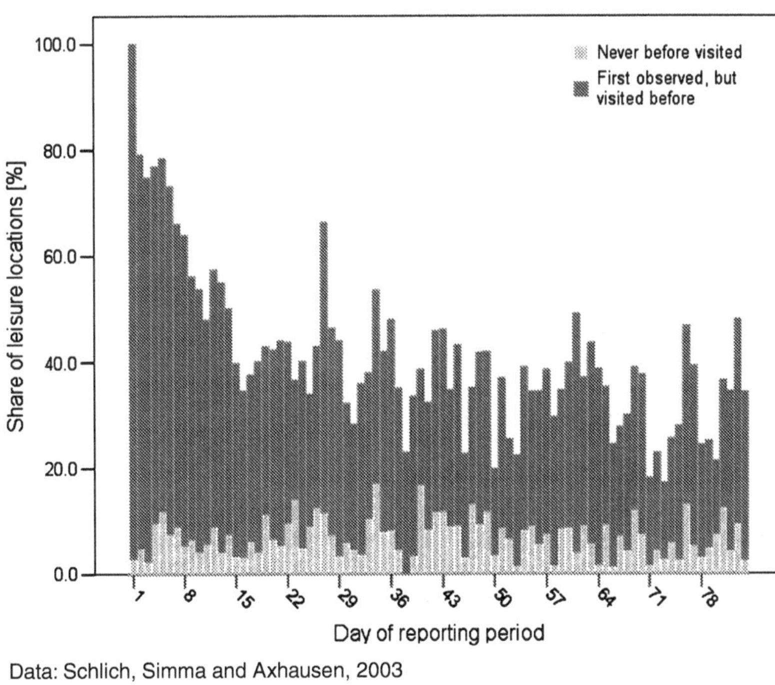

Source: Schlich *et al.* (2003).

Figure 3: Share of Never Before Observed and Never Before Visited Leisure Locations over a 12-Week Period

This pattern of an on-going search for new locations is consistent across a range of six long-duration datasets, which have been analysed by Schönfelder and Axhausen (2003a, 2003b, 2004) and later Schönfelder *et al.* (2006). It would be interesting to know whether this search for the new is equally pronounced among the other dimension of travel behaviour: activity type, routes or participants in the activities undertaken.

DIGRESSION 1: ACTIVITY SPACE AND INNOVATION

The observed constant rates of innovation with regards to visited locations raises the question of how much space, or how many activity locations have to be within reach for a traveller to satisfy her or his need for the new and different. The traveller's mental map is likely to be substantially larger, as it includes second-hand information. One possibility is to restrict the measurement to the observed locations and search for a measure that at least approximates the likely underlying mental map. Any systematic measure based on these locations observed could be called *activity space*. Schönfelder and Axhausen (2003a, 2003b) propose a number of approaches for this purpose, partially derived from the ecological research on animal home ranges. The most satisfying are the simple to calculate and robust size of the 95% confidence ellipse (see Figure 4a) and the shortest path network (Figure 4b), which requires substantially more background data in the form of coded road and public transport networks and more computation. It is defined as the length of the set of links, which are part of the shortest paths between the locations visited by a person, in the sequence observed[7].

The 95% confidence ellipse rests on strong parametric assumptions and imposes a particular geometric form. It also generalises and covers parts of the city or urbanised landscape, which the traveller is very unlikely to know. It is therefore not fully suitable as a proxy measure of the underlying mental map. The shortest path network requires fewer parametric assumptions and only considers observed locations. The shortest path network should also be a better approximation, as it only includes parts of the city, of which the traveller has first-hand experience; assuming that we observe the traveller long enough, we can hope that he or she starts to exhaust the locations which form part of the mental map. Any innovation is then likely to add only marginal amounts to its size.

The analysis in the papers cited above shows that the activity spaces follow a left-skewed distribution, that the size is not strongly tied to the travellers' socio-demographics, but that it is dependent on the total number of trips observed. This weak link between the socio-demographics and the patterns of trip making was also found by Schlich (2004) who likewise worked with long-duration diaries. The activity

[7] If GPS records are available, the computed shortest paths may be replaced by the observed routes.

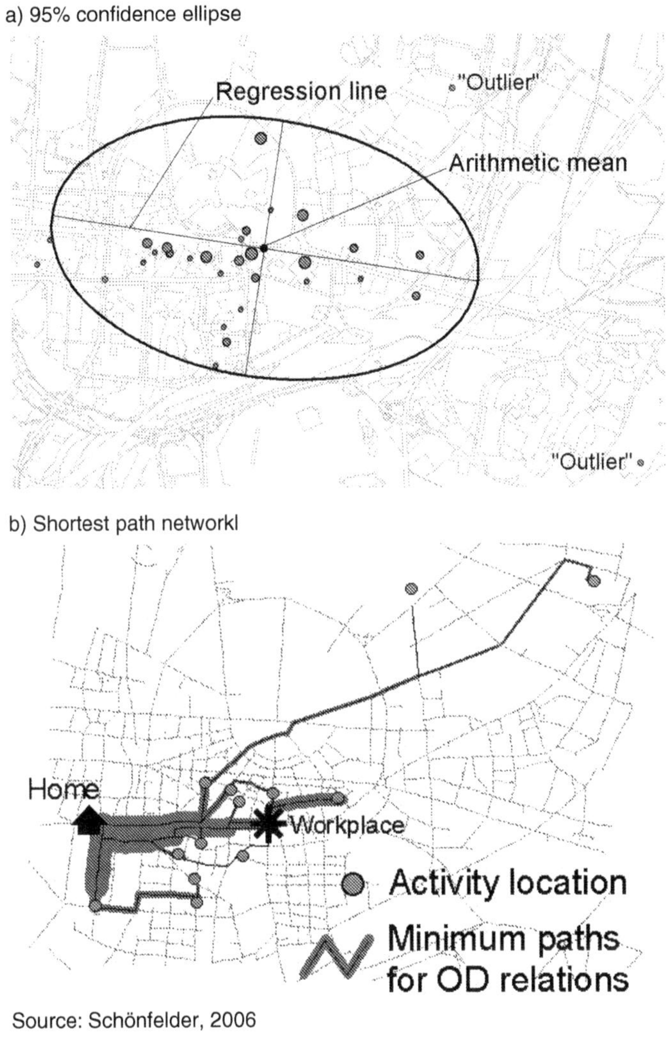

Source: Schönfelder (2006).

Figure 4: Examples of Activity Space Measures

level and variety-seeking seem to be independent of the usual explanatory variables. An interesting question to ask is then to the extent to which travellers self-select as residents into locations allowing them to satisfy this need with lower generalised cost for high(er) quality activities.

DIGRESSION 2: CONSTRUCTION OF THE CHOICE SETS

Travel behaviour research divides decision-making into two parts: construction of the choice set and the choice itself. The bulk of the research during the last three decades has gone into modelling the second part with continuously more complex structures, mostly of the general extreme value type (see, e.g., Ben-Akiva and Lerman, 1985; Train, 2003; Bhat, 2006; Daly and Bierlaire, 2006). The problem of the choice set formation has been acknowledged, but essentially ignored in practice, as it tied up with the unobservable traveller's *personal world* and the equally difficult to observe choice situation of the specific case. Generally, it is not known when the details of a particular activity and movement are decided upon, under what time pressure and with whom, if anybody, the choice was negotiated. The work of Doherty and collaborators (e.g., Doherty and Miller, 2000) has shed light on the lead times of the choices, but not yet on the other aspects. The large share of observed choices labelled impulsive, spontaneous or last minute in his surveys, is noticeable through out, and indicates that travellers have often little time or take little time to decide.

Figure 5 proposes a structure for the choice process. It highlights, that the choice among the alternatives constructed by the traveller is only the last of a series of choices. The first is the meta-decision that a decision is required. The person must have detected, that his or her situation has changed and that the current behaviour might produce an unsatisfactory outcome. Examples in the travel context would be the onset of congestion on

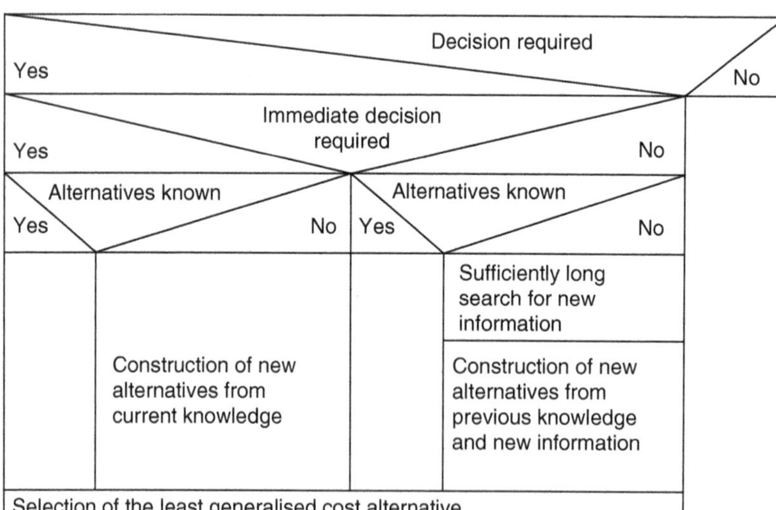

Figure 5: Simplified Model of the Choice Process

a planned route, the unproductive meeting or activity, the realisation that on-street parking is not readily available or the rapid technical deterioration of the personal car. The time available for that choice must be defined. Again, in the transport context, this might be the driving time until the next junction or exit, or the time until the end of the leasing period for the current car. The traveller needs to assess whether he or she has enough alternatives to choose from or to establish or construct new ones. Hägerstrand (1970) identified the physical, social and legal constraints that will limit the alternatives to those compatible with time-space regime of the environment, the prior commitments to others and the locations reachable with available mobility tools[8]. Depending on the time pressure, there might be time for an extensive search for information, say in the case of a new work location or a new residence, or just enough time to recall earlier trips and their routes that involve the upcoming exit ramp. The end of the extensive search has also to be decided upon (see Varian, 1999, for the economics of search). Only now can the traveller decide. Sometimes no alternative is found, so no change in behaviour is possible; in other cases the number of alternatives is large, as for example in the case of holiday destinations or after searching various airline websites for the routings of a multi-destination round-the-world journey.

A PROPOSED CONCEPTUAL FRAMEWORK

The view formulated above sketches the traveller as a *network actor* in a double sense, a person who negotiates both the physical space of locations and infrastructure networks and the social space of his or her social networks, ranging from work, school, clubs, church, neighbourhood to friendship and family. Through investment in his or her mobility and networking tools, the traveller reduces the variable generalised costs of travel, either through faster speeds or higher comfort or lower variable expenditures. Positioning himself or herself in an environment which matches the desired profile for the rate of innovation he or she can pursue, the projects envisaged and the commitments accepted. This is a profoundly dynamic understanding of the traveller, as just about all elements are subject to on-going change: the person learns and thereby updates his or her personal world, mental map and activity repertoire. The social networks grow and change through the actions of other members of the networks, which in turn will change the joint activity space, as new locations are added by new network members. Finally, one would expect that the traveller will generally seek to improve his or her situation, or at least to stabilise it. Offer (2006) discusses the problem of the choice of time horizon for this improvement.

Such a dynamic view of the individual requires a division of the modelled processes into short-term and long(er)-term ones. The perspective changes from the descriptive one

[8] The partial relaxation of the social constraints through the general availability of mobile phones was 30 years in the future when Hägerstrand's (1970) seminal paper was published. He could also not foresee the attendant tolerance for the announced delay and real-time rescheduling.

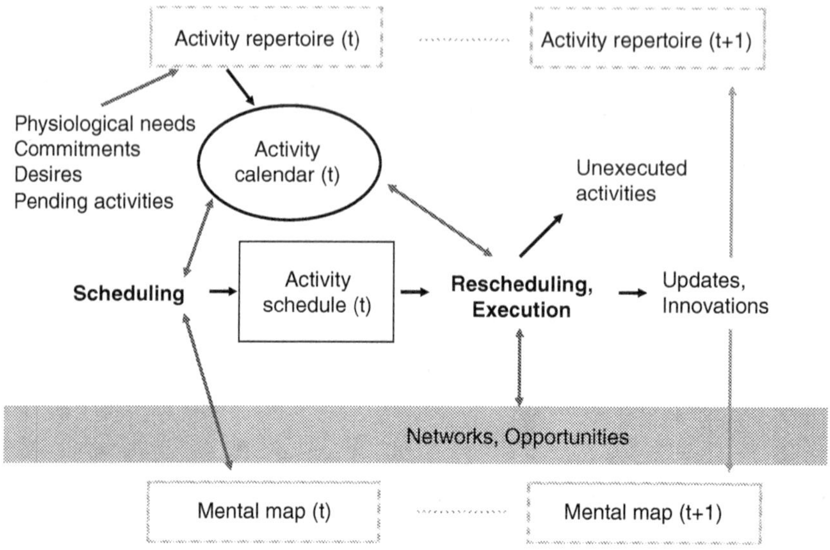

Figure 6: Modelling the Individual's Day-to-Day Dynamics

taken so far to an operational one. One task of travel behaviour research is to predict changes due to policy intervention or exogenous changes, such as immigration, population growth or decline, changing prices, etc. One possibility for such a division is suggested in Figures 6 and 7. The understanding referred to here does not require an *equilibrium*[9], but assumes a willingness on the part of the traveller to improve his or her situation incrementally. While behaviourally appealing, such an approach might still need equilibrating mechanisms to produce consistent results in application contexts.

The central process for the short term (Figure 6) is the formulation of schedules as the complete description of a day (number, type and sequence of activities, their durations and locations, modes and routes, finally group size and composition for travel and activities). It is assumed here that the scheduler draws from an *activity calendar* that lists the activities or, more generally, activity types that the traveller has to accomplish due to project engagements, commitments, physiological needs or desires. This list reflects the person's *activity repertoire* that can be expanded through interaction with others and the environment. As people generally aim to improve and possibly even optimise their schedules, they will draw on their mental maps to reduce the effort and uncertainties of the day. We know that travellers, as a rule, do not fully book their days, but leave slots for the unexpected and the unplanned. In a simulation framework, it might be necessary

[9] Equilibria in the travel behaviour context are defined as situations, in which the expected generalised costs on which choices are based are equal to those encountered by the modelled agents during the execution of these choices. There are no inconsistencies left.

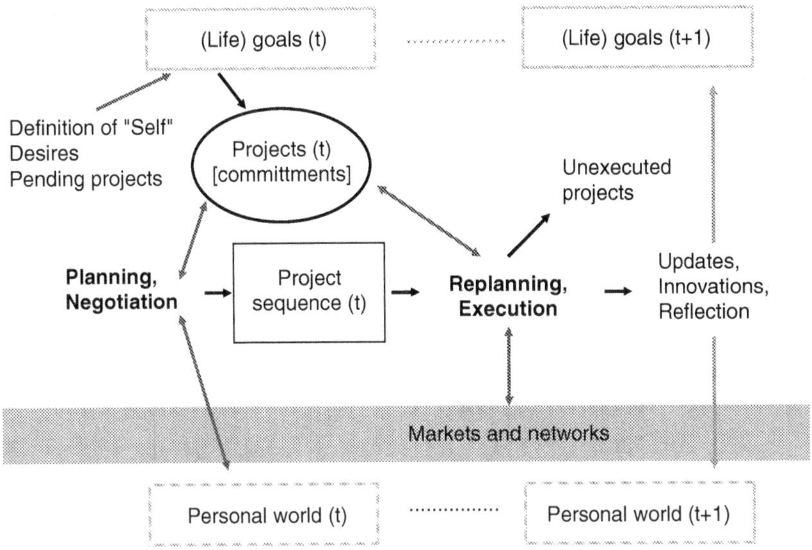

Figure 7: Modelling the Individual's Longer-Term Dynamics

for the sake of computational convenience to impose the assumption that the current day has been fully allocated by some arbitrary point of time in the previous night.

The execution of the schedule requires interaction with others in the infrastructure networks and in activity locations, such as shops, cinemas or other persons' homes. In some cases, the resulting congestion or the failure of an activity opportunity to deliver the expected service or good forces travellers to adjust and to reschedule. At the end of the day, travellers will have updated their knowledge about the elements in their activity repertoires and mental maps. They may have developed new solutions to the fulfilment of their needs by innovating, here trying new routes, modes or locations, by drawing on new information, by expanding their *expectation space* or by accepting solutions proposed by others on the basis of their knowledge. The expectation space is a third view of the personal world. It is the set of simple rules or heuristics that people develop about the environment, and their generalisations about the organisation of space. Examples are the heuristics of how to detect the onset of congestion, how to find a gasoline station in an otherwise unknown part of town or expectations about the store composition of any local shopping mall.

Longer-term processes (see Figure 7), which structure the shorter-term ones, revolve around the projects which a traveller formulates to translate his or her life goals and his or her understanding of himself or herself into reality (Nuttin, 1984; Carver and

Scheier, 1998). In any one period these projects need to be sequenced to provide a reasonable load and prioritisation. This planning requires negotiation with others, as many projects will depend on synchronisation with, input of, presence of or permission of others. Again, the interaction with still others in the markets and networks during execution will require adjustments and changes, including the abandonment of certain projects or project elements. The experience will update and expand the individual's *personal world* but also shape the set of life goals pursued in the next time interval.

The impact on travel behaviour modelling of such a reframing of the task would be profound. It would move the centre of attention away from the idea of equilibrium and towards concepts such as innovation, solution generation, life goals and commitments to people and ideas. These concepts are not unknown to activity-based travel behaviour analysis (see, e.g., Gärling *et al.*, 1998), but they would need to be moved centre-stage.

OUTLOOK

This chapter has tried to provide concepts enabling a discussion of the dynamics of travel behaviour. At the same time, it highlights the need for a view of travellers as network actors constantly aiming to stabilise or improve their situation. The model system sketched in the last section seems feasible to apply with today's computers and software tools. Between feasible and operational is a large gap, however, which needs to be bridged by appropriate research. The major topics of such research would need to be the description of social networks and their impact on travel decisions, the modelling of the mental maps and activity repertoires of travellers, the identification of the solution strategies applied by individuals when they schedule their day and with it the days of the persons they interact with and contact. Last but not least, all results of the research need to be translated into computational models, which can support the decision-making of firms and governments.

ACKNOWLEDGEMENTS

This chapter draws on Axhausen (2006) and discussions during my lecturing and supervision. I gratefully acknowledge the input from those taking part.

REFERENCES

Axhausen, K.W. (2000). Definition of Movement and Activity for Transport Modelling. In: *Handbook of Transport Modelling* (D.A. Hensher and K.J. Button, eds), Vol. 1, Elsevier, Oxford, 271–284.

Axhausen, K.W. (2003). Definitions and Measurement Problems. In: *Capturing Long Distance Travel* (K.W. Axhausen, J.L. Madre, J.W. Polak, and P. Toint, eds), Research Science Press, Baldock, 8–25.

Axhausen, K.W. (2005). *Social Factor in Future Travel: A Qualitative Assessment* (*Arbeitsbericht Verkehrs- und Raumplanung, 294*), IVT, ETH Zürich, Zürich.

Axhausen, K.W. (2006a). *Social Networks, Mobility Biographies and Travel: The Survey Challenges* (*Arbeitsbericht Verkehrs- und Raumplanung, 343*), IVT, ETH Zürich, Zürich.

Axhausen, K.W. (ed.) (2006b). *Moving through Nets: The Physical and Social Dimensions of Travel*, Pergamon, Oxford.

Axhausen, K.W. and L. Hurni (eds) (2005). *Zeitkarten Schweiz 1950–2000*, IVT and IKA, ETH Zürich, Zürich.

Axhausen, K.W., M. Löchl, R. Schlich, T. Buhl, and P. Widmer (2006). Fatigue in Long-Duration Travel Diaries, *Transportation*, in press.

Axhausen, K.W., A. Zimmermann, S. Schönfelder, G. Rindsfüser, and T. Haupt (2002). Observing the Rhythms of Daily Life: A Six-Week Travel Diary, *Transportation*, **29**, 95–124.

Becker, G.S. (1976). *The Economic Approach to Human Behavior*, University of Chicago Press, Chicago.

Ben-Akiva, M.E. and S.R. Lerman (1985). *Discrete Choice Analysis: Theory and Application to Travel Demand*, MIT Press, Cambridge.

Bhat, C.R. (2006). Econometric Choice Formulations: Alternative Model Structures, Estimation Techniques and Emerging Directions. In: *Moving Through Nets: The Physical and Social Dimensions of Travel* (K.W. Axhausen, ed.), Elsevier, Oxford, 47–82.

Carver, C.S. and M.F. Scheier (1998). *On the Self-Regulation of Behavior*, Cambridge University Press, Cambridge.

Daly, A. and M. Bierlaire (2006). A General and Operational Representation of Generalised Extreme Value Models, *Transportation Research B*, **40**, 285–305.

DeSerpa, A. (1971). A Theory of Economics of Time, *The Economic Journal*, **81**, 828–846.

Doherty, S.T. and E.J. Miller (2000). A Computerized Household Activity Scheduling Survey, *Transportation*, **27**, 75–97.

Gärling, T., T. Laitila, and K. Westin (eds) (1998). *Theoretical Foundations of Travel Choice Modeling*, Pergamon, Oxford.

Hägerstrand, T. (1970). What about People in Regional Science? *Papers of the Regional Science Association*, **24** (1), 7–21.

Hensher, D.A. (1977). *Value of Business Travel Time*, Pergamon, Oxford.

Hensher, D.A. (ed.) (2001). *The Leading Edge in Travel Behaviour Research*, Pergamon, Oxford.

Hensher, D.A. and P.R. Stopher (eds) (1979). *Behavioural Travel Modelling*, Croom Helm, London.

International Association for Travel Behaviour (ed.) (1989). *Travel Behaviour Research*, Avebury, Aldershot, UK.

Jara-Díaz, S.R. and C. Guevara (2003). Behind the Subjective Value of Travel Time Savings: The Perception of Work, Leisure and Travel from a Joint Mode Choice–Activity Model, *Journal of Transport Economics and Policy*, 37, 29–46.

Kaufmann, V. (2002). *Re-Thinking Mobility: Contemporary Sociology*, Ashgate, Aldershot, UK.

Larsen, J., K.W. Axhausen, and J. Urry (2006). Geographies of Social Networks: Meetings, Travel and Communication amongst Youngish People, *Mobilities*, 1(2), 261–283.

Löchl, M., S. Schönfelder, R. Schlich, T. Buhl, P. Widmer, and K.W. Axhausen (2005). *Stabilität des Verkehrsverhaltens*, Final Report, SVI 2001/514. *Schriftenreihe, 1120*, Bundesamt für Strassen, UVEK, Bern.

Lyons, G., J. Jain, and D. Holley (2006). *The Use of Travel Time by Rail Passengers in Great Britain*, Working Paper, UWE, Bristol.

Madre, J.-L., K.W. Axhausen, and W. Brög (2006). Immobility in Travel Diary Surveys: An Overview, *Transportation*, in press.

Mahmassani, H. (ed.) (2002). *In Perpetual Motion: Travel Behaviour Research Opportunities and Application Challenges*, Pergamon, Oxford.

Nuttin, J. (1984). *Motivation, Planning and Action: A Relational Theory of Behaviour Dynamics*, Leuven University Press, Leuven.

Offer, A. (2006). *The Challenge of Affluence: Self-Control and Well-Being in the United States and Britain since 1950*, Oxford University Press, Oxford.

Ortuzar, J. de Dios, D.A. Hensher, and S. Jara-Diaz (eds) (1998). *Travel Behaviour Research: An Update*, Pergamon, Oxford.

Ortuzar, J. de Dios and L.G. Willumsen (2001). *Modelling Transport*, Wiley, Chichester.

Rijkswaterstaat (ed.) (1986). *Behavioural Research for Transport Policy*, VNU Science Press, Utrecht.

Schlich, R. (2004). *Verhaltenshomogene Gruppen in Längsschnitterhebungen*, Doctoral Dissertation, ETH Zürich, Zürich.

Schlich, R., S. Schönfelder, S. Hanson, and K.W. Axhausen (2004). The Structures of Leisure Travel: Temporal and Spatial Variability, *Transport Reviews*, **24**, 219–228.

Schlich, R., A. Simma, and K.W. Axhausen (2003). Determinanten des Freizeitverkehrs: Modellierung und empirische Befunde, Research project SVI 2000/443. *Schriftenreihe, 1071*, Bundesamt für Strassen, UVEK, Bern.

Schönfelder, S. (2006). *Urban rythms: Modelling the Rythms of Individual Travel Behaviour*, Department Bau, Umwelt und Geomatik, ETH Zürich, Zürich.

Schönfelder, S. and K.W. Axhausen (2003a). Activity Spaces: Measures of Social Exclusion? *Transport Policy*, **10**, 273–286.

Schönfelder, S. and K.W. Axhausen (2003b). On the Variability of Human Activity Spaces. In: *The Real and Virtual Worlds of Spatial Planning* (M. Koll-Schretzenmayr, M. Keiner, and G. Nussbaumer, eds), Springer, Heidelberg, 237–262.

Schönfelder, S. and K.W. Axhausen (2004). *Structure and Innovation of Human Activity Spaces* (*Arbeitsberichte Verkehrs- und Raumplanung, 258*), IVT, ETH Zürich, Zürich.

Schönfelder, S., H. Li, R. Guensler, J. Ogle, and K.W. Axhausen (2006). *Analysis of Commute Atlanta Instrumented Vehicle GPS Data: Destination Choice Behavior and Activity Spaces* (*Arbeitsberichte Verkehr und Raumplanung, 350*), IVT, ETH Zürich, Zürich.

Seligman, A.B. (2000). *The Problem of Trust*, Princeton University Press, Princeton.

Stopher, P.R. and M.E.H. Lee-Gosselin (eds) (1997). *Understanding Travel Behavior in an Era of Change*, Pergamon, Oxford.

Stopher, P.R. and A.H. Meyburg (eds) (1976). *Behavioral Travel-Demand Models*, Lexington Books, Lexington.

Stopher, P.R., A.H. Meyburg, and W. Brög (eds) (1981). *New Horizons in Travel-Behavior Research*, Lexington Books, Lexington.

Train, K. (2003). *Discrete Choice Methods with Simulation*, Cambridge University Press, Cambridge.

TRB (ed.) (1974). *Behavioral Demand Modeling and Valuation of Travel Time*, Special Report 149, Transportation Research Board, Washington, DC.

TRB (ed.) (1983). *Travel Analysis Methods for the 1980s*, Special Report 201, Transportation Research Board, Washington, DC.

Urry, J. (2000). *Sociology beyond Society: Mobilities for the 21st Century*, Routledge, London.

Varian, H.R. (1999). Economics and Search, Note Published as www.sims.berkeley.edu/~hal/Papers/sigir/sigir.html (March 2006).

Xiang, L. and J.W. Polak (2005). *On the Prevalence of Expected Utility and Non-Expected Utility Decision Rules in Travellers: Choice amongst Uncertain Prospects*, Paper presented the 45th Congress of the European Regional Science Association, Amsterdam, August.

Threats from Car Traffic to the Quality of Urban Life
Tommy Gärling and Linda Steg (Editors)

10

DETERMINANTS OF CAR DEPENDENCE

Stephen Stradling, Transport Research Institute, Napier University, Edinburgh, UK

ABSTRACT

This chapter looks at how car dependence varies in Scotland, a highly motorised country. An overview of the factors influencing car dependence and travel and transport choices more generally is developed. Car-dependent places, car-dependent trips and car-dependent persons are distinguished. It is shown that most motorists are already multi-modal transport users with prior experience of other modes. The prospects for modal shift from die-hard drivers, complacent car users, malcontented motorists and aspiring environmentalists are given along with the different policy pulls and pushes required to impact on their car use.

INTRODUCTION

As Part 1 of this volume has shown, at both an individual and collective level the disbenefits of car travel and threats to the quality of life from car traffic are increasingly apparent (see RAC, 1995; Engwicht, 1998; Adams, 2000; Litman, 1999; Newman and Kenworthy, 1999; Semlyen, 2000; Goodwin, 2001; Gärling *et al.*, 2002; Stradling, 2002a, 2002b; Sloman, 2003). Yet car ownership continues to rise despite a growing policy focus on reducing car dependency and achieving modal shift. To understand how such efforts might be made more effective a detailed knowledge is needed of transport behaviours and the opportunities that facilitate or constrain them, the patterns of lifestyle obligations that drive them and the personal preferences and inclinations that underpin them.

Using findings from a number of recent studies of car users in Scotland, this chapter looks at how car ownership, car use and the prospects for modal shift from car to more sustainable modes varies across locations, trips and population segments in a highly motorised country and how this knowledge could assist the management of future travel demand.

CAR-DEPENDENT PLACES, TRIPS AND PEOPLE

Car Access

Scotland, like other developed countries, is currently a car-dependent society. Figures from the 2001 Census show Scotland with a population of just over 5 million people in 2.2 million households with 2.0 million cars or vans owned or available for use by these households. Table 1 shows the percentage of households with none, one, and two or more cars or vans available. Figures are given for Scotland as a whole and for each of the eight Scottish parliamentary regions, in descending order of the number of cars or vans as a proportion of the number of households recorded for each region. This final column highlights the contrast in car availability between the more rural (e.g., Highlands and Islands) and the more urban areas of Scotland (e.g., Glasgow).

Those with access to a car in their household tend to make more trips. Ormston *et al.* (2004) asked 1,024 respondents from four travel-to-work areas in both urban and rural parts of Scotland how many trips they make in a typical week. Responses ranged from 0 to 113, with a mode of 14. After square root transformation to counter the skewness of the distribution, analysis showed a significant effect for number of cars available to the household (Median trips per person per week: no cars in household 16 trips; one car 18 trips; two or more cars 22 trips).

Data from the 2002 Scottish Social Attitudes survey (Anderson and Stradling, 2004) showed that three-quarters of Scottish adults now live in households which own or have regular use of at least one car and that above six in ten adults (63%) say they currently drive.

There is, however, considerable variation in patterns of car access and use across different sections of the population. Tables 1 and 2 show that people living in Scotland's rural areas

Table 1: Percentage of Households with Cars or Vans Available in Scotland 2001

Area	*% with None*	*% With One*	*% With Two or More*	*Cars per 100 Households*
Scotland	34	43	23	93
Highlands and Islands	26	49	26	106
Mid Scotland and Fife	27	46	27	106
South of Scotland	28	46	27	105
North East Scotland	31	44	26	101
West of Scotland	33	42	25	96
Central Scotland	34	43	23	94
Lothians	36	44	20	88
Glasgow	55	35	10	57

Source: Adapted from Scottish Census data Table KS17.

Table 2: Population Access to Cars and Current Drivers in Scotland 2002

[N = 1,665]	*% Resident in Household with Car*	*% Current Drivers*
All	75	63
Males	77	73
Females	73	54
Age		
18–24	66	41
25–39	79	71
40–64	82	72
65+	57	42
Urban/rural area		
Accessible urban areas	62	60
Rural and remote urban	87	75

are more likely than those in urban areas to be drivers and to have household access to a vehicle. Table 2 shows that while equivalent proportions of male (77%) and female (73%) Scottish adults live in households with access to a car, three-quarters of males (73%) but only half of females (54%) currently drive. Those in the youngest and oldest age groups are less likely to live in households with access to a vehicle or to be current drivers.

Car access and use are also strongly patterned by income. Figure 1 shows the distribution of car access by household income, using data from the Scottish Household Survey (SHS) for 1999–2003 (Stradling *et al.*, 2005). Car ownership is, *inter alia*, a status marker and this is signalled in the cost of cars. Access to a car varies substantially with household income.

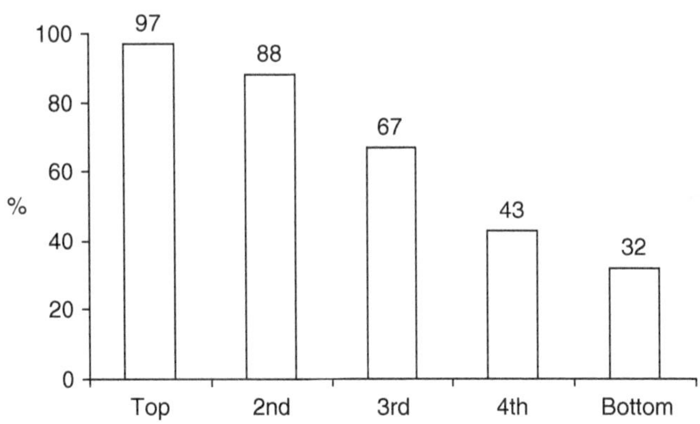

Figure 1: Percentage of Respondents from Households with Access to a Car for Annual Household Income Quintiles

From Origins to Destinations

Car use also varies with journey purpose. The SHS travel diary recorded details of almost 27,000 reported journeys on the day before interview, with interviews conducted on all days of the week. Table 3 uses SHS figures (Scottish Executive, 2004a) to show the percentage of previous-day journeys for core "lifestyle maintenance" activities.

Travel joins up the places where people go to lead their lives and meet their obligations (Stradling *et al.*, 2000; Stradling, 2002c) and Table 3 gives a snapshot of the quotidian round of daily life for Scottish adults. Two-thirds of journeys were for attending place of work, restocking larder or wardrobe, social network maintenance and escorting others less able to make their way alone.

Table 4, using data from the same source (Scottish Executive, 2004a), shows the percentage of journeys for each activity that were undertaken by car, whether as driver or passenger. Business trips are the most car-dependent journeys, with trips by adults for education and for eating/drinking at the opposite end of the scale.

Mode Substitution

Farrington *et al.* (1998, p. 3) deemed as structurally dependent on the car "those who are dependent . . . because there are no viable alternatives" and as consciously dependent on the car "those who rely on their vehicle but could realistically undertake their journeys by alternative modes". The former are unable to switch modes, the latter unwilling.

Table 3: Percentage of Journeys by Adults (16+) on Previous Day by Trip Purpose

[N = 26,944]	*% of Journeys*
Commuting	24
Shopping	23
Visiting friends or relatives	12
Escort	8
Sport/entertainment	6
Other personal business	6
Holiday/day trip	4
Business	4
Eating/drinking	3
Education	3
Visit hospital or other health	3
Other or not recorded	4

Source: Adapted from Scottish Executive (2004a, Table 4).

Table 4: Percentage of journeys by Adults (16+) as Car Driver, Passenger or Both by Trip Purpose

[N = 26,944]	*Driver*	*Passenger*	*Both*
Business	77	6	83
Escort	67	9	76
Other personal business	60	16	76
Visiting friends or relatives	51	22	73
Commuting	60	11	71
Sport/entertainment	50	21	71
Visit hospital or other health	42	29	71
Holiday/day trip	40	23	63
Shopping	46	16	62
Eating/drinking	24	27	51
Education	26	10	36
Other or not recorded	39	17	56

Source: Data from Scottish Executive (2004a, Table 4).

Able to Use Other Modes

In two studies of travel awareness (NFO World Group and Napier University Transport Research Institute, 2001, 2003), respondents were asked how often they undertook various lifestyle maintenance activities and, for those they undertook, how often they used various travel modes, including car, to access these activities. Those who undertook each activity by car were then asked whether it would be practical for them to use each of four alternative modes (bus, walk, train, cycle) for that activity. Table 5 combines data from the two studies and shows, for a set of trip-types currently undertaken by car, the percentage of drivers who say they always do it by car and the percentage of drivers who undertake such activity who say they could do so by each of four other modes (some respondents indicated it would be practical for them to use more than one alternative mode). The activities are arranged in descending order of the percent saying that "None of these" would be a practical alternative for them. Table 5 shows that while 57% of those who currently commute by car say they could not do the journey otherwise, 43% thus could commute other than by car. Of the non-commute activities, "Supermarket shopping" was the most car-dependent trip, but even for this activity only three-quarters (72%) report that they "Always" did it by car. Evenings out for leisure purposes – a category covering a wide range of possibilities, some local, some distant – was the least car-dependent activity on this measure, with only one-third (35%) of drivers saying that they always did this by car. Half of drivers (47%) who escorted children to or from school always did this by car; but further analysis of the data shows that 10% of respondents who were drivers and who escorted children to school never did so by car.

Mackett and Ahern (2000, p. 23) noted that ". . . many households in Britain are car dependent, that is, their whole lifestyle depends upon having a car available to undertake

Table 5: Percentage of Drivers Who Always Do Each Activity by Car and Percentage of Drivers Doing Each Activity Who Say It Would be Practical to Use Each of Four Other Modes of Transport

[N = 392 – 1,598]	*% Always by Car*	*Bus*	*Walk*	*Train*	*Cycle*	*None of These*
Travel to work[a]	100	25	14	9	9	57
Supermarket shopping	72	29	19	<1	2	56
Go away for a weekend	67	20	<1	38	<1	52
Take children to leisure activities[b]	59	30	25	3	2	50
Leisure activities at the weekend	58	31	22	12	8	47
Evenings out for leisure purposes	35	36	22	9	8	43
Visit friends or relatives	53	33	37	11	6	36
Town centre shopping	48	47	17	10	1	35
Take children to/from school[b]	47	13	54	0	2	35

Source: Data from NFO World Group and Napier University Transport Research Institute (2001, 2003).

[a] Respondents who travel to work by car.

[b] Respondents with children in the household.

their range of activities", but these figures show that the converse does not hold, that at least for the common, core, current lifestyle activities of Table 5, having a car available does not invariably result in it being used to meet an activity-generated transport need, whether because of within-household competition for the car, green motives, or other reasons.

Stradling (2003) using this dataset showed that of persons from noncar-owning households who do each of these activities, between 14% (take children to/from school) and 57% (evenings out for leisure purposes) undertake them by car at least some of the time. Indeed, small but finite numbers, between <1% for evenings out and 7% for weekends away, say they manage to always do these things by car, despite not having a car available in their household. The reach of the car stretches beyond car owners. Even some persons without cars depend on the car, and on the good offices of others with access to a car, to serve part of their transport needs.

Use of Other Modes

Table 6 shows the frequency of use of personal motorised, public transport (PT) and self-propelled modes by a sample of Edinburgh adults living close to an urban bus corridor.

Table 6: Rated Frequency of Use of Different Travel Modes by Adult Residents Living Close to an Edinburgh Quality Bus Corridor

[N = 1,016]	*Most Days*	*Most Working Days*	*Once or Twice a Week*	*About Once a Fortnight*	*About Once a Month*	*Several Times a Year*	*About Once a Year or Less*	*Never*
Personal motorised transport								
Car as driver	33	5	16	2	2	5	2	35.2
Passenger in car with member of household driving	8	2	25	8	4	8	4	40.9
Passenger in car with friend or relative driving	2	1	16	15	15	31	8	12.0
Passenger in car with colleague driving	0	1	5	3	7	15	10	58.1
Moped			0				0	99.6
Motorbike	0	0	0	0	0	1	0	97.8
Public transport								
Bus	25	15	29	11	7	9	1	1.7
Airport bus	0		0	0	2	22	25	50.6
Night bus	0	0	2	2	4	7	14	71.1
Taxi	1	1	13	16	17	36	11	6.1
Train	0	0	2	6	11	38	28	14.0
Self-propelled transport								
Bicycle	3	2	6	3	3	9	6	68.8
Walk (more than 5 minutes from house)	69	7	19	2	1	1	0	0.9

Note: Zero signifies greater than zero but not more than 0.5%. Empty cells signify zero.

Table 7: Extent and Statistically Significant Differences in Frequent Mode Use by Gender, Age Band and Household Income Band amongst Residents Living Close to an Edinburgh Quality Bus Corridor

[N = 1,016]	*Percent Using Mode Once a Week or More*	*Gender Differences in Frequency of Usage*	*Age Group Mode is Most Frequently Used by*	*Household Income Differences in Frequency of Usage*
Walk (more than 5 minutes from house)	95		17–24	
Bus	69		45+	
Car as driver	54	M > F	35+	Hi > Med > Lo
Passenger in car with member of household driving	35	F > M	45+	Hi > Med > Lo
Passenger in car with friend or relative driving	19	F > M	55–64	Lo > Med > Hi
Taxi	15		17–34	Hi > Med, Lo
Bicycle	11	M > F	25–44	Med > Hi, Lo
Passenger in car with colleague driving	6		25–54	Hi > Med > Lo
Night Bus	2		17–24	
Train	2		17–34	
Motorbike	<1			
Airport Bus	<1		25–34	

Note: Empty cells indicate that differences were not statistically significant. > indicates significantly greater use of mode at 5% significance level.

When the number of modes that each respondent used more often than "Never" is computed (Stradling, 2004), only 3% of respondents are mono-modal, saying they only ever used one of the twelve listed modes of travel. Two respondents claim to use all twelve. Four out of five respondents (79%) use five or more modes. Half (51%) had used seven or more. Most respondents were multi-modal transport users.

But even amongst this urban sample showing relatively high mobility and transport accessibility, there is substantial variation by age, and some variation by household income and by gender. Table 7 shows the proportion of all respondents in the study reporting using each mode frequently (once a week or more often) and those modes on which there were statistically significant differences by sex, age band and household income (here divided into high-, medium- and low-income bands). For example, one in five (19%) had travelled as a passenger in a car with a friend or relative driving once a week or more often and this proportion was higher for females than males (F > M), was highest in the 55–64 age group, and those in the lowest third for household income had done it more often, on average, than those in the middle income group, who had done it more than those in the highest income group (Lo > Med > Hi).

Table 8: Frequency of Use of Different Modes of Transport by Scottish Car Drivers

[N = 1,220] Row per cent	*Most Days*	*Once or Twice a Week*	*Once a Fortnight*	*Once a Month*	*Several Times a Year*	*Once a Year or Less*	*Never*
Car/van driver	80	16	1	1	1	1	None
Car/van passenger	12	34	9	10	12	5	20
Motorbike	1	1	0	0	1	1	96
Bus	5	15	6	7	15	13	44
Train	2	3	4	5	23	19	44
Taxi	1	8	9	13	26	12	32
Bicycle	2	5	3	5	9	4	73
Walking for at least 10 minutes	55	23	5	3	3	1	9

Source: Data from NFO World Group and Napier University Transport Research Institute (2001, 2003).

Using data from two large surveys of Scottish adults (NFO World Group and Napier University Transport Research Institute, 2001, 2003), Table 8 shows that multi-mode use also applies to car drivers. Over half the drivers (56%) had also used bus and train. A quarter (24%) cycled several times a year or more often and only one in eleven car drivers (9%) said that they never walk for at least 10 minutes. This, of course, has implications for the application of demand management measures requiring car drivers to relinquish some car use in favour of increased use of other, more sustainable, modes. Many drivers will have prior experience of such modes. Familiarity with alternatives should reduce uncertainty and hence anxiety about changing travel patterns and transport choices.

The number of modes that each respondent indicated they used more often than "Never" was computed (Stradling, 2005). Only 14 of 1,220 car drivers (1.1%) said they only used one mode of travel. One per cent used all eight and 60 per cent used five or more modes. The number of non-car modes (bus, train, taxi, bicycle, walking for at least 10 minutes) was also computed. Only 3% of car drivers say they never used such modes. Twenty-six per cent of drivers used two or more such modes at least weekly; 53% used two or more monthly or more often.

Car drivers used on average only around one more mode than non-drivers (mean for drivers 4.83 and for non-drivers 4.01) and for both groups the average number of non-car modes was close to 3 (mean for drivers 2.98 and for non-drivers 3.18) (Stradling, 2005).

Driver Types

Dudleston *et al.* (2005) found that 77% of drivers and 85% of non-drivers in a large Scotland-wide sample agree that "people should be encouraged to walk, cycle and use public transport more". In addition, 39% of drivers agree that "reducing my car use would make me feel good"; 44% feel "travelling by car can be stressful", but 43% believe that "driving my car is too convenient to give up for the sake of the environment". Using cluster analysis of attitudinal items four driver types were identified:

- Die-hard drivers – comprising 26% of Scottish drivers (20% of Scottish adults);
- Complacent car users – 28% of drivers (21% of adults);
- Malcontented motorists – 24% of drivers (18% of adults);
- Aspiring environmentalists – 24% of drivers (18% of adults).

While the average annual car mileage for the four groups was similar, the segments are differentiated by the extent to which they exhibit attachment to the car, are willing to consider alternative modes, are already multi-modal, feel willing and able to reduce their car use, and are aware of transport and environmental issues.

Die-hard drivers (DHD) like driving and would use the bus only if they had to. Almost none of them believe that higher motoring taxes should be introduced for the sake of the environment and there is overwhelming support for more road building to reduce congestion. There are slightly more males than females in this group. *Car complacents* (CC) are less attached to their cars but currently see no reason to change. They generally do not consider using transport modes other than the car and faced with a journey to make will commonly just reach for their car keys. *Malcontented motorists* (MM) find that current conditions on the road such as congestion and the behaviour of other drivers make driving stressful and they would like to reduce their car use but cannot see how. They say that being able to reduce their car use would make them feel good, but feel there are no practical alternatives for the journeys they have to make. They were over-represented in accessible rural areas of Scotland. *Aspiring environmentalists* (AE) are actively trying to reduce their car use, already use many other modes and are driven by an awareness of environmental issues and a sense of responsibility for their contribution to planetary degradation. Table 9 shows level of mode use and a number of the car use and environmental attitude statements on which the four driver types differed.

Thus to various extents and for a range of reasons, many drivers in Scotland are currently ready to reduce their car use. Even many AE, who already use more modes ever and frequently, like travelling in a car, further testifying to the attractiveness of the automobile; and many DHD, though not as many as among the other segments, say that "being environmentally responsible is important to me". However, the groups differ sharply on the other attitude items covering car use ("I find car driving can be stressful sometimes"), the consequences of cutting car use ("Reducing my car use would make me feel good"), the ease of cutting car use ("It would be easy for me to reduce my car use"), the compulsion to cut car use ("Environmental threats such as global warming have been exaggerated") and equity issues over who should bear the cost ("For the sake of the environment, car users should pay higher taxes", "I would be willing to pay higher taxes on car use if I knew the revenue would be used to support public transport").

Table 9: Mode Use and Attitudes of Die-Hard Drivers(DHD), Car Complacents (CC), Malcontented Motorists (MM) and Aspiring Environmentalists (AE)

	DHD	*CC*	*MM*	*AE*
Annual car mileage: % over 10,000 miles per annum	24	22	26	25
Number of modes (of eight) used ever	4.0	4.5	4.4	5.2
Number of modes (of eight) used once a month or more often	2.9	3.3	3.3	3.9
Percent agree				
I like travelling in a car	98%	82%	82%	73%
I find car driving can be stressful sometimes	25%	28%	66%	67%
Reducing my car use would make me feel good	5%	21%	65%	78%
It would be easy for me to reduce my car use	17%	15%	21%	47%
Being environmentally responsible is important to me	61%	76%	85%	89%
Environmental threats such as global warming have been exaggerated	39%	19%	20%	9%
People should be allowed to use their cars as much as they like, even if it causes damage to the environment	48%	13%	19%	7%
For the sake of the environment, car users should pay higher taxes	4%	5%	17%	39%
I would be willing to pay higher taxes on car use if I knew the revenue would be used to support public transport	11%	9%	38%	46%
It is important to build more roads to reduce congestion	72%	23%	60%	30%

Applying Prochaska and DiClemente's (1992) stages-of-change approach to readiness for cutting car use, drivers were identified as being in the pre-contemplation stage (I'm not even thinking about changing), the contemplation stage (I'm thinking about it but haven't tried to change yet), the action phase (I'm trying to change, even though it's not easy), or the maintenance stage (I've made the change, now I must keep it up). Table 10 shows the figures at each stage for each driver type.

Table 11 summarises suggestions from Dudleston *et al.* (2005) as to how susceptible each driver type would be to policy measures to encourage reductions in car use. DHD are both unwilling and less able to cut car use while the AE are willing, more able, and generally already engaged in change. Table 11 suggests that it may be most productive to (a) encourage those who already use alternative modes (e.g., aspiring environmentalists) to use them a little more, (b) encourage those keen to reduce their car travel (e.g., malcontented motorists) to begin to experiment with alternative modes by providing assistance, assurance and encouragement via travel blending, Travel Smart, Indimark or

Table 10: Percentage of Car Driver Types at Each Stage of Change

Percent Agree	*DHD*	*CC*	*MM*	*AE*
Pre-contemplation I have not tried to reduce the amount I use my car over the past 12 months and I am not thinking of doing so in the next 6 months	88	57	40	15
Contemplation I have not tried to reduce the amount I use my car over the last 12 months, but I am thinking of doing so over the next 6 months	6	18	13	20
Action I have already tried to reduce my car use in small ways over the last 12 months and I am planning to use my car less over the next 6 months	3	14	20	33
Action I have tried to use my car less over the last 12 months and I will be trying to reduce it even more over the next 12 months	1	5	13	17
Maintenance I have already reduced my car use as much as I can and I am now trying to keep it that way	3	6	14	15

similar social marketing procedures (see Thøgersen, Chapter 20 in this volume) and (c) endeavour to raise the level of travel awareness of those with currently unconsidered potential for mode switching (e.g., car complacents). Table 11 thus comprises a framework that could be used to design travel awareness campaigns. This will involve focusing on the particular frustrations and (mis)perceptions of each group through targeted messages and service improvements. For example, it is suggested that the MM should respond to promotional messages which remind them of the frustrations encountered with current levels of congestion together with messages which tell them of the relatively relaxing qualities of public transport and which reinforce their environmental imperatives. The AE should require less persuasion to use alternatives but should be kept informed of the non-car travel opportunities available to them.

CONCLUSIONS

Organisms maximise under constraint (Dunbar, 2001) and were the automobile an organism we would deem it as having been remarkably successful in carving out an environmental niche and in adapting the behaviour of its host to its requirements. In little over a century cars have colonised the planet. Future historians may well characterise the twentieth century as the century of the car, during which around one billion cars were manufactured (Urry, 1999) of which over half a billion (500 million: Shove, 1998)

Table 11: Potential Interventions to Influence Mode Split for Each Driver Type

	DHD	*CC*	*MM*	*AE*
Willingness to use car less	Very low	Low	Very high	Very high
Perceived ability to use car less	Low	High	Very low	High
Environmental imperative	Very low	Low	High	Very high
Potential "switchability"	Very low	Medium	High	Very high
Drivers to change	None	• Acknowledgement of existing alternatives • Not particularly passionate about car use	• Frustrated by congestion and stress of driving • High environmental imperative and travel awareness	• High environmental imperative and travel awareness • Positive attitudes towards PT • Experience of PT and bicycle use • Sympathy for car restraint
Obstacles to change	• High psychological and actual dependence on the car • Passionate about car use and the right to drive • Do not feel personally responsible • Do not believe others will change their behaviour too	• Low-travel awareness • Low-environmental imperative	• Perceived lack of alternatives • Do not believe other people will change their behaviour too	High travel demand
Policy options	• Weaken stereotypical image of PT users • Hard "push" measures (non-fiscal)	• Raise awareness of negative effects of car use • Raise awareness of monetary costs of car use • Promote positive qualities of PT	Promote messages which reinforce: • Moral obligation • Positive qualities of PT and negative aspects of the car	• Improve PT and cycling provision • Promotion of alternatives • Disincentives to car use

Note: PT, Public transport.

are currently occupying the streets, garages, car parks and grass verges of the world. And they have the potential for even further growth. As Adams (2000) points out, global population growth is currently increasing even faster than car ownership and, he asks, "What would be the result should China and the rest of the Third World sustain their growth rates in motorization and succeed in their aspirations to catch up with the developed world?" (Adams, 2000, p. 109). Where there are few cars today there is aspiration for many cars tomorrow. And in the motorised world, where there are many cars today, ". . . no country has yet achieved a lasting and large-scale downturn in the total volume of traffic" (Goodwin, 2001).

Changing individual travel behaviour will not be easy. In car-dependent places the infrastructure maintains and reproduces the continued use of the car – "The whole country is geared for the car" one respondent complained (interviewed in Stradling *et al.*, 1998, 1999). Land use planning decisions over the location of origins (e.g., homes) and destinations (e.g., work, school, retail and other restorative opportunities) may even be seen as requiring car travel – "Nice house on an estate, but the nearest shop is four miles away, the school is three-quarters of a mile away; the nearest pub is certainly a car drive" (respondent interviewed in Mitchell and Lawson, 1998). And many appreciate the autonomy as well as the mobility that the car conveys – "I just like driving . . . I only go places when I can drive"; "One of the reasons I like driving is because I'm in control" (respondents interviewed in Stradling *et al.*, 1998, 1999).

This chapter has endeavoured to distinguish car-dependent places, car-dependent trips and car-dependent persons. A new out-of-town retail park with arterial access, no or little bus provision and free parking is a car-dependent place, and is so as a result of land use planning decisions. Supermarket shopping was the most car-dependent journey type and evenings out for leisure purposes the least in the study cited here (Table 5). Travel to work is typically a car-dependent trip and 43% of those who commute by car say it would not be practical for them to travel to work by any other mode (Table 5). But there are many who are able – but currently unwilling – to change: just under a third of working adults in Scotland say it would be practical for them to travel to work by bus compared with the 12% who currently do so, 22% say it would be practical to walk to work compared with the 11% who presently do so, currently just 2% cycle to work although a further 8% say it would be practical for them to do so, 11% say it would be practical for them to travel to work by train but only 3% currently do (Dudleston *et al.*, 2005).

The commercial marketing literature indicates that targeting is essential for any realistic marketing campaign (again, see Thøgersen, Chapter 20 in this volume). Walking is the single most practical alternative for taking children to and from school. This kind of journey could be targeted to encourage modal shift away from car use towards walking. Making the journey by bus was the most frequently endorsed alternative form of transport for visiting friends and relatives, town centre shopping and evenings out for leisure purposes. Targeting these journey types and improving services to make them more convenient would potentially encourage a modal shift from car use to bus use. Thirty-nine per cent of those who go away for weekends by car say that the train provides a practical alternative to the car. Train operators in Scotland could focus on this journey type to encourage greater train use.

This chapter has also shown that most people, including car drivers, are multi-modal travellers, using more than one transport mode at different times to meet their transport needs. As multi-modal travellers, car drivers thus already have some familiarity with alternative modes. Overall, only 11% of car drivers in Scotland indicated that they could not practically use a bus, train, walk or cycle for any of their journeys and are structurally car dependent – there are no viable alternatives. Seven per cent were consciously car dependent – they could realistically undertake *all* the trip types they were questioned about other than by car, but do not. These two figures establish the ends of the potential modal shift distribution, those who cannot and those who will not cut car use. The segmentation analysis into driver types provides additional fine detail in mapping out the full terrain.

The DHD like driving and are resistant to reducing their car use. The CC are less attached to their cars but currently see no reason to change. The MM find that current conditions make driving stressful and would like to reduce their car use but cannot see how. The AE are actively trying to reduce their car use, already use more modes and are driven by an awareness of environmental issues and a sense of responsibility for their contribution to planetary degradation.

The Scottish Executive's transport white paper (Scottish Executive, 2004b) noted that individual's travel choices determine demand for transport. Individual travel and transport decisions – whether and where to travel, and by what transport mode – are driven by the interaction of three broad factors: the individual's perceptions of their obligations (What journeys do I have to make?), opportunities (How could I make these journeys?) and inclinations (How would I like to make these journeys?).

Current lifestyle patterns generate travel needs. Travel behaviour researchers (e.g., Axhausen and Gärling, 1992; Vilhelmson, Chapter 8 in this volume) refer to these as derived travel demand. What they derive from are a person's present formal and informal social and personal obligations. Persons with jobs are currently generally obliged to attend their place of work in order to discharge that obligation; parents of school age children are obliged to contrive their safe and timely arrival at school. Larders and wardrobes need to be stocked so retail outlets and cash machines must be visited and, with the consumer acting as the final link in the retail distribution chain, purchases transported home. Relatives and friends need to be visited, leisure opportunities attended. Transport joins up the places where people go to lead their lives (Stradling *et al.*, 2000) and meet their obligations to self and others (Stradling, 2002c). Which transport mode is chosen to meet obligation access needs will depend first on which modes are available or, rather, which are perceived as available by the potential user – a bus route or timetable not known about will not find a place in the individual's decision set – and second on which modes they are more inclined to use, which they judge attractive by virtue of, amongst other factors, not making excessive demands on their personal resources of money, time, and physical, cognitive and affective effort (Stradling *et al.*, 2000; Stradling, 2002c).

To reduce car use and provoke modal shift to more sustainable modes of travel, should we be tough on car dependence or tough on the causes of car dependence? Car dependence can be reduced by modifying the opportunities for travel through improving the availability and accessibility of alternative modes; by modifying the lifestyle patterns that generate obligations to travel from current origins to present destinations; or by modifying the inclinations and preferences towards travel by alternative modes, for example by marketing public transport (Stradling, 2002c) or de-marketing the car (Wright and Egan, 2000).

The challenge for travel demand management policy is to put in place a mix of measures that reduces transport-related disadvantage and exclusion – drivers travel more, older persons and those in the lower social classes travel less – whilst conserving the planet's natural resources. Other northern European states spend more than the UK on facilitating walking and cycling and on subsidising public transport with lower fares and higher levels of service, service integration and integrated multi-modal ticketing (CfIT, 2001; Colin Buchanan and Partners, 2003).

A combination of pull and push travel demand management measures (Steg and Vlek, 1997) is required, making car use less attractive and alternatives to car use more attractive. "Spending money on public transport investment and subsidy appears to bring about consistent year on year increases in public transport patronage. Where modal shift from car to public transport (and cycling and walking) is sought, however, parking restraint and road space reallocation are required" (Colin Buchanan and Partners, 2003, p. 75).

Increasing the opportunities for using more sustainable modes attracts the willing making more of them able; decreasing the opportunities for car use encourages the currently able but unwilling to rethink their individual travel choices.

REFERENCES

Adams, J. (2000). Hypermobility, *Prospect*, 50 (March), 27–31.

Anderson, S. and S.G. Stradling (2004) *Attitudes to Car Use and Modal Shift in Scotland*, Scottish Executive Social Research, Edinburgh.

Axhausen, K. and T. Gärling (1992). Activity-Based Approaches to Travel Analysis: Conceptual Frameworks, Models and Research Problems, *Transport Reviews*, **12**, 324–341.

CfIT (2001). *European Best Practice in Delivering Integrated Transport*, Report to Commission for Integrated Transport, London.

Colin Buchanan and Partners (2003). *Transferability of Best Practice in Transport Policy Delivery: Final Report*, Scottish Executive, Edinburgh.

Dudleston, A., E. Hewitt, S.G. Stradling, and J. Anable (2005). *Public Perceptions of Travel Awareness – Phase Three*, Scottish Executive, Edinburgh.

Dunbar, R. (2001). Darwinising Ourselves, *The Psychologist*, **14** (8), 420–421.

Engwicht, D. (1998). *Street Reclaiming: Creating Liveable Streets and Vibrant Communities*, Annandale: Pluto Press Australia.

Farrington, J., D. Gray, S. Martin, and D. Roberts (1998). *Car Dependence in Rural Scotland*, Development Department Research Programme Research Findings No. 53, The Scottish Office, Edinburgh.

Gärling, T., A. Gärling, and P. Loukopoulos (2002). Forecasting Psychological Consequences of Car Use Reduction: A Challenge to an Environmental Psychology of Transportation, *Applied Psychology: An International Review*, **51**, 90–106.

Goodwin, P.B. (2001). Traffic Reduction. In: *Handbook of Transport Systems and Traffic Control* (K. Button and D. Hensher, eds), Elsevier, Oxford, 21–32.

Litman, T. (1999) *The Costs of Automobile Dependency and the Benefits of Balanced Transport*, Victoria Transport Policy Institute (www.vtpi.org/autodep.html).

Mackett, R.L. and A. Ahern (2000). *Potential for Mode Transfer of Short Trips: Report on the Analysis of the Survey Results*, Report to the Department of the Environment, Transport and the Regions, Centre for Transport Studies, University College London, September.

Mitchell, C.G.B. and S. Lawson (1998). *The Great British Motorist 1998*, AA Report, Automobile Association, Basingstoke.

NFO World Group and Napier University Transport Research Institute (2001). *Public Perceptions of Travel Awareness*, Final Report to Central Research Unit, Scottish Executive, Edinburgh.

NFO World Group and Napier University Transport Research Institute (2003). *Public Perceptions of Travel Awareness – Stage 2*, Final Report to Central Research Unit, Scottish Executive, Edinburgh.

Newman, P. and J. Kenworthy (1999). *Sustainability and Cities: Overcoming Automobile Dependence*, Island Press, Washington, DC.

Ormston, R., S.G. Stradling, T. Rye, J. Cooper, K. Hamilton, I.O. York, P. Emmerson, and C. Vance (2004). *Integrated Ticketing*, Scottish Executive Social Research, Edinburgh.

Prochaska, J.O. and C.C. DiClemente (1992). Stages of Change in the Modification of Problem Behaviors, *Progress in Behavior Modification*, 28, 184–218.

RAC (1995). *Car Dependence*, RAC Foundation for Motoring and the Environment, London.

Scottish Executive (2004a). *Scottish Household Survey Travel Diary Results for 2002*, Scottish Executive Statistical Bulletin Transport Series Trn/2004/4, Scottish Executive National Statistics, Edinburgh, May.

Scottish Executive (2004b). *Scotland's Transport Future. The Transport White Paper – June 2004*, Scottish Executive, Edinburgh.

Semlyen, A. (2000). *Cutting Your Car Use. Save Money, Be Healthy, Be Green!* Green Books, Dartington Totnes, UK.

Shove, E. (1998). *Consuming Automobility*, SceneSusTech Discussion Paper, Department of Sociology, Lancaster University, Lancaster, UK.

Sloman, L. (2003). *Less Traffic Where People Live: How Local Transport Schemes Can Help Cut Traffic*, Transport for Quality of Life, Machynlleth.

Steg, L. and C. Vlek (1997). The Role of Problem Awareness in Willingness-to-Change Car Use and in Evaluating Relevant Policy Measures. In: *Traffic and Transport Psychology* (T. Rothengatter and E.C. Vaya, eds), Pergamon, Oxford, 465–475.

Stradling, S.G. (2002a). Combating Car Dependence. In: *Behavioural Research in Road Safety XII* (G.B. Grayson, ed.), Transport Research Laboratory, Crowthorne, 174–197.

Stradling, S.G. (2002b). *Persuading People Out of Their Cars*, Inaugural Lecture, Napier University, 27 March.

Stradling, S.G. (2002c). Transport User Needs and Marketing Public Transport, *Municipal Engineer*, **151** (1), 23–28.

Stradling, S.G. (2003). Reducing Car Dependence. In: *Integrated Futures and Transport Choices* (J. Hine and J. Preston, eds), Ashgate, Aldershot, UK, 100–115.

Stradling, S.G. (2004). Changing Individual Travel Behaviour: From Policy to Perceived Behavioural Control. Paper presented at the 3rd International Conference on Traffic and Transport Psychology, Nottingham, September.

Stradling, S.G. (2005). Readiness for Modal Shift in Scotland, *Scottish Geographical Journal*, **120**, 265–275.

Stradling, S.G., M. Carreno, N. Ferguson, T. Rye, D. Halden, P. Davidson, J. Anable, S. Hope, B. Alder, T. Ryley, and M. Wigan (2005). *Scottish Household Survey Analytical Topic Report: Accessibility and Transport*, Scottish Executive Social Research, Edinburgh.

Stradling, S.G., M.L. Meadows, and S. Beatty (1998). Psychological Benefits and Disbenefits of Driving. In: *Behavioural Research in Road Safety VIII* (G.B.Grayson, ed.), Transport Research Laboratory, Crowthorne, UK.

Stradling, S.G., M.L. Meadows, and S. Beatty (1999). *Factors Affecting Car Use Choices*, Report to the Department of the Environment, Transport and the Regions, Transport Research Institute, Napier University, Edinburgh.

Stradling, S.G., M.L. Meadows, and S. Beatty (2000). Helping Drivers Out of Their Cars: Integrating Transport Policy and Social Psychology, *Transport Policy*, 7, 207–215.

Urry, J. (1999). *Automobility, Car Culture, and Weightless Travel*, Unpublished paper, Department of Sociology, Lancaster University, Lancaster, UK (http://www.lancaster.ac.uk/sociology/soc008ju.html).

Wright, C. and J. Egan (2000). De-Marketing the Car, *Transport Policy*, 7, 287–294.

Threats from Car Traffic to the Quality of Urban Life
Tommy Gärling and Linda Steg (Editors)

11

Instrumental Motives for Private Car Use

Cecilia Jakobsson, Department of Psychology, Göteborg University, Göteborg, Sweden

ABSTRACT

This chapter focuses on instrumental motives for car use. These motives are related to the perception of the car as a fast, convenient, and affordable tool for urban travel. It is in general viewed as superior to alternative travel modes for reaching destinations where everyday activities such as work, maintenance, and leisure can be carried out. An analysis of the role of instrumental motives relative to other internal and external motivating factors is presented. That instrumental motives may counteract car use reduction, is emphasized.

INTRODUCTION

A main argument made in this chapter is that for maintaining a lifestyle that involves many activities at different places; it is important that travel is fast, convenient, and affordable. Due to advancements in automobile technology, massive investments in road infrastructure, and spread of affluence, no other travel mode may in these respects compete with the private car for urban travel. Instrumental reasons are therefore likely to be a main motive for purchasing as well as using private cars. In support of the importance of this instrumental-tool function of the car, it has been found that when explaining why people prefer to use the car even for short distances where walking or cycling might be a realistic option, they state reasons such as convenience and alleviation of time pressure (Mackett, 2003). In a study by Jakobsson *et al.* (2002), the major reasons for using the car included statements like "public transport services are too inaccessible and takes too long " or "I need the car when I go shopping, because it is convenient not to have to carry the groceries." Interestingly, these comments were often followed by or expressed together with feelings of being too dependent on the car and of frustration due to lack of alternatives. Thus, a conflict was apparent within many individuals. There is clearly a need to further investigate the motives of car use, disentangling the existence of possible complex relationships and sometimes conflicts between reasons or motives. One conflict is between the instrumental value of the car for getting around in daily life on the one hand, and a desire for a more sustainable and safe environment on the other hand. Because of the positive outcomes of car use in terms of attainment of goals related to a particular lifestyle, one may furthermore claim that instrumental motives are intimately linked to symbolic and affective motives (such as feelings of freedom, independence, power, status, or privacy) discussed elsewhere (Steg *et al.*, 2001; Steg, 2005) as well as in other chapters in this volume (see Gatersleben, Chapter 12).

A possible caveat is that the stated instrumental reasons are justifications – something that is socially acceptable to report – rather than having a major influence on the motivation to choose the car over other travel modes. In this chapter it will be taken for granted that instrumental motives exist. Such instrumental motives for car use will first be defined and then discussed. In real life they are likely to interact with a number of other motives. An analysis will then be presented, attempting to disentangle when instrumental motives seem to play a more decisive role relative to other motivating factors, both external and internal to the individual.

Another aim of the chapter is to show that people may fail to respond to transport policies targeting reduced car use, because instrumental motives are in conflict with such a reduction. If society wants people to break car-use dependencies, it is important then to take instrumental motives seriously because, at least in car users' minds, they are barriers to change.

INSTRUMENTAL MOTIVES FOR CAR USE

Fulfilling biological needs, social obligations, and personal desires require that people move from one place to another in the environment to perform goal-directed behaviours such as work, maintenance activities (e.g., shopping), and various leisure activities (Gärling and Garvill, 1993). That demand for travel is derived from this requirement, it constitutes a basic tenet of the activity-based approach to travel behaviour (e.g., Jones *et al.*, 1983; Recker *et al.*, 1986; Kitamura, 1988; Axhausen and Gärling, 1992; Ettema and Timmermans, 1997; Bhat and Koppelman, 1999). The instrumental motives for car use are thus related to engagement in activities that are separate from travel *per se*. The value of the tool or instrument that the car constitutes stems from the satisfaction derived from the engagement in the activities. Still, when alternatives exist, some tools are better than others. Speed, convenience or flexibility, and affordability are factors that favour the car over other travel modes in an urban environment.

Some factors, external to the individual, that have been shown to influence car use are income, occupation, gender, number of children in the household, residential area, and access to public transportation (Lu and Pas, 1999). The activity-based approach is an important point of departure for explaining that household car use depends on such factors. The location of the home, the workplace, and the other activity nodes (as well as the accessibility and quality of transport alternatives) are assumed to determine the frequency and the length of car trips. Less is however known about how these factors interact with subjective interpretations of the objective circumstances and expectations from others as reflected in norms, gender roles, and social roles. Their interaction with instrumental motives is discussed in the section following the next section.

CAR USE AND MOTIVATION

The study of motivation as a determinant of behaviour has a long history in psychology, although focus has shifted over the years between "hot" perspectives, represented by constructs such as emotion, desire, drive, and goal-orientation, and "cold" cognitive perspectives, entailing constructs like information processing, attention, memory, and feedback (Eagly and Chaiken, 1993). Motivational concepts were, for example, central to the behaviourist tradition, which claimed that human behaviour is shaped by incentives and reinforcements. In line with this tradition reinforcement analysis has been applied to travel behaviour by, for instance, Geller *et al.* (1982). This analysis posits that travel behaviour is foremost shaped by consequences that can be experienced directly in time and space, for example, speed. Also in line with the behaviourist tradition, positive consequences, that is, rewards are taken to be more reinforcing than negative consequences (punishments) that will only suppress behaviour temporarily. This is illustrated in Table 1 (which is based on a theoretical model presented in Geller *et al.*, 1982) that shows hypothetical direct and delayed positive and negative consequences of car use and public transport. The dominance of positive direct consequences of using the car is apparent as well as several possible conflicts that may influence motivation to choose another travel mode. Note that the consequences listed are mainly but not exclusively related to instrumental motives.

Influences on motivation can be divided into factors that are internal or external to the individual. The former may be referred to as psychological and the latter as situational or environmental. The idea is that motivation for choosing car or public transport may be induced both by external factors, such as economic incentives and disincentives, legislation, available infrastructure, and social norms and internal factors, such as intentions, attitudes, and personal norms.

Governments may use information, regulations or economic incentives and disincentives in order to influence car users' motivation. This chapter will not analyse the effectiveness of such different strategies (but see Part 3 of this volume). However, one may generally conclude that external influences such as costs or regulations targeting behaviour

Table 1: Positive and Negative Direct and Delayed Consequences of Private Car and Public Transport

		Direct	*Delayed*
Private car	+	Fast	
		Comfortable	
		Prestigious	
		Private	
		Flexible	
		Free choice of route	
		Possible to carry cargo etc.	
		Predictable	
		Payment later	
		Pleasure of driving	
	–	Congestion	Environmental deterioration
		Difficult to park	Fuel cost
			Maintenance cost
Public transport	+	Time to read	Reduce environmental deterioration
		Possibility to rest/sleep	
		No worries about parking	
	–	Pay a fare	
		Uncomfortable	
		Crowded	
		Not private	
		Exposed to weather conditions	
		Limited possibility to carry cargo	
		Inflexible	
		Low prestige	
		Long travel time	
		Low possibility to choose route	

Source: Adapted from Jakobsson (2004).

may be less effective in strengthening motivation, because individuals feeling forced to do something are deprived of a valued freedom of choice. They are still forced to change their behaviour. In contrast, strategies targeting people's knowledge and attitudes may have weak behavioural effects, at least in the short term. In the longer term, however, motivation for using alternatives to the car may be strengthened by, for example, information campaigns and social modelling affecting knowledge and attitudes.

Internal factors, such as beliefs, values, attitudes, personal norms, and problem awareness have also been shown to influence car use (Garvill *et al.*, 2000; Steg *et al.*, 2001; Nordlund and Garvill, 2003). Together with norms, which may be external (social norms) or internal (personal norms) to the individual, people's attitudes are probably the most studied psychological determinant of car use (Verplanken *et al.*, 1994; Gärling *et al.*, 1998; Tertoolen *et al.*, 1998; Garvill *et al.*, 2000; Thorpe *et al.*, 2000; Gärling *et al.*, 2003). There are several definitions of attitude. One useful definition was proposed by Eagly and Chaiken (1993). It is broad and therefore applicable to attitudes towards the environment and the car use. They define attitude as a psychological inner state that is expressed by evaluating (overtly or covertly, affectively or cognitively, behaviourally or not) a phenomenon, a behaviour, or an object. The evaluation can be placed on a continuum from negative to positive. In line with this, an attitude towards car use would consist of several components: a cognitive component, which reflects knowledge of and reasoning about, for example, the positive and negative effects caused by car use; an emotional component involving positive and negative feelings towards the car use and the environment; and a behavioural component reflecting whether or not the person has an intention to act in line with his or her feelings and knowledge.

One of the most influential theories that takes both internal and external factors into account is the Theory of Planned Behaviour, or TPB (Ajzen, 1985, 1991). In the TPB, motivation to perform a behaviour is assumed to be related to attitudes (positive or negative) towards the behaviour, and subjective norms, which is based on judgements of significant others' (e.g., family and friends) opinion regarding the behaviour, and the individual's desire to comply or defy the norm. The TPB differs from its predecessor, the Theory of Reasoned Action (Fishbein and Ajzen, 1975), by the inclusion of perceived behavioural control as a determinant of intention. If individuals perceive a low control over barriers in the environment, it will affect the relationship between attitude and behaviour negatively. Perceived control may be strengthened if the individual plans how to carry out the behaviour (Gärling and Fujii, 2002). The attitude towards behaviour is dependent on the individual's perceptions and salient beliefs about the consequences of the behaviour as well as the evaluation of these consequences.

To sum up, the motivation to use the car stems from a number of sources, some of them external to the individual and some others internal or psychological. Since the focus of this chapter is on the instrumental motives for car use, we need to show that the car is primarily valued, because it is a tool for the individual to engage in essential activities. However, this will certainly vary for different individuals, some people have a stronger car use habit (choosing the car routinely with very little cognitive effort) or stronger preference for the car (affective and symbolic reasons may be more prominent than instrumental reasons), but it may also vary depending on the situation and the purpose of the trip. This will be further discussed in the next section, which is an attempt to illuminate the interaction between different motives and between situational factors and motives, as well as the effect of varying levels of awareness when processing the information underlying a decision to use the car or not.

IMPULSIVE AND COMPULSIVE CAR USE

Instrumental motives for car use may be more significant under certain conditions. For example, in situations that call for flexibility and speed, these motives also become more important. Evidence for the importance of trip-type on which factors are more influential was provided by Anable and Gatersleben (2005). They found that car users rated instrumental factors (convenience) as more important for work trips and affective factors more important for leisure trips (see also Gatersleben, Chapter 12 in this volume). It is thus plausible to assume that there are different processes that dominate the choice of using the car depending on the situation and the type of trip (Jakobsson, 2004). This assumption is also based on a research on habits and car use (see Fujii and Gärling, Chapter 13 in this volume). It has been found that depending on the frequency of the trip and the degree of intention versus automatic decision-making regarding trips, a distinction can be made between habitual trips and planned trips. Decision-making related to trips may thus range from being at a high level of awareness and deliberation, for instance, when a trip with low frequency (e.g., a weekend vacation trip) is planned, to being at a low level of awareness, automatic and habitual, for instance, for high frequency trips (e.g., daily work trips). When a behaviour becomes habitual, no intention is formed, resulting in less conscious, more automatic processing of information (Verplanken *et al.*, 1994; Ouelette and Wood, 1998).

In addition, when situational factors dominate, car users are likely to act on impulses. It has been shown that impulsive behaviour is less influenced by intentions since these are not elaborate or specific (Gollwitzer and Bargh, 1996). In case of impulsive behaviour, intentions may also be formed close in time to their implementation, because of influences from situational factors (Gärling *et al.*, 1998). Thus, a car trip could then be said to be impulsive if an intention is formed very close in time to or simultaneously with the behaviour. Impulses are related to fast, automatic information processing, and are probably more influenced by affective motives, originating within the individual rather than by instrumental motives triggered by situational factors.

A distinction needs to be made between impulsive and compulsive car use. Compulsive car use is associated with instrumental motives, external factors, and loss of control due to situational factors. Important situational factors preventing car users from implementing their intentions can be labelled "unexpected events." Some of the trips in a field study were described by the respondents to be compulsive rather than impulsive (Jakobsson *et al.*, 2002; Jakobsson, 2004). Compulsive trips made car users not to follow a plan recorded in a prospective car log. The reasons given, included unexpected needs to travel by car, for instance, time pressure, medical emergencies, bad weather, and/or forgetting to shop for items. Jakobsson (2004) investigated different trip purposes in order to examine the differences in control individuals exerted over trip-making (operationalised as degree of planning in advance), depending on the type of trip ranging from work trips (assumed to be habitual) to leisure and shopping trips (assumed to be impulsive). Results revealed that many more trips were executed by households than they had planned. As was expected, larger discrepancies were found for shopping and chauffeuring trips (escorting other people), whereas the discrepancy was smaller for work trips. Furthermore, it was found that the number of shopping trips more strongly exceeded the planned number if the household had more than one car, that the number of chauffeuring trips more strongly exceeded the planned number if the household had children, and that the number of leisure trips more strongly exceeded the planned number if the household had a higher income. Questionnaires distributed at the end of data collection revealed the participants' stated reasons for the observed discrepancies. Activities that had been suppressed due to, for example, illness or weather conditions were the main reason for not making a planned car trip. The reason for making additional trips by car was primarily unplanned activities, most often shopping, chauffeuring, and personal services, or switching travel mode to car from other modes due to weather conditions or time pressure. In conclusion, the discrepancies were produced to a large extent by events outside the individuals' control justifying that the trips are termed compulsive.

THE CAR-USE REDUCTION PROCESS AND BARRIERS TO CHANGE

External changes may motivate car users to set car-use reduction goals, at least if these changes are forced. Economic disincentives may be forceful if being difficult to evade. Legislation is one way to prevent evasion. If the costs are experienced to be too high, then this may also have the effect of forcing change, but at the same time there may be counterproductive changes in internal factors leading to less motivation or negative influences. Baron and Jurney (1993) have shown that norms such as fairness and infringement on personal freedom prevent people from voting for a coerced reform, despite being in favour of the long-term consequences. This may explain that a generally negative attitude towards different measures designed to reduce car use can coexist with a high awareness of environmental problems due to car use, and an understanding of the need to deal with these problems. In a similar vein, it has been shown how economic incentives may lead to elimination or "crowding out" of preexisting intrinsic motivation to act in a socially responsible way (Frey, 1993; Deci *et al.*, 1999; Frey and Jegen, 2001). The individual feels bribed when the control shifts from inside to the outside situation, and the reaction to this is rejection rather than acceptance of a proposed measure.

Common to environmental and social psychology, is also a game-theoretical or social-dilemma approach to the decision-making surrounding car use (Gärling *et al.*, 2002). Households are faced with choices reflecting a conflict between the individual's interest and the collective's or society's interest (Dawes, 1980; Messick and Brewer, 1983; Ostrom, 1998). A reduction of car use involves individual sacrifices in order to achieve a collective benefit (e.g., reduced air pollution). In such a social dilemma, it is always beneficial for the individual to make a selfish choice. Yet, if a majority chooses to act in this way, the outcome is worse for everyone than if the majority chooses to act in the collective interest. In other words, it is necessary to make choices that are regarded as personal sacrifices (for example, to reduce car use) in order to achieve an improved outcome for everyone. This insight is probably not always present in car users.

When considering car use, the negative consequences for the environment will occur in the future or may be seen as caused by someone else. Positive consequences, on the other hand, for example, fast transportation, are more easily associated with car use and directly experienced (Everett and Watson, 1987). There are also negative consequences of car use that affect the individual directly. In urban areas these mainly consist of lack of parking spaces and congestion leading to prolonged travel times. An additional direct individual cost in the form of economic disincentives may outweigh the benefits of car use and make alternatives more attractive.

Everett and Watson (1987), and Garvill (1999) posit two reasons as to why car use fails to be reduced despite the negative environmental effects it causes. At the same time as these are reasons why car use reduction fails, they are reasons why car use is sustained and growing. Firstly, the positive effects favour the individual, while the negative effects accrue mainly to the society. Secondly, the positive effects materialize closer in time than the costs. As referred to the above, Table 1 summarizes some possible positive and negative consequences of car use and compares these to benefits and costs of public transport. The table includes some of the most important barriers to motivate car-use reduction, and most of them are instrumental in nature. A majority of car users will be trapped in this conflict, unwilling to set and implement a car-use reduction goal. Yet, symbolic-affective factors may also play a role. If car users set a goal, there are still more obstacles for them to face when the car-use reduction goal or intention shall be implemented. Here, primarily, situational factors play a role.

Finally, one major barrier that is commonly mentioned by car users is the lack of viable alternatives to the car. This perception, in combination with distance to work and other destinations has been shown to constitute a main reason not to reduce car use even when a willingness to do so exists (Kingham *et al.*, 2001).

Given that there are alternatives to the car as well as motivation to reduce car use, there is still one additional step before the actual implementation of car-use reduction may occur. After having set a car-use reduction goal, to attain the goal households must form a plan for how to do this (Gollwitzer, 1998; Gärling and Fujii, 2002). To break habits or resist impulses so that a car-use reduction goal can be reached, it is conjectured that planning is an important factor, since it facilitates the performance of a behaviour by increasing strength of motivation and enhancing memory as well as coordinating concurrent, possibly conflicting goals and plans. In support of this conjecture, inducing planning in households has proved to be an effective measure by which to reduce car use (Gärling *et al.*, 1998; Jakobsson *et al.*, 2002; Garvill *et al.*, 2003). Such a plan consists of sets of predetermined choices contingent on specified conditions. For instance, one may have one plan for reducing work trips, another for shopping trips, and a third for leisure trips.

Even if planning prepares car users for a reduction in car use, there may still be discrepancies between the intentions formed in the planning phase and the implementation of these plans. As previously discussed in the section on impulsive and compulsive car use, the reasons for these discrepancies can be found both within the individual as well as in the environment. In the TPB (Ajzen, 1985, 1991), perceived behavioural control refers to the degree of control a person believes he or she has over events related to the success of performing a behaviour or reaching a goal. These control beliefs include skills, knowledge, needed resources, and opportunities. They are assumed to influence behaviour through intentions. An issue is, whether perceived control corresponds to actual control (Ajzen, 1991). In particular, the factors related to available opportunities and resources may be misjudged.

In the research on car-use reduction, the terms error of omission and error of commission have been used when car users fail to implement an intention concerning how much to use the car (Fujii and Gärling, 2003). Errors of commission refer to a failure to reduce the use of the car despite having stated one would. It may be due to the fact that people are unrealistic, and do not take into account concurrent conflicting plans or that they tend to change their minds due to weak or unstable intentions.

The consequences of a car-use reduction are evaluated in parallel to the implementation of the car-use reduction goal. A basic principle is that human behaviour is goal-directed and controlled by negative feedback (Carver and Scheier, 1982, 1998). Negative feedback informs of a discrepancy between the current state and a goal, and if any discrepancy is registered, how much and in which direction behaviour needs to be changed. A prerequisite for the effectiveness of negative feedback is that the reduction goal is specific. A comparison would otherwise be difficult. Transparency, immediacy, and high frequency of feedback also increase the likelihood of a change. If an implemented chosen behavioural change fails to achieve the car-use reduction goal, one may try an alternative behavioural change. However, this outcome may also cause a change in the reduction goal or that it is abandoned. This is assumed to be a more frequent occurrence when the reduction goal is unrealistic or the commitment is low.

Even if the reduction goal is attained, there may be negative side effects making the behavioural change unstable. Sacrifices may, for example, be experienced more negatively than expected. The reduction goal is also assumed to be subordinate to other important goals like comfort and wellbeing. If the evaluation shows that the reduction goal is in conflict with any goal that is perceived to be more important, it may be changed or abandoned.

CONCLUSIONS

This chapter has highlighted the tool function of the car referred to as instrumental motives for car use. The role these motives play, relative to other internal as well as external motivating factors has been analysed. When identifying motives, instrumental or others, which affect car use, one needs to take into account the different attributes of different types of trips depending, for example, on the type of activity engagement that motivates them. Furthermore, the car use reduction process was discussed. It is conceptualised as consisting of the stages of motivation, planning, and implementation. Instrumental-related factors, such as perceptions of the convenience of the car, lack of alternatives to driving to work or other frequently visited destinations, and a tight time schedule for activities play an important role in all these stages.

Is the car used more than needed from an instrumental point of view? This question has not yet received a definite answer. More research is needed for this, focusing on both the goal (activity participation) of travel as well as how the interaction of internal factors with situational factors determines the attractiveness of car use, relative to alternative travel modes.

ACKNOWLEDGEMENTS

Financial support for the preparation of this chapter was obtained through grants from the Swedish Agency for Innovation Systems (#2002-00434) and the Swedish Research Council for Environment, Agricultural Sciences, and Spatial Planning (#25.9/2001-1763). I thank the editors Tommy Gärling and Linda Steg for their valuable comments.

REFERENCES

Ajzen, I. (1985). From Intentions to Actions: A Theory of Planned Behavior. In: *Action-Control: From Cognition to Behavior* (J. Kuhl and J. Beckman, eds), Springer, Heidelberg, 11–39.

Ajzen, I. (1991). The Theory of Planned Behavior, *Organizational Behavior and Human Decision Processes*, **50**, 179–211.

Anable, J. and B. Gatersleben (2005). All Work and No Play? The Role of Instrumental and Affective Factors in Work and Leisure Journeys by Different Travel Modes, *Transportation Research Part A: Policy and Practice*, **39**, 163–181.

Axhausen, K. and T. Gärling (1992). Activity-Based Approaches to Travel Analysis: Conceptual Frameworks, Models and Research problems, *Transport Reviews*, **12**, 324–341.

Baron, J. and J. Jurney (1993). Norms Against Voting for Coerced Reform, *Journal of Personality and Social Psychology*, **64**, 347–355.

Bhat, C.R. and F.S. Koppelman, (1999). A Retrospective and Prospective Survey of Time-Use Research, *Transportation*, **26**, 119–139.

Carver, C.S. and M.F. Scheier (1982). Control Theory: A Useful Conceptual Framework for Personality-Social, Clinical, and Health Psychology, *Psychological Bulletin*, **92**, 111–135.

Carver, C.S. and M.F. Scheier (1998). *On The Self-Regulation of Behavior*, Cambridge University Press, Cambridge.

Dawes, R.M. (1980). Social dilemmas, *Annual Review of Psychology*, **31**, 169–193.

Deci, E.L., R. Koestner, and R.M. Ryan (1999). A Meta-Analytic Review of Experiments Examining the Effects of Extrinsic Rewards on Intrinsic Motivation, *Psychological Bulletin*, **125**, 627–668.

Eagly, A.H. and S. Chaiken (1993). *The Psychology of Attitudes*, Harcourt Brace Jovanowich College Publishers, Fort Worth.

Ettema, D.F. and H.J.P. Timmermans (eds) (1997). Theories and Models of Activity Patterns. In: *Activity-Based Approaches to Travel Analysis*, Pergamon, Oxford, 1–36.

Everett, P.B. and B.G. Watson (1987). Psychological Contributions to Transportation. In: *Handbook of Environmental Psychology*, Vol. 2 (D. Stokols and I. Altman, eds), John Wiley, New York, 987–1008.

Fishbein, M. and I. Ajzen (1975). *Belief, Attitude, Intention and Behavior: An Introduction to Theory and Research*, Addison-Wesley, Reading, MA.

Frey, B.S. (1993). Motivation as a Limit to Pricing, *Journal of Economic Psychology*, **14**, 635–664.

Frey, B.S. and R. Jegen (2001). Motivation Crowding Theory, *Journal of Economic Surveys*, **15**, 589–611.

Fujii, S. and T. Gärling (2003). Application of Attitude Theory for Improved Predictive Accuracy of Stated Preference Methods in Travel Demand Analysis, *Transportation Research A: Policy and Practice*, **37**, 389–402.

Gärling, T., A. Biel, and M. Gustafsson (2002). The Human Interdependence Paradigm and Its Application in Environmental Psychology. In: *Handbook of Environmental Psychology* (R. Bechtel and A. Churchman, eds), Wiley, New York, 85–94.

Gärling, T. and S. Fujii (2002). Structural Equation Modeling of Determinants of Planning, *Scandinavian Journal of Psychology*, **43**, 1–8.

Gärling, T. and J. Garvill (1993). Psychological Explanations of Participation in Everyday Activities. In: *Behavior and Environment: Psychological and Geographical Approaches* (T. Gärling and R.G. Golledge, eds), Elsevier/North-Holland, Amsterdam, 270–297.

Gärling, T., R. Gillholm, and A. Gärling (1998). Reintroducing Attitude Theory in Travel Behavior Research: The Validity of an Interactive Interview Procedure to Predict Car Use, *Transportation*, **25**, 129–146.

Gärling, T., P. Loukopoulos, and M. Lee-Gosselin (2003). Public Attitudes. In: *Handbook of Transportation. Volume 4: Transportation and the Environment* (K. Button and D.A. Hensher, eds), Elsevier, Amsterdam, 725–737.

Garvill, J. (1999). Choice of Transportation Mode: Factors Influencing Drivers' Willingness to Reduce Personal Car Use and Support Car Regulations. In: *Resolving Social Dilemmas: Dynamic, Structural, and Intergroup Aspects* (M. Foddy, M. Smithson, S. Schneider and M. Hogg, eds), Psychology Press, Philadelphia, PA, US, 263–279.

Garvill, J., A. Marell, and A. Nordlund (2000). Effects of Values, Environmental Concern, Personal Norm, Attitude, and Habit on Car Use, *International Journal of Psychology*, **35**, 83–84.

Garvill, J., A. Marell, and A. Nordlund (2003). Effects of Increased Awareness on Choice of Travel Mode, *Transportation*, **30**, 63–79.

Geller, E.S., R.A. Winett, P.B. Everett, and R.C. Winkler (1982). *Preserving the Environment: New Strategies for Behavior Change*, Pergamon Press, New York.

Gollwitzer, P.M. (1998). Metacognition in Action: The Importance of Implementation Intentions, *Personality and Social Psychology Review*, **2**, 124–136.

Gollwitzer, P.M., and J.A. Bargh (1996). *The Psychology of Action: Linking Cognition and Motivation to Behavior*, Guilford Press, New York.

Jakobsson, C. (2004). Accuracy of Household Planning of Car Use: Comparing Prospective to Actual Car Logs, *Transportation Research Part F: Traffic Psychology and Behaviour*, **7**, 31–42.

Jakobsson, C., S. Fujji, and T. Gärling (2002). Effects of Economic Disincentives on Private Car Use, *Transportation*, **29**, 349–370.

Jones, P., M.C. Dix, M. I. Clarke, and I. G. Heggie (1983). *Understanding Travel Behavior*, Gower, Aldershot, UK.

Kingham, S., J. Dickinson, and S. Copsey (2001). Travelling to Work: Will People Move Out of Their Cars? *Transport Policy*, **8**, 151–160.

Kitamura, R. (1988). An Evaluation of Activity-Based Travel Analysis, *Transportation*, **15**, 9–34.

Lu, X., and E.I. Pas (1999). Socio-Demographics, Activity Participation and Travel Behavior, *Transportation Research Part A: Policy and Practice*, **33**, 1–18.

Mackett, R.L. (2003). Why Do People Use Their Cars for Short Trips? *Transportation*, **30**, 329–349.

Messick, D.M., and M.B. Brewer (1983). Solving Social Dilemmas: A Review, *Review of Personality and Social Psychology*, **4**, 11–44.

Nordlund, A.M. and J. Garvill (2003). Effects of Values, Problem Awareness, and Personal Norm on Willingness to Reduce Personal Car Use, *Journal of Environmental Psychology*, **23**, 339–347.

Ostrom, E. (1998). A Behavioral Approach to the Rational Choice Theory of Collective Action, *American Political Science Review*, **92**, 1–22.

Ouelette, J.A., and W. Wood (1998). Habit and Intention in Everyday Life: The Multiple Processes by Which Past Behavior Predicts Future Behavior, *Psychological Bulletin*, **124**, 54–74.

Recker, W.W., M.G. McNally, and G.S. Roth (1986). A Model of Complex Travel Behavior: Theoretical Development, *Transportation Research A*, **20**, 307–318.

Steg, L. (2005). Car use: Lust and Must. Instrumental, Symbolic and Affective Motives for Car Use, *Transportation Research Part A: Policy and Practice*, **39**, 147–162.

Steg, L., K. Geurs, and M. Ras (2001). The Effects of Motivational Factors on Car Use: A Multidisciplinary Modelling Approach, *Transportation Research Part A: Policy and Practice*, **35**, 789–806.

Steg, L., C. Vlek, and G. Slotegraaf (2001). Instrumental-Reasoned and Symbolic-Affective Motives for Using a Motor Car, *Transportation Research Part F: Traffic Psychology and Behaviour*, 4, 151–169.

Tertoolen, G., van D. Kreveld, and B. Verstraten (1998). Psychological Resistance Against Attempts to Reduce Private Car Use, *Transportation Research Part A: Policy and Practice*, 32, 171–181.

Thorpe, N., P. Hills, and S. Jaensirisak (2000). Public Attitudes to TDM Measures: A Comparative Study, *Transport Policy*, 7, 243–257.

Verplanken, B., H. Aarts, Van A. Knippenberg, and Van C. Knippenberg (1994). Attitude Versus General Habit: Antecedents of Travel Mode Choice, *Journal of Applied Social Psychology*, 24, 285–300.

Tommy Gärling and Linda Steg (Editors)

12

AFFECTIVE AND SYMBOLIC ASPECTS OF CAR USE

Birgitta Gatersleben, Department of Psychology, University of Surrey, Guildford, UK

ABSTRACT

Cars are attractive not only because of their instrumental advantages such as flexibility, cost and speed, but also because the car can influence mood (e.g., is thrilling, exciting and relaxing) and because it can provide an important social symbol to express status and power. This chapter gives an overview of the limited empirical research that examines the affective and symbolic aspects of car possession and use. Most of this exploratory research is not theory driven and the chapter hopes to provide more insight into definitions and operationalisations of the relevant concepts, which may aid future research and theory development.

INTRODUCTION

> 'Cars are above all machines that move people, but they do so in many senses of the word'. (Sheller, 2004, p. 221)

From a very early age children learn to like cars (Stokes and Hallett, 1992). The average toy shop these days will provide an enormous array of transport-related toys such as cars, fire engines, planes, petrol stations etc. The number of brightly coloured cars available for babies and toddlers is overwhelming. Television programmes for young children often feature a 'cute' car with a pleasant personality. In films for teenagers, cars play an important role in terms of a symbol for independence and freedom. 'When cars are presented to children, they are nearly always presented in a positive way' (Stokes and Hallett, 1992, p. 173).

This analysis by Stokes and Hallett (1992) clearly suggests that the car is not merely a useful object that can help us get from A to B in a comfortable, fast manner. In fact,

Stokes and Hallett (1992) suggest that the psychological reasons for having and using a car (e.g., habit, emotional attachment) are more important than the practical ones (e.g., convenience, speed, cost). There is something special about a car. Not only do people use it to drive to places, they also dream, flirt, relax, listen to music, sing, chat, sleep; and some people even get buried in a car. Macintyre *et al.* (1998) and Hiscock *et al.* (2002) have shown that, controlling for age and social class, those who have access to a car are, on average, of better health. Moreover, the car can provide psycho-social benefits which can be health promoting. Gärling *et al.* (2002) suggest that a reduction of car use can have severe negative long-term psychological consequences if people fail to adapt properly. It can, for instance, lead to reduced well-being due to limited participation, negative effects on family functioning and increased stress due to longer travel time and distance and a loss of control.

The thought that cars have more than instrumental value has been discussed by many scholars (e.g., Flink, 1975; Sachs, 1984; Marsh and Collett, 1986; Goodwin, 1995; Diekstra and Kroon, 2003). However, only in recent years have these views been supported by empirical data. This chapter gives an overview of this limited research, which examines the views of individual car users and non-car users on the instrumental, affective and symbolic aspects of car possession and use. Most of this research is explorative and it originates in a variety of social science disciplines. There is therefore little consistency in the definitions and operationalisations of the relevant concepts. This chapter examines the findings with a view to provide a clearer insight into the different concepts, which may aid future research.

CONCEPTUALISATION OF INSTRUMENTAL, AFFECTIVE AND SYMBOLIC ASPECTS

Steg *et al.* (2001) were the first to specifically examine the relative importance of instrumental, affective and symbolic aspects in relation to car possession and use. Instrumental aspects, such as costs and flexibility relate to the general instrumental function of the car. Emotional aspects refer to the emotions evoked by travelling, such as feeling stressed, excited, pleasant or bored as formulated in theories on affect (e.g., Mehrabian, 1980; Russell, 1980, 2003). Symbolic aspects are related to people's desire to express their social identity and status (Schlenker, 1982).

Instrumental aspects are the focus of most travel behaviour research to date. This research has shown that the car is generally rated more positively than other modes of transport (especially public transport) on practical and convenience-based aspects, such as the accessibility, reliability, safety of transport, travel speed and cost. These aspects are also positively related to car use and negatively to willingness to reduce car use. On the negative side, cars are rated lower on aspects such as the health and fitness effects as well as their environmental effects (e.g., Verplanken *et al.*, 1994; Nilsson and Küller, 2000; Bamberg and Schmidt, 2001; Fujii *et al.*, 2001; Hunecke *et al.*, 2001; Joireman *et al.*, 2001; Matthies *et al.*, 2002).

The majority of the studies conducted, which examine the affective experience of travel, focus on the negative experience of stress (e.g., Koslowsky and Krausz, 1993; Koslowsky *et al.*, 1995; Evans *et al.*, 2002; Wener *et al.*, 2004) or aggression (e.g., Parkinson, 2001). These studies generally conclude that driving is stressful and that aggression is relatively more likely while driving than during other activities. Very little research has examined positive emotions related to travel behaviour such as the feeling of thrill when driving a car. However, for some people driving is pleasant and the car can provide a place to relax and unwind after a day of work (e.g., Steg *et al.*, 2001; Gatersleben and Uzzell, 2007; Mann and Abraham, 2006). Steg *et al.* (2001; Steg, 2005) and Gatersleben and Uzzell (2007) suggest that in examining the emotional aspects of car use it may be useful to refer to the theory of affect proposed by Russell and colleagues (e.g., Russell, 1980, 2003). According to this theory, two independent dimensions underlie all emotional states, emotional dispositions and affective appraisals. The pleasure–displeasure dimension reflects the degree to which a person feels happy or satisfied. The activation–deactivation dimension is a combination of activity (excited versus calm) and alertness (awake versus sleepy). Mehrabian and Russell (1974; Russell and Mehrabian, 1977; Mehrabian, 1980), distinguish three rather than two independent dimensions: pleasure, arousal (activation) and dominance. The dominance/submissive dimension reflects the extent to which a person feels in control, free and unrestricted. In terms of car use this distinction is particularly interesting. As this chapter will demonstrate dominance defined as autonomy, control or freedom features frequently as an affective or symbolic aspect.

Dittmar (1992) suggests that material possessions can have symbolic meaning by enabling people to express their social and personal identity. People have a psychological need to gain other peoples' approval. One way to ensure the approval of others is to manipulate and adapt one's outward actions, through the process of 'self presentation' or 'impression management' (Schlenker, 1982). Material possessions are especially useful for manipulating the impression that we make on others because they are stereotypical descriptors of socio-economic and social-cultural groups which are commonly understood by all, and as such they are instant symbols of identity, social standing, attitudes and beliefs (Dittmar, 1994). The private car in particular could have strong symbolic appeal, because, through years of persistent media advertisement, it is now a symbol that denotes social status, confidence, power and competence (Marsh and Collett, 1986; Stokes and Hallett, 1992; Stradling *et al.*, 1999; Hiscock *et al.*, 2002).

Another element of the symbolic value of the car is psychological attachment or the extent to which the car becomes part of someone's identity (Goodwin, 1997; Steg, 2005; Mann and Abraham, 2006). Belk (1988) demonstrated that people perceive a whole range of material objects, including the car as part of the self. Belk (1991) refers to Sartre (1943) who suggested that there are three ways in which possessions may become part of the self: through mastery and control of the object, by creating an object and through getting to know the object. Particularly, the first and latter are significant in the use of a motor car. Identity process theory (Breakwell, 1986) suggests that an individual will engage in a variety of behavioural and psychological coping strategies if their identity is threatened. Dittmar (1992) and Maguire (1980) showed that such reactions can be found when people experience unintended loss of their material possessions. Some of the severe reactions that can occur, for instance, at the threat of rising fuel prices (panic buying, demonstrations) suggest that at least for some people car dependence is very deeply rooted (e.g., Goodwin, 1995; Sheller, 2004).

REVIEW OF PREVIOUS RESEARCH

Overview of Research Findings

Ellaway *et al.* (2003) showed that the possession and use of a car can have positive effects on health and well-being. They conducted a large-scale survey study and showed that, controlling for age and social class, respondents who had access to a car scored higher on self-esteem and mastery than those who did not. Those who had access to a car also believed that the car provided more protection, autonomy and prestige, and regular drivers were more positive about cars than passengers.

Hiscock *et al.* (2002) suggest that access to a car might have such beneficial effects on people because it provides ontological security (a long-term tendency to believe that things are reliable and secure as opposed to threatening). Ontological security is important from an early age as it enables the development of self-identity by enhancing a sense of protection, autonomy and prestige. Hiscock *et al.* (2002) suggest that (like homes) 'cars may provide enhanced ontological security to those who have access to them and this may be partly why car owners appear to be so attached to their cars and to have better health' (Hiscock *et al.*, 2002, p. 121). An in-depth interview study was conducted with people who were 'at the margins of car ownership' so that they would be in a good position to convey the relative costs and benefits of the car. Respondents were interviewed about the car and public transport use. The study revealed that the car was seen to provide these psycho-social benefits to a greater extent than public transport. The car provided protection from the social environment in terms of undesirable social contact, privacy, personal space and violence and from the physical environment in terms of the weather. Some respondents believed it provided protection against accidents because in a car they were in control. The car was perceived to provide autonomy as convenience (being able to get somewhere immediately, quickly and frequently), autonomy as choice (to carry things, to go places) and autonomy as reliability (over journey times, frequency). One respondent, for instance, indicated 'it isn't easy getting all the buses and trains . . . whereas with a car you can just take off and go' (Hiscock *et al.*, 2002, p. 127).

Finally, it was found that respondents also believed that the car could provide prestige in terms of it being a symbol of income, having an exciting life, masculinity, being a valued member of society (unlike cyclists or users of public transport) and in terms of being able to enhance self-worth. One respondent said 'if he had a Rolls Royce . . . they might give him more esteem, more prestige' (Hiscock *et al.*, 2002, p. 130).

Lupton (2002) asked 77 residents of Sydney (drivers and non-drivers) to talk about the positive and negative aspects of driving. She found very similar results. The majority of respondents indicated that they enjoyed driving and mainly for reasons of convenience, independence and particularly freedom: 'driving first of all is like a sense of freedom' and 'I love driving because I can feel the car. You know I can control it' (Lupton, 2002, p. 280). The worst aspect of driving was related to a restriction of this freedom. Lupton (2002) found that although control was the most important positive aspect of the car, some respondents also referred to the car as a place to relax, wind down and enjoy some solitude or time away from the rest of the world: 'it's a release sometimes . . . get onto the road . . . and chill out' (Lupton, 2002, p. 280).

Mann and Abraham (2006) examined the driving experiences of commuters by means of in-depth interviews (see also Mann, 2006). Using Interpretative Phenomenological Analysis (IPA) they distinguished five major themes: affect-utility integration, journey-based affect, autonomy, personal space and identity. The first theme will be discussed later in this chapter. In relation to journey-based affect, Mann and Abraham (2006) found that although this differed between respondents (mainly car drivers) most respondents had both positive and negative feelings about their journeys. Car-based journeys, however, were rated more positively (more comfortable, relaxing and enjoyable) than journeys by public transport. As in the previous studies, control or autonomy was an important theme distinguished by Mann and Abraham (2006). Control was perceived as a positive aspect of the car in contrast to public transport. However, respondents indicated that traffic congestion and parking problems can lead to a loss of control. Although, many respondents felt that they could regain this control by route planning. Some respondents also referred to public transport in terms of control mainly by discussing how it can allow freedom from responsibility. Mann and Abraham (2006) also found references to personal space in terms of being alone without intrusion and a feeling of ownership of space. The final theme distinguished by Mann and Abraham (2006) is called car ownership and identity. This theme mainly referred to people's attachment to the car, that is, respondents expressing a feeling that they cannot do without a car (due to commitments such as work) so that car ownership is not a matter of choice. Some respondents also referred to the car as a status symbol, for instance, by expressing a desire to own a more expensive car or by expressing an anti-status by being proud of not owning a 'big' car.

The studies discussed so far have used qualitative research methods in the form of in-depth interviews to explore how people feel and think about cars. Jensen (1999) compared both qualitative interviews and a quantitative telephone survey. Jensen (1999) aimed to examine 'what it is in people's lives, in the daily life of the family and the individual that has made and continues to make them transport themselves so often and so far' (Jensen, 1999, p. 20). The in-depth interviews with a random sample of Danish drivers and non-drivers revealed six types of transport users: three car-user groups (passionate car drivers, everyday car drivers and leisure time car drivers) and three non-car driver groups (cyclists/public transport users of the hearth, cyclists/public transport users of convenience and cyclists/public transport users of necessity). The passionate car drivers value the car for affective and symbolic reasons. This type of driver 'loves to drive it [the car]', 'doesn't get stressed while driving', and 'sees his or her car as a symbol of freedom'. They believe that 'the make of the car says something about one's own personality' and that 'certain makes reflect success' (Jensen, 1999, p. 21). The everyday car driver is mainly described in instrumental terms: 'it is the easiest, the quickest and often the cheapest' (Jensen, 1999, p. 21). However, references are also made to affective aspects: 'it would be boring and hard to live without' (Jensen, 1999, p. 22), 'he or she likes to drive even though it can be stressful . . .' (Jensen, 1999, p. 22), including dominance-related aspects: (Mehrabian and Russell, 1974; Russell and Mehrabian, 1977; Mehrabian, 1980) 'the car gives a sense of independence'. Everyday car drivers, however, do acknowledge negative aspects of cars as well. The leisure driver considers cars useful and comfortable but not particularly pleasurable and they have a positive attitude towards alternatives. Although status and power played a role in the qualitative interview studies, very few people in the quantitative telephone survey agreed with statements referring to the car as a symbol of status, power or personality. Yet, they did agree that the car is a symbol of freedom and independence. This suggests that although status and power may play a role, it is easily rejected when respondents are directly confronted with it, perhaps because it is perceived to be socially undesirable. This would suggest that less obtrusive research techniques may be more applicable when studying the symbolic aspects of car use.

Steg *et al.* (2001) developed three different methodologies to examine the instrumental-reasoned and the symbolic-affective motives for car use: a similarity sorting, a Q-sort and a semantic differential method. A list with 60 attractive and unattractive aspects of cars was developed. In the semantic differential task, respondents (all car users) were asked to rate the car on these aspects. The aspects were also used to develop a set of 32 short scenarios, which describe a situation involving the car. These scenarios were sorted by the respondents, first on similarity and then on attractiveness. In the semantic differential task, respondents based their attractiveness judgements on instrumental aspects such as availability, flexibility, carrying capacity, convenience and time on the positive side and financial and environmental costs on the negative side. Symbolic aspects did not feature at all in this task and affect in terms of stress and driving pleasure did not feature highly, although control and freedom did, be it in more instrumental terms. In the attractiveness card sorts, however, respondents relied much more on affective judgements. In fact the most attractive scenario was: 'Finally, you bought your dream car. Your neighbour watches in admiration when you are testing your new car.

The car holds the road well' (Steg *et al.*, 2001, p. 167). The fourth most important scenario was: 'You pick up your son from school. You hear him saying to his friend that your car is much better and stronger than the car of his father. You decide not to go home right away but to make a short drive first' (Steg *et al.*, 2001, p. 167). Clearly, this study suggests that findings can be very different depending on the methodology that is being used.

However, there is some evidence that the importance of non-instrumental aspects can also be found in quantitative studies. Stradling *et al.* (1999), for instance, conducted a large-scale survey among nearly 800 British drivers. Amongst others, these respondents were asked about their agreement with 25 statements on the cost and benefits of driving. In relation to affect it was found that almost all respondents agreed that the car provides a sense of freedom. About half also agreed that it provides a feeling of being in control and of self-confidence, while about half agreed that it is stressful. Most drivers indicated that they enjoyed driving (one-third enjoyed it a lot).

Anable (2005) conducted a large-scale quantitative survey among leisure travellers, which included attitude statements on the instrumental, affective and symbolic aspects of the car. Most of the car users in her study reported that they did not particularly enjoy driving but felt that they could not do without a car. However, 20% of the respondents did report that they find driving very pleasant and that they would not want to give it up.

Steg (2003) asked a random sample of 1803 Dutch households to judge both car and public transport on various instrumental, affective and social-symbolic aspects. She found that overall the car was rated more positively than public transport on all aspects (except safety), even by infrequent drivers. The car scored, especially, more positively not only on instrumental aspects such as convenience, independence, flexibility, comfort, speed and reliability, but also on pleasure and status. Moreover, frequent car users were more positive than infrequent car users about all of these aspects except status. Surprisingly, in light of the studies described above, the car and public transport differed less with respect to freedom and control.

A study by Steg (2005) assessed evaluations of 32 positive aspects of car use, clustered into three factors (symbolic-affective, instrumental and independence). She found that frequent drivers were more positive about the symbolic-affective aspects of the car, but not necessarily about the instrumental aspects or independence. Using a different measure of instrumental, affective and symbolic aspects she also found that those who drive more are more likely to say that the car has symbolic value (e.g., expresses status and identity). A second study showed that car use for the daily commute is positively related to descriptive norms (familiar others drive), injunctive norms (familiar others feel I should drive), affect (less stress) and self-presentation (e.g., driving suits me better). Evaluations of the instrumental aspects of car use did not add to explaining the variance in reported car use.

Gatersleben and Uzzell (2007) showed that although the daily commute can be stressful, it can also be boring, relaxing and exciting. The main source of positive affect for all commuters is natural scenery (see also Parsons *et al.*, 1998). For car users, music and looking at or talking to other people were also sources of positive affect and some car users also mentioned flexibility and control. However, for car users (and users of public transport) a loss of control due to delays and congestion was by far the most important source of negative affect.

Altogether these studies suggest that at least for some people the possession and use of a car is related to different positive as well as negative affective and symbolic aspects. Although the importance of these aspects is more profound in qualitative studies, some quantitative studies have also found that affective and symbolic aspects play a role in people's driving experiences and may be related to driving frequency. It is difficult, however, to draw any causal conclusions on the basis of these mainly exploratory studies. For instance, it is not clear whether car dependency results from instrumental or psychological variables or vice versa. Moreover, the studies provide only limited insight into the extent to which affective and symbolic appraisals of the car may help to predict car use and mode choice. Finally, the relationship between instrumental, affective and symbolic aspects is often unclear.

The Relationship between Instrumental, Affective and Symbolic Aspects

Mann and Abraham (2006) suggest that individuals do not make clear distinctions between affective and instrumental aspects of the car (Mann, 2006, p. 135; Mann and Abraham, 2006). Instrumental aspects such as time and reliability were often viewed as causes of important affective consequences. Respondents expressed boredom and frustration in relation to reliability and travel time, guilt in terms of environmental pollution and anger in relation to the cost of public transport. This has also been acknowledged by Steg (2004) who showed that affect may result from instrumental as well as symbolic aspects. Gatersleben and Uzzell (2007) also showed that affective experiences often have an instrumental basis according to respondents. Steg (2005) found three dimensions underlying judgements about the attractive aspects of car use: symbolic affective (e.g., prestige, kick of driving, driving is relaxing), instrumental (e.g., safe, comfortable) and independence (e.g., availability, independence, route choice). General driving pleasure ('driving is enjoyable') fell within the instrumental factor. These findings suggest that these aspects cannot be seen as independent. In their earlier paper (partly based on the same data), Steg *et al.* (2001) described very similar results. In another study by Steg (2003), respondents were asked to evaluate the car and public transport on 17 instrumental, affective and symbolic aspects. Again, a factor analysis revealed no clear distinction between instrumental and affective aspects. Both for car and public transport, judgements of pleasure appeared to correlate with judgements of variety and cosiness; judgements of dominance and control were related to security and speed. For cars, arousal was related to sexiness and status. The latter suggests that symbolic aspects are more strongly related to positive arousal whereas instrumental aspects are more strongly related to pleasure. This is supported by Hiscock *et al.* (2002) who showed that respondents who saw the car as a symbol of status, also saw it as a symbol of having an exciting life.

It appears that affective appraisals of the car are related to instrumental and symbolic functions of the car. The relationship between instrumental and symbolic aspects was also found in some of the studies, especially the interview studies where respondents referred to functional properties of the car, which allowed them to express personal values and status. One of the respondents in Jensen (1999) for instance, indicated that 'the make of the car says something about one's own personality' (Jensen, 1999, p. 21). The two scenarios in the study by Steg *et al.* (2001) described above referred to 'new', 'big' cars, which are admired by others. Mann and Abraham (2006) indicated that some respondents expressed a sense of pride in not owning a powerful new car, which they called an expression of anti-status. Steg *et al.* (2001) also found that one of their car scenarios which clustered with other status scenarios referred to such a situation: 'Your colleagues jeer at you and your ordinary car. You don't care, it brings you everywhere, and you have a nice shelter against rain. That's all you need, isn't it?' (Steg *et al.*, 2001, p. 166).

Taken together, these exploratory studies suggest that the instrumental properties of a car may influence the affective and symbolic reactions to – and experiences with – this mode. But, as mentioned before, the studies do not allow us to draw any causal conclusions. In order to better understand the interrelationships between the instrumental, affective and symbolic evaluations of the car as well as their relationship with general attitudes, intentions and behaviours it is necessary to conduct theory-driven research, which uses clear measures and operationalisations of these different concepts. Dittmar's (1992, 1994, 1996) model of material possessions may provide a starting point for this work. Dittmar suggests that possessions fulfil two functions: instrumental (the car is fast, cheap, convenient) and symbolic (e.g., the car allows me to express my individual and group identity). The studies presented in this paper suggest that the extent to which the car fulfils these two functions influences our affective experiences with this mode (e.g., makes me feel relaxed, stressed, excited, powerful, free, in control, bored).

The Automobile: Control, Freedom, Independence and Autonomy

The car is different from other modes of transport not because it provides mobility but because it provides autonomous mobility (Flink, 1975). This autonomy is often described as one of the most important aspects of the car (e.g., Flink, 1975; Marsh and Collett, 1986; Goodwin, 1995; Diekstra and Kroon, 2003).

The studies described here asked people for their own thoughts and feelings; the concept of control or autonomy in these studies therefore refers to control in affective terms (feeling in control) rather than in behavioural terms (being in control). As mentioned above, Mehrabian distinguished three (rather than two) independent dimensions that underlie all emotions: pleasure, arousal (or activation) and dominance (e.g., Mehrabian and Russell, 1974; Russell and Mehrabian, 1977; Mehrabian, 1980). The latter dimension refers to the extent to which people feel in control, free and unrestricted. The studies presented in this chapter suggest that at least for the study of car use it may indeed be useful to distinguish dominance as an affective factor in addition to pleasure and arousal (or activation).

In all studies described above, autonomy, control, independence and/or freedom appeared to be one of the most, if not the most important positive aspect of the car (Stradling *et al.*, 1999; Hiscock *et al.*, 2002; Lupton, 2002; Steg, 2003; Mann and Abraham, 2006). For instance, Hiscock *et al.* (2002) suggest that 'convenience provided respondents with feelings of control over their lives' (Hiscock *et al.*, 2002, p. 126) and 'unreliability [of public transport] meant people felt less in control of their lives' (Hiscock *et al.*, 2002, p. 129). Being able to control undesirable social contacts mainly exists when respondents discuss control over the indoor environment and particularly when people talk about the difference between public transport and the private car (Hiscock, 2002; Lupton, 2002; Mann and Abraham, 2006).

Several of the studies discussed also refer to control in symbolic terms. In these cases, references are often still made to affect or emotion but with a more relational connotation, e.g., feeling powerful, or stronger than others. Steg *et al.* (2001) found three dimensions underlying the Q-sort task. The first dimension is of particular interest here as it distinguishes situations in which the driver is in control versus those where he or she is not in control. Being in control in this study was mainly related to scenarios describing symbolic and affective aspects, such as the proud son and dream car episodes described above. Jensen (1999) found in both her qualitative and the quantitative studies that most people agreed that the car is an important symbol of freedom and independence.

The studies suggest that control or dominance has a strong instrumental basis. In most of the studies, references were made to control in terms of being able to choose travel time, mode and route (e.g., Stradling *et al.*, 1999; Hiscock, 2002; Lupton, 2002; Steg, 2003; Mann and Abraham, 2006). As mentioned before, Hiscock (2002) found a link between convenience and reliability and feelings of control over ones life. One of the respondents of Mann (2006) said, 'I'd be annoyed [waiting for buses] because I wouldn't be in control' (Mann, 2006, p. 149).

The studies described in this chapter suggest that more than one dimension might be underlying the control or dominance dimension. Not only does it involve instrumental, symbolic and affective aspects, Jensen (1999) also suggests that control, independence and freedom are not the same concepts. According to Jensen (1999), it is important to think about what a person is assumed to be in control of (one's life or one's travel route). Freedom should be contrasted from force and can thus be defined as the ability to choose. Independence is related to freedom but it is not the same: 'it is the "nature" of the concept that it must be independence from something or in relation to something' (Jensen, 1999, p. 31). In a car you can be independent of others; you can be yourself. But, in contrast to public transport, a car also provides independence from time schedules. The factor analysis by Steg (2003) on instrumental, affective and symbolic aspects of the car also suggested that independence and freedom/control may be different factors. Independence was related to convenience, comfort and flexibility. Freedom and control were related to each other and to security and travel speed. This was the case both for the car and for public transport.

The above analysis of the studies reveals that control, freedom and independence are very important advantages of the car according to individual car users and non-users. Theoretically, it may be assumed that these aspects are related to one underlying affective control or dominance dimension, but the studies show some conflicting evidence. As this aspect or these aspects appear to play an important role in the evaluation of travel experiences and perhaps in travel mode choice it seems worthwhile to conduct more theory-driven research, which develops clear definitions and operationalisations of the concept(s). The research findings reported here suggest that there is more than one dimension underlying the control concept, which may be related in different ways to instrumental and symbolic aspects of car use.

CONCLUSIONS

Although the studies discussed in this chapter may have raised more questions than answers, they have undoubtedly cast more light onto the relative importance of instrumental, affective and symbolic aspects for travel experiences and driving frequency. Perhaps, one of the most interesting conclusions that can be drawn from these studies is that it is not easy, if not impossible, to ignore the psychological, affective and symbolic aspects of car use and their relationship with instrumental aspects.

For most people the car provides the most flexible and convenient mode of transport. But at least for some the car also has more deeply rooted psychological significance. Anable (2005) and Jensen (1999) both showed that some people attach great symbolic and affective value to their car. Steg *et al.* (2001) showed that the use of indirect non-obtrusive research methods may even reveal that affective and symbolic aspects may be more important than survey studies would suggest. Social desirability may prevent study participants in survey research from expressing what the car really means to them.

The studies discussed in this chapter reveal that a great deal of work is yet to be done. But before this work is conducted, it is essential to develop clearer definitions and operationalisations of the key concepts: instrumental, affective and symbolic aspects of car use. The theoretical basis of the research to date is not strong, and more theory-driven research is necessary (Steg, 2005). Owing to the exploratory nature of most studies the relevant concepts are not clearly defined, which makes interpretation of the data difficult and generalisation of the findings impossible.

It is undoubtedly true that the car is often chosen as a mode of transport over other modes for instrumental reasons. The psychological relationship between people and their car, however, will probably play a very important part when trying to persuade people to reduce their car use for reasons of environmental sustainability and when predicting social and psychological reactions to policy initiatives, which aim to bring about such reductions. Gärling *et al.* (2002) suggest that a reduction in car use may lead to a loss of control, which might have serious psychological consequences and they refer to Murphy (1988) by suggesting that it 'might be better to never have had control than to have lost it' (Murphy, 1988, p. 100).

REFERENCES

Anable, J. (2005). 'Complacent Car Addicts' or 'Aspiring Environmentalists'? Identifying Travel Behaviour Segments Using Attitude Theory, *Transport Policy*, **12**, 65–78.

Bamberg, S. and P. Schmidt (2001). Theory-Driven Evaluation of an Intervention to Reduce the Private Car-Use, *Journal of Applied Social Psychology*, **31** (6), 1300–1329.

Belk, R.W. (1988). Possessions and the Extended Self, *Journal of Consumer Research*, **15**, 139–168.

Belk, R.W. (1991). The Ineludable Mysteries of Possessions. In: *To Have Possessions: A Handbook on Ownership and Poverty* (Special Issue) (F.W. Rudmin, ed.), *Journal of Social Behavior and Personality*, **6** (6), 17–55.

Breakwel, G.M. (1986). *Coping with Threatened Identities*, Methuen, London.

Diekstra, R.F.W. and M. Kroon (2003). Cars and Behaviour. In: *Sustainable Transport. Planning for Walking and Cycling in Urban Environments* (R. Tolley, ed.), Woodland Publishing, Cambridge, 252–264.

Dittmar, H. (1992). *The Social Psychology of Material Possessions: To Have is To Be*, Hemel Hempstead, UK.

Dittmar, H. (1994). Material Possessions as Stereotypes: Material Images of Different Socio-economic Groups, *Journal of Economic Psychology*, **15**, 561–585.

Dittmar, H. (1996). The Social Psychology of Economic and Consumer Behaviour. In: *Applied Social Psychology* (G.R. Semin and K. Fiedler, eds), Sage, London, 145–172.

Ellaway, A., S. Macintyre, R. Hiscock, and A. Kearns (2003). In the Driving Seat: Psychosocial Benefits from Private Motor Vehicle Transport Compared to Public Transport, *Transportation Research, Part F*, **6**, 217–231.

Evans, G., R. Wener, and D. Phillips (2002). The Morning Rush Hour: Predictability and Commuter Stress, *Environment and Behavior*, **34** (4), 521–530.

Flink, J.J. (1975). *The Car Culture*, MIT Press, Massachusetts.

Fujii, S., T. Gärling, and R. Kitamura (2001). Changes in Drivers' Perceptions and Use of Public Transport During a Freeway Closure: Effects of Temporary Structural Change on Cooperation in a Real-life Social Dilemma, *Environment and Behavior*, 33 (6), 796–808.

Gärling, T., A. Gärling, and P. Loukopoulos (2002). Forecasting Psychological Consequences of Car Use Reduction: A Challenge to and Environmental Psychology of Transportation, *Applied Psychology: An International Review*, **51** (1), 90–106.

Gatersleben, B. and D. Uzzell (2007). Affective Appraisals of the Daily Commute: Comparing Perceptions of Drivers, Cyclists, Walkers and Users of Public Transport, *Environment and Behavior*.

Goodwin, P.B. (ed.) (1995). *Car Dependence*, RAC Foundation for Motoring and the Environment, London.

Goodwin, P.B. (1997). Mobility and Car Dependence. In: *Traffic and Transport Psychology. Theory and Application* (J.A. Rothengatter and E. Carbonell Vaya, eds), Pergamon, Oxford, 449–464.

Hiscock, R., S. Macintyre, A. Kearns, and A. Ellaway (2002). Means of Transport and Ontological Security: Do Cars Provide Psycho-Social Benefits to their Users? *Transportation Research, Part D*, 7, 119–135.

Hunecke, M., A. Blöbaum, E. Matthies and R. Höger (2001). Responsibility and Environment: Ecological Norm Orientation and External Factors in the Domain of Travel Mode Choice Behavior, *Environment and Behavior*, 33 (6), 830–852.

Jensen, M. (1999). Passion and Heart in Transport – A Sociological Analysis on Transport Behaviour, *Transport Policy*, **6**, 19–33.

Joireman, J., P. Van Lange, M. Van Vugt, A. Wood, T. Leestand, and C. Lambert (2001). Structural Solutions to Social Dilemmas: A Field Study on Commuters' Willingness to Fund Improvements in Public Transit, *Journal of Applied Social Psychology*, **31** (3), 504–526.

Koslowsky, M., A. Kluger, and M. Reich (1995). *Commuting Stress: Causes, Effects and Methods of Coping*, Plenum Press, New York.

Koslowsky, M. and M. Krausz (1993). On the Relationship Between Commuting, Stress Symptoms, and Attitudinal Measures: A LISREL Application, *Journal of Applied Behavioral Science*, **29** (4), 485–492.

Lupton, D. (2002). Road Rage: Drivers' Understandings and Experiences, *Journal of Sociology*, **38** (3), 275–290.

Macintyre, S., A. Ellaway, G. Der, G. Ford, and K. Hunt (1998). Do Housing Tenure and Car Access Predict Health because they are Simply Markers of Income and Self Esteem? A Scottish Study, *Journal of Epidemiology and Community Health*, **52** (10), 657–663.

Maguire, M. (1980). The Impact of Burglary upon Victims, *British Journal of Criminology*, **20**, 261–275.

Mann, E. (2006). *Cognitive and Affective Antecedents of Commuter Transport Mode Choices*, Unpublished Doctoral Dissertation, University of Sussex, UK.

Mann, E. and C. Abraham (2006, in press). The Role of Affect in UK Commuters' Travel Mode Choices: An Interpretative Phenomenological Analysis, *British Journal of Psychology*, **97** (2), 155–176.

Marsh, P. and P. Collett (1986). *Driving Passion: The Psychology of the Car*, Jonathan Cape, London.

Matthies, E., S. Kuhn, and C.A. Klöckner (2002). Travel Mode Choice of Women: The Result of Limitation, Ecological Norm or Weak Habit? *Environment and Behavior*, **34** (2), 163–177.

Mehrabian, A. (1980). *Basic Dimensions for a General Psychological Theory*, Oelschlager, Gunn & Hain, Cambridge, MA.

Mehrabian, A. and J. Russell (1974). *An Approach to Environmental Psychology*, MIT Press, Cambridge, MA.

Murphy, L.R. (1988). Workplace Interventions for Stress Reduction and Prevention. In: *Causes, Coping and Consequences of Stress at Work* (C.L. Cooper and R. Payne, eds), Wiley, Chichester, 301–342.

Nilsson, M. and R. Külller (2000). Travel Behaviour and Environmental Concern, *Transportation Research Part D*, **5**, 211–234.

Parkinson, B. (2001). Anger On and Off the Road, *British Journal of Psychology*, **92**, 507–526.

Parsons, R., C.G. Tassinary, R.S. Ulrich, M.R. Hebl and M. Grossman-Alexander (1998). The View from the Road: Implications for Stress Recovery and Immunization, *Journal of Environmental Psychology*, **18**, 113–139.

Russell, J.A. (1980). A Circumplex Model of Affect, *Journal of Personality and Social Psychology*, **39**, 1161–1178.

Russell, J.A. (2003). Core Affect and the Psychological Construction of Emotion, *Psychological Review*, **110** (1), 145–172.

Russell, J. and A. Mehrabian (1977) Evidence of a Three-Factor Theory of Emotions, *Journal of Research in Personality*, **11**, 273–294.

Sachs, W. (1984). *Die Liebe zum Automobil. Ein Rückblick in die Geschichte unserer Wünsche*, Rohwolt, Reirkbeck bei Hamburg.

Sartre, J. (1943). *Being and Nothingness: A Phenomenological Essay on Ontology*, Philosophical Library, New York.

Schlenker, B.R. (1982). Translating Actions into Attitudes: An Identity-Analytic Approach to the Explanation of Social Conduct. In: *Advances in Experimental Social Psychology* (L. Berkowitz, ed.), Academic Press, New York, 194–248.

Sheller, M. (2004). Automotive Emotions. Feeling the Car, *Theory Culture and Society*, **21** (4/5), 221–242.

Steg, L. (2003). Can Public Transport Compete with the Private Car? *IATSS Research*, **27** (2), 27–36.

Steg, L. (2004). Car Use Lust and Must. In: *Traffic and Transport Psychology* (T. Rothegatter and R.D. Huguenin, eds), Elsevier, Oxford, 443–452.

Steg, L. (2005). Car Use: Lust and Must. Instrumental, Symbolic and Affective Motives for Car Use, *Transportation Research, Part A*, **39**, 147–162.

Steg, L., C. Vlek, and G. Slotegraaf (2001). Instrumental-Reasoned and Symbolic-Affective Motives for using a Motor Car, *Transportation Research, Part F*, **4**, 151–169.

Stokes, G. and Hallett, S. (1992). The Role of Advertising and the Car, *Transport Reviews*, **12** (2), 171–183.

Stradling, S.G., M.L. Meadows, and S. Beatty (1999). *Factors Affecting Car Use Choices*, Transport Research Institute, Napier University, Edinburgh.

Verplanken, B., H. Aarts, A. Van Knippenberg, and C. Van Knippenberg (1994). Attitude versus General Habit: Antecedents of Travel Mode Choice, *Journal of Applied Social Psychology*, **24** (4), 285–300.

Wener, R., G. Evans, D. Philips, and N. Nadler (2004). Running for the 7:45. The Effects of Public Transit Improvements on Commuter Stress, *Transportation*, **30** (2), 203–220.

Threats from Car Traffic to the Quality of Urban Life
Tommy Gärling and Linda Steg (Editors)

13

ROLE AND ACQUISITION OF CAR-USE HABIT

Satoshi Fujii, Department of Civil Engineering, Tokyo Institute of Technology, Tokyo, Japan
Tommy Gärling, Department of Psychology, Göteborg University, Göteborg, Sweden

ABSTRACT

It has been assumed that car-use habits make people automatically choose to use the car instead of alternative travel modes such as public transport, bicycle, or walking. Consistency between intention and behavior is thus reduced in travel mode choice. In this chapter, we assume that car-use habits reflect script-based choice. Based on this assumption, we describe a method for measuring the strength of car-use habit. Further, we discuss how a car-use habit is acquired, its impact on travel mode choice, and how to break it.

INTRODUCTION

Many individuals repeat the same trips by car almost every day, for instance, when commuting to work (e.g., Kitamura, 1990; Golob *et al.*, 1997). Is such frequent car use a habit? In everyday language it may be referred to as such. But what *is* a habit? Is it something else than repeated behavior? If it is something else, what are the implications? In this chapter, we argue that habits are less controlled by conscious decisions (e.g., Ronis *et al.*, 1989; Bargh, 1997). It is further argued that when car use is habitual, it is not determined by intention, attitude, subjective norm, and perceived control in the way it is posited to be in the theory of planned behavior (TPB) (Ajzen, 1985, 1991). Furthermore, only observing that a behavior is repeated is not sufficient evidence for that it is habitual.

In this chapter, we first present a definition of car-use habits and how they can be measured. The process of acquiring a car-use habit is then analyzed. Following this, we highlight several impacts that car-use habits may have on travel behavior. A consequence is that the prevalence of car-use habits needs to be taken into account when designing and implementing policy measures targeted at reduced or changed car use. In the final section, we discuss policy measures for how to break car-use habits.

DEFINITION AND MEASUREMENT OF CAR-USE HABIT

Definition of Car-Use Habit

Oulette and Wood (1998) refer to a habit as a tendency to repeat past behavior in a stable context. Aarts *et al.* (1998) and Verplanken and Aarts (1999) refer to a habit as "goal-directed automaticity" in implementing a behavior. It is assumed that a habit is automatically triggered by situational cues, implying that any deliberation of alternatives are reduced or eliminated. For example, a habitual driver is unlikely to think of public transport as an alternative (Verplanken *et al.*, 1998).

As Ajzen (2002) notes, there is abundant evidence that attitudes, intentions, and simple acts can be activated automatically. Yet, most routines in daily life such as, for instance, car-use habits are *semi-automatical*, implying that they consist of both automatic and nonautomatic or controlled phases. In our view, the important issue in understanding repeated car use is whether the choice to drive is made automatically, not whether the consequent behavior (driving) is automatically executed. Yet, as explained below, the evaluation of the *outcome* of the behavior (e.g., failure to achieve the goal of being in time at the destination) is still important.

Ajzen (2002) suggests that repeating the same choice over and over in the same context may lead to an *intention* being automatically activated without deliberation. A related way of conceptualizing "goal-directed automaticity" in travel mode choice would be to assume that acquired *scripts* (Schank and Abelson, 1977) influence predecisional information search (Verplanken *et al.*, 1997) as well as choice (Gärling *et al.*, 2001; Fujii and Gärling, 2003). If travel choice alternatives are conceptualized as sets of attribute levels (e.g., Ben-Akiva and Lerman, 1985), after repeatedly making the same choice, existing covariation between attribute levels may be discovered and utilized. Thus, some attributes may become proxies for other attributes. Then it is no longer necessary to search information on all attributes, because the same choice can be made on the basis of a subset of attribute information. This is how we define *script-based choice* (Svenson, 1990). It is implied that the people have a simplified, prototypical representation of the choice alternatives stored in memory, called a *script*, that can be retrieved if only a subset of the information is available. Thus, the assumption is that a car-use habit is a tendency to retrieve a script to use the car when facing choices of travel mode. Such retrieval will in general alleviate a person from extensive deliberation preceding the travel mode choice.

Our proposed definition goes beyond simply assuming that habit is automatically triggered by situational cues (Aarts *et al.*, 1998). We assume instead that a frequently applied decision rule is retrieved from memory and again applied. A knowledge structure is thus believed to exist in memory (cf. Aarts and Dijksterhuis, 2000b) that makes this possible. Since minimal information would be sufficient for retrieval of the script, acquisition and application of the choice do not necessarily assume a stable context. We define strength of a car-use habit as the ease with which car-use scripts are retrieved. Thus, the strength of car-use habit (H) is

$$H = \sum_{i=1}^{n} E_i \tag{1}$$

where n denotes the number of car-use scripts stored in memory, E_i ease of retrieval script i from memory. Habit strength increases if car-use scripts are stored in memory that correspond to different situations (if n is large), for instance, work commute, pick up child, shopping, and so forth. The strength of the car-use habit would also increase if the ease with which each script is retrieved increases (if E_i increases). Ease of retrieval is assumed to depend on the absence of alternative scripts. For instance, people may have two scripts for how to travel to a restaurant for dinner. One entails driving the own car. The other one entails walking. In this case, the retrieval of the car-use script is less easy than if only one car-use script existed, and it is less likely that the car is used.

A car-use habit should increase in strength with frequency of car use. However, frequent car use does not necessarily lead to the acquisition of a car-use habit. This is because car use is influenced by other factors such as, for instance, situational constraints and intentions (Ajzen, 2002). It follows that the frequency of car use is not a perfect measure of a car-use habit. An alternative measure is therefore needed.

Measurement of Car-Use Habits

A method for measuring car-use habits may be based on observations of predecisional information processing. Verplanken *et al.* (1997) using the information-board technique (Payne *et al.*, 1993) found that, when making travel mode choice, frequent car drivers acquired less information about alternative travel modes than nonfrequent car drivers did. Based on this and similar findings, the *response-frequency measure* was developed (Verplanken *et al.*, 1998) in order to measure the habit strength in travel mode choice. As the example shown in Table 1, several different everyday situations are presented asking respondents to respond to each as quickly as possible with the first mode that comes to mind. The assumption is that such generic or prototypical situations trigger a script-based choice. Furthermore, if respondents respond fast, the role of deliberate intention would be minimized. Given that these conditions are fulfilled, the response-frequency measure should be a relatively pure measure of habit defined as

Table 1: The Response-Frequency Measure of Strength of Car-Use Habit

Respond* quickly *to each by selecting the first travel mode that comes to mind				
Visiting home of a friend	❑ auto	❑ train	❑ bicycle	
	❑ bus	❑ walk	❑ motorbike	❑ other
Shopping clothes	❑ auto	❑ train	❑ bicycle	
	❑ bus	❑ walk	❑ motorbike	❑ other
Watching movies	❑ auto	❑ train	❑ bicycle	
	❑ bus	❑ walk	❑ motorbike	❑ other
Going out for lunch	❑ auto	❑ train	❑ bicycle	
	❑ bus	❑ walk	❑ motorbike	❑ other
Going back to your home	❑ auto	❑ train	❑ bicycle	
	❑ bus	❑ walk	❑ motorbike	❑ other
Going to a playland	❑ auto	❑ train	❑ bicycle	
	❑ bus	❑ walk	❑ motorbike	❑ other
Going on a date	❑ auto	❑ train	❑ bicycle	
	❑ bus	❑ walk	❑ motorbike	❑ other
Going out for dinner	❑ auto	❑ train	❑ bicycle	
	❑ bus	❑ walk	❑ motorbike	❑ other
Barbecuing with friends	❑ auto	❑ train	❑ bicycle	
	❑ bus	❑ walk	❑ motorbike	❑ other
Going for bathing in the sea	❑ auto	❑ train	❑ bicycle	
	❑ bus	❑ walk	❑ motorbike	❑ other
Going for skiing	❑ auto	❑ train	❑ bicycle	
	❑ bus	❑ walk	❑ motorbike	❑ other
Commuting to school	❑ auto	❑ train	❑ bicycle	
	❑ bus	❑ walk	❑ motorbike	❑ other
Going to a convenience store	❑ auto	❑ train	❑ bicycle	
	❑ bus	❑ walk	❑ motorbike	❑ other
Going to a hospital	❑ auto	❑ train	❑ bicycle	
	❑ bus	❑ walk	❑ motorbike	❑ other
Going to a book store	❑ auto	❑ train	❑ bicycle	
	❑ bus	❑ walk	❑ motorbike	❑ other

Source: Adapted from Fujii and Kitamura (2003).

script-based choice, and of the *strength* of car-use habit as related to the frequency with which the car is chosen.

The response-frequency measure has been used in several studies to measure the strength of car-use habits (Verplanken *et al.*, 1994, 1997; Aarts *et al.*, 1998; Aarts and Dijksterhuis, 2000a; Gärling *et al.*, 2001; Fujii and Gärling, 2003; Fujii and Kitamura, 2003). The measure has been shown to have acceptable reliability. Witnessing to its validity, a modest positive correlation has been found with self-reported frequency of car use, and the measure has also been found to correlate with a positive attitude toward car use.

Verplanken and Orbell (2003) noted that the response-frequency measure is limited to multiple-choice situations. They therefore developed another self-report measure consisting

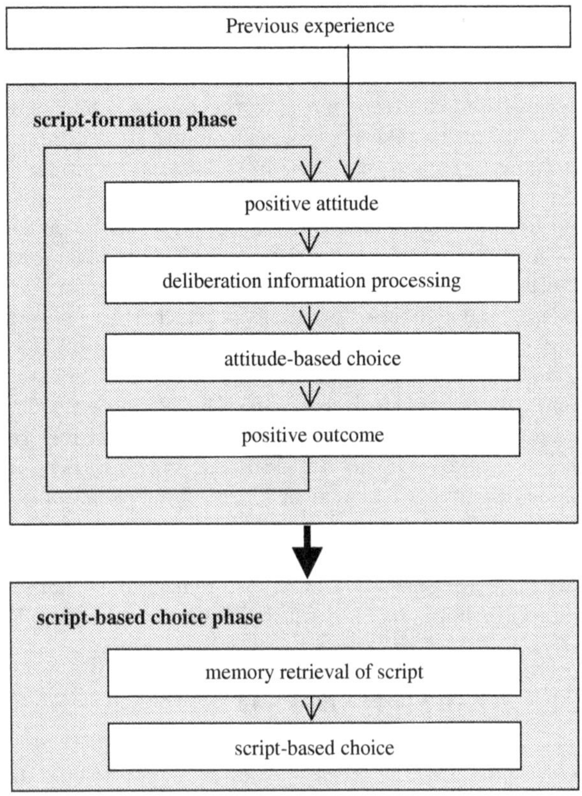

Source: Adapted from Gärling *et al.* (2001).

Figure 1: A Process Model of Acquisition of Script-Based Choice

of questions tapping participants' meta-cognitive knowledge related to automaticity of the behavior. This measure has not been used in studies of car-use habits.

ACQUISITION OF CAR-USE HABIT

Gärling *et al.* (2001) proposed a process model of the acquisition of script-based choices of car use. As Figure 1 shows, they assumed that a positive attitude motivates car use and is further strengthened by a positive outcome of the choice to use the car. After repeated such deliberate choices, script-based choice would develop so that a script accessible in memory makes possible to infer information of the choice alternatives.

This would gradually reduce the need to search information. The model was confirmed by correlational analyses as well as experiments with nondrivers as participants (Gärling *et al.*, 2001; Gärling, 2004). Fujii and Gärling (2003) also found that forced repetitious choices lead to script-based choices in a real-world context. Students who after graduation moved to new places to take up jobs were found to develop car-use habits, as assessed by the response-frequency measure, after starting to commute by car.

If car users attend to limited situational cues, many situations become more similar. For example, a car commuter may develop a script "to use the car when travelling from home in the morning" instead of a script "to use the car when travel from home in the morning to *work*". In this way, the number of situations to which a car-use script applies would increase, thus further strengthening the car-use habit (cf. equation (1)). Eventually, this generalization of the script may result in a script-based choice to always use the car when traveling from home. An example is the experiment by Gärling *et al.* (2001) showing that after having acquired a car-use habit when the destination was farther away than walking distance, decreases in distance were not noted so that car was chosen for trips to destinations within walking distance. A requirement for the development of such a maladaptive choice is thus that evaluations of outcomes are distorted. There is abundant evidence for this, either because feedback is delayed or nontransparent (Brehmer, 1995), or that the feedback is misinterpreted, frequently in a way so that it confirms the positive attitude motivating the choice (Einhorn and Hogarth, 1978; Klayman and Ha, 1987).

IMPACTS OF CAR-USE HABITS ON TRAVEL BEHAVIOR

An observable consequence of the acquisition of a car-use habit is an increase in the frequency of car use. A car-use habit also has other impacts on travel behavior. In this section, we describe several such impacts.

Interference with Intentions and Other Psychological Factors

Theories that have been frequently used to model the determinants of behavior are the theory of reasoned action (TRA) (Fishbein and Ajzen, 1975) and its successor the TPB (Ajzen, 1985, 1991). This latter theory assumes that intentions are formed after deliberation that takes into account factors such as attitude, subjective norm, and perceived control. Intention is the proximal cause of behavior. In a meta-analysis of studies reporting behavior and intention consistency, Sheppard *et al.* (1988) found an average correlation between intention and behavior of 0.53 (corresponding to a hit rate of about 75%). Ajzen (2002) notes several factors that may account for the inconsistency between intention and behavior. The most important ones related to travel behavior were identified by Fujii and Gärling (2003) and Gärling *et al.* (1998). One is that people sometimes engage in "impulsive behavior," another one is that a habit has been acquired.

Triandis (1977) proposed that the following relations hold between habit (H), intention (I), and behavior (B):

$$B = (w_h H + w_i I) \times F \quad w_h + w_i = 1 \tag{2}$$

where F denotes the situational constraints and w_h and w_i are the weights that determine the relative impact of habit and intention. Since the sum of these weights is fixed to 1, the model posits that the role of intention decreases when the strength of habit increases, and the reverse. Ajzen (2002) questioned such a reciprocal relationship. However, this criticism relied on a change of definition from intention being the result of deliberation to being *spontaneous* (in the case of a habit). Whether being described as a "spontaneous intention" or as script-based choice, as we assume, does not seem to make much difference.

As noted, the direct consequence of the acquisition of a car-use habit is that car use is automatically chosen. Thus, habitual drivers facing a travel mode choice would choose the car by automatically retrieving a car-use script. Note that such retrieval may be interfered with by constraints that eliminate the car as the most attractive alternative. Intentions not to use the car (or to use an alternative travel mode such as public transport) would also override the habit. Still, since scripts to choose the car is *automatically* retrieved from memory based on minimal information from the environment, intentions and constraints may have no or marginal effects. It has been argued that support for this proposition is that past behavior better predicts subsequent behavior than intention (e.g., Oulette and Wood, 1998). Ajzen (2002) rightly argues that it is also necessary to show that some independent measure of habit (e.g., the response-frequency measure) accounts for the variance that intention does not accounted for. Gärling *et al.* (2001) showed that the response-frequency measure of car-use habit actually did this.

According to TRA and TPB (Fishbein and Ajzen, 1975; Ajzen, 1985, 1991), intention is influenced by attitude, subjective norm, and perceived control. Therefore, if habit reduces the role of intention, it is implied that the roles of the determinants of intention are also reduced. As exemplified above, the condition necessary for this to occur would be that there are situational or environmental changes (shorter distances) that are not perceived after that a choice (to drive the car) had been made deliberately. As noted, this may be accounted for by the fact that the effect of feedback on choice is weak.

Releasing Cognitive Resources

Since habits are implemented automatically, people are alleviated of cognitive load (Baumeister *et al.*, 1998), thus freeing their minds for other tasks. Deliberating a travel mode choice every time would impose unwanted load on cognitive processing in everyday life (Garvill *et al.*, 2003). Thus, people who have developed a car-use habit can allocate cognitive resources to other daily tasks.

Some travel choices are complex entailing choices of activity or errand and duration, destination, departure time, mode, and route (Gärling *et al.*, 1997). Retrieval of scripts enables people to overcome their cognitive limitations in making such complex choices (Gärling, 2004). However, they may be unable to evaluate whether the choices they make attain their goals, for instance, to minimize travel distance.

Minimization of Information Acquisition

A consequence of automatic choice owing to the acquisition of a car-use habit is that information acquisition is reduced. This also implies alleviation of cognitive load. Therefore, a car-use habit may be adaptive as long as the circumstances do not change in important respects. However, it may not lead to the goals people have under other, changed circumstances. Thus, even though car-use habits (like habits in general) help people to adapt to complex conditions, sometimes they make them behave irrationally (i.e., not attaining their goals).

Increased Difficulty of Changing Travel Behavior

Even if large budgets are allocated to constructing or improving infrastructure for public transport to improve alternative, more sustainable travel modes, such improvements may not be noted by car users with a strong car-use habit. In other words, the transport policy may affect only those who have not yet developed a car-use habit since they are the only ones likely to notice the change. Therefore, in order to change the behavior of those with a strong car-use habit, infrastructure investments would not alone be sufficient. As will be discussed further in the next section, methods to modify travel behavior need to be designed and implemented in ways that take this into account.

Summary

To summarize, if people develop a car-use habit (1) they would choose the car without deliberation or automatically; (2) intentions and other psychological determinants such as attitude, subjective norm, and perceived control would guide their behavior to a lesser extent; and (3) they would not acquire information about alternative travel modes. Once a car-use habit has been developed, it seems to follow that it must be broken in order to achieve a change from car use to alternative, more sustainable travel modes.

METHODS OF BREAKING CAR-USE HABITS

Common wisdom tells that methods of changing behavior entail "money," "power," or "words." "Money" refers to economic factors including money as well as goods and services that can be traded in markets. "Power" refers to physical power (barriers) as well as political power (regulation). "Words" refer to various types of communication. Methods of all these types have been used to change travel behavior (see Chapter 15 by Loukopoulos in this volume). The question we raise in this section is whether any or all of them can be used to break a car-use habit?

Changes in Payoffs and Regulation

There are two types of methods to change payoffs for breaking a car-use habit, pull measures, and push measures (Vlek and Michon, 1992; Steg, 2003). Pull measures increase benefits from using other travel modes than car. Examples include increases of service level of public transport (such as reduction of travel time, increases in the number of seats, or improved air-conditioning), rebates on fares for public transport, or construction of new bicycle and pedestrian roads. As already noted, such measures alone are not likely to be effective in breaking car-use habits.

Push measures imply decreasing benefits associated with car use. Examples are road pricing and reduction of road capacity. Push measures also include coercive regulations such as prohibition of car use. These measures are likely to be effective in changing car use. Are they also effective in breaking a car-use habit? In several studies, we found that temporary changes in payoffs and regulations that forced car users to use alternative travel modes induced lasting changes in car use (Fujii *et al.*, 2001; Fujii and Gärling, 2003, 2005; Fujii and Kitamura, 2003). The impacts of temporary changes were particularly strong for habitual car users who had little or no previous experience of using other travel modes. Thus, temporarily changing payoffs and regulations appear to break car-use habits through affecting habitual car users' perceptions of and attitudes toward the alternative travel modes.

Even though changes in payoffs are able to break car-use habits, there are factors that impede their implementation (low public acceptance of push measures, see Chapter 17 by Gärling and Loukopoulos in this volume, the monetary costs for pull measures, and as argued above, that habitual drivers are unaware of pull measures) as well as yielding unintended side effects such as eroding intrinsic motivation (Frey, 1993; Deci *et al.*, 1999).

Communication Measures

There are two types of communication measures that potentially can break a car-use habit: individualized communication and mass communication. Individualized communication includes personal conversation, workshops, education, and travel feedback programs (TFP) (see Fujii and Taniguchi, 2005) such as individualized marketing (Brög *et al.*, 2003), travel blending (Rose and Ampt, 2001), or personal journey plans (Jones, 2003) (see also Chapter 20 by Thøgersen in this volume). Unlike mass communication, individualized communication provides car users with information or messages that are customized based on their current travel behavior or their attitudes toward their travel behavior. Individualized communication is more likely to be accepted by habitual car drivers than mass communication. Car users cannot as easily ignore individualized communication as they can ignore mass communication if it is well designed to make them feel it is socially desirable to respond and easy to do. This would certainly apply to habitual as much as to nonhabitual car users. For example, a telephone call from the city government with a request to answer a few simple questions, made at the beginning of a program of individualized communication, would more likely to be attended to than comparable messages on TV or in newspapers. Thus, individualized communication has the potential of being more effective in breaking a car-use habit than mass communication.

The main reason why communication would be effective in breaking car-use habit is that it directly affects psychological factors that are determinants of the *behavioral change*. This is illustrated in Figure 2 that integrates TPB (Ajzen, 1985, 1991), norm activation theory (Schwartz, 1977), the theory of implementation intention (Gollwitzer, 1993, 1996; Gollwitzer and Brandstätter, 1997), and the process model of acquisition of script-based choice (Gärling *et al.*, 2001). As can be seen, intention ("I will change to use bus instead of car") is influenced by attitude toward change ("I like to change to use bus instead of car"), perceived control ("I think it is not difficult to change to use bus instead of car"), subjective norm ("I think others expect that I change to use bus instead of

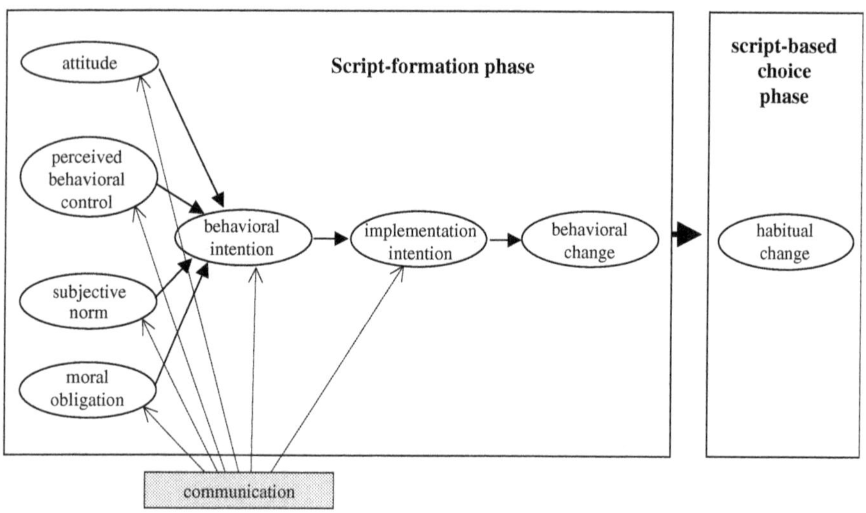

Figure 2: A Process Model of the Effects of Individualized Communication on Change of Habit

car"), and moral obligation ("I feel a moral obligation to change to use bus instead of car"). These determinants of intention to change would be more strongly changed through *individualized* communication. Attitudes toward behavioral change may become more positive by messages informing about the positive aspects of the change of habit (such as reducing health hazards and the prevalence of accidents). Perceived control may be enhanced by information about how to use alternative travel modes. Subjective norm may be enhanced by providing feedback about others' attitudes toward behavioral change. Moral obligation may be indirectly activated by messages informing about the negative consequences of car use (cf. Taniguchi and Fujii, 2004).

An intention may also need to be directly strengthened by using persuasive messages in order to make habitual car users thoroughly deliberate their travel mode choice. Otherwise, the car-use habit may not be broken. In a similar vein, Verplanken and Faess (1999) showed that even though people form intentions, this does not prevent them from reverting to old habits. Another way of accomplishing the same thing would be to make habitual car users form an intention to implement choices of alternative travel modes. Implementation intentions are formed through planning how to implement behavior (Gollwitzer, 1993, 1996; Gollwitzer and Brandstätter, 1997). Individualized advices about this may therefore be designed. Taniguchi and Fujii (2004) observed positive such effects on changes from habitual car use to other travel modes.

The cost per driver or household of individualized communication is in general higher than the cost of mass communication. Still, the cost for individualized communication is likely to be much less than that for infrastructure investment such as railway or bicycle road construction (Brög *et al.*, 2003). Furthermore, unlike push measures including changing payoffs and regulations, policymakers would not meet with public opposition when they implement communication measures.

Combination of Measures

A TFP (cf. Fujii and Taniguchi, 2005) includes many types of information and messages in a single program. The tenet is that it is necessary to combine different pull and push measures. As an example, a TFP recently implemented to promote public transport three months after building a public transport system resulted in the number of passengers increasing by about 50% (Taniguchi and Fujii, 2007). Yet, a TFP only uses communication type measures. It would be even more effective to include pull measures such as infrastructure investment in more sustainable travel modes. Such measures are sometimes necessary because the availability of attractive alternatives is essential. These measures may also be effective if habitual car users through successful communication learn about alternative travel modes. In a similar vein, push measures such as regulation or road pricing may not be sufficient to break car-use habits because these induce "rat running" (Emmerink *et al.*, 1995), that is, that drivers continue to use the car during periods or in areas without any regulations or pricing. In many cases, such "rat running" is less attractive than choosing some alternative travel mode. If communication informs habitual car users of how to use the alternative travel mode, they are likely to abandon their car-use habit.

Although a single experience of forced use of other travel modes is an important trigger, it may still not be sufficient for breaking car-use habits. As noted above, a single experience of using an alternative travel mode may lead to its repeated use by frequent car drivers. A precondition is likely to be that the outcome is positive (Gärling *et al.*, 2001; see Figure 1) and that this fact is clearly communicated to habitual car users. If people use alternative travel modes (such as bus or train) repeatedly, they may thus acquire a new habit (a bus- or train-use habit). The development of such a new habit would further weaken the incompatible old car-use habit.

SUMMARY AND CONCLUSIONS

In this chapter, we first discussed the nature and measurement of car-use habits. We equated car-use habit with script-based choice. We assumed that habitual strength is related to the number of car-use scripts stored in memory that are retrieved in different situations as well as to the ease with which each is retrieved. The latter in turn depends on the absence of scripts of using alternative travel modes. Following Gärling *et al.* (2001), we analyzed the process of acquiring a car-use habit. We then described several impacts of car-use habit on travel behavior. People with a car-use habit are likely to choose car use automatically (without much deliberation). This will reduce the role of intention and therefore also the influences of determinants of intention such as attitude, subjective norm, and perceived control. A car-use habit will also minimize acquisition of information about alternative travel modes. At the same time, cognitive resources are freed for other tasks. On balance, the difficulty to change car use is likely to increase with the strength of a car-use habit. Finally, we discussed three types of measures targeting changes in car use, that is, changes in payoffs, regulations, and communication measures. As we noted, all these measures entail elements that may be effective in breaking car-use habits. But changes in payoffs that reduce benefits from using the car and regulations of car use are not likely to be accepted by the public, and changes in payoffs that increase payoffs of using alternative travel modes may be too costly. In addition, changes in payoffs and regulations may erode intrinsic motivation to use other travel modes than car such as public transport, bicycling, and walking.

Still, we may be able to avoid all these implementation difficulties by using communication measures, particularly *individualized* communication such as TFP that are much more likely than other communication measures to reach habitual car users. It is argued that these measures should target psychological factors such as attitudes, moral obligations, intentions, and implementation intentions. Furthermore, the effectiveness of communication measures in breaking car-use habits would be enhanced by combining them with the other types of measures, that is, changes in payoff and regulation.

If the goal is a sustainable society in the future, we cannot simply continue to ignore the serious negative consequences of private car use. A thorough scientific understanding of car-use habits is essential for designing policy measures to change car use. All available such measures would be made more effective by taking into account the frequent prevalence of car-use habit.

REFERENCES

Aarts, H. and A. Dijksterhuis (2000a). The Automatic Activation of Goal-Directed Behavior: The Case of Travel Habit, *Journal of Environmental Psychology*, 20, 75–82.

Aarts, H. and A. Dijksterhuis (2000b). Habit as Knowledge Structure: Automaticity in Goal-Directed Behavior, *Journal of Personality and Social Psychology*, 78, 53–63.

Aarts, H., B. Verplanken, and K. van Knippenberg (1998). Predicting Behavior from Actions in the Past: Repeated Decision Making or a Matter of Habit? *Journal of Applied Social Psychology*, 28, 1355–1374.

Ajzen, I. (1985). From Intentions to Actions: A Theory of Planned Behavior. In: *Action Control: From Cognition to Behavior* (J. Kuhl and J. Beckmann, eds), Springer, Heidelberg, 11–39.

Ajzen, I. (1991). The Theory of Planned Behavior, *Organizational Behavior and Human Decision Processes*, 50, 179–211.

Ajzen, I. (2002). Residual Effects of Past on Later Behavior: Habituation and Reasoned Action Perspectives, *Personality and Social Psychology Review*, 6, 107–122.

Bargh, J.A. (1997). The Automaticity of Everyday Life. In: *The Automaticity of Everyday Life: Advances in Social Cognition* (R.S. Wyer Jr., ed.), Vol. 10, Erlbaum, Mahwah, NJ, 1–61.

Baumeister, R.F., E. Bratslavsky, M. Muraven, and D.M. Tice (1998). Ego Depletion: Is the Active Self a Limited Resource? *Journal of Personality and Social Psychology*, 74, 1252–1265.

Ben-Akiva, M. and S.R. Lerman (1985). *Discrete Choice Analysis*, MIT Press, Cambridge, MA.

Brehmer, B. (1995). Feedback Delays in Complex Dynamic Decision Making. In: *Complex Decision Making: The European Perspective* (P.A. Frensch and J. Funke, eds), Erlbaum, Hillsdale, NJ, 103–130.

Brög, W., E. Erl, and N. Mense (2003). Individualised Marketing Changing Travel Behaviour for a Better Environment, Paper presented at the TRIP Research

Conference: The Economic and Environmental Consequences of Regulating Traffic, Denmark, Hillerød, February.

Deci, E.L., R. Koestner, and R.M. Ryan (1999). A Meta-Analytic Review of Experiments Examining the Effects of Extrinsic Rewards on Intrinsic Motivation, *Psychological Bulletin*, **125**, 627–668.

Einhorn, H.J. and R.M. Hogarth (1978). Confidence in Judgment: Persistence of the Illusion of Validity, *Psychological Review*, **85**, 396–416.

Emmerink, R.H.M., P. Nijkamp, and P. Rietveld (1995). Is Congestion Pricing a First-Best Strategy in Transport Policy? A Critical Review of Arguments, *Environment and Planning B*, **22**, 581–602.

Fishbein, M. and I. Ajzen (1975). *Belief, Attitude, Intention, and Behavior: An Introduction to Theory and Research*, Addison-Wesley, Reading, MA.

Frey, B.S. (1993). Motivation as a Limit to Pricing, *Journal of Economic Psychology*, **14**, 635–664.

Fujii, S. and T. Gärling (2003). Development of Script-Based Travel Mode Choice after Forced Change, *Transportation Research F*, **6**, 117–124.

Fujii, S. and T. Gärling (2005). Temporary Structural Change: A Strategy to Break Car-Use Habit and Promote Public Transport. In: *Traffic and Transport Psychology* (G. Underwood, ed.), Amsterdam, Elsevier, 585–592.

Fujii, S., T. Gärling, and R. Kitamura (2001). Changes in Drivers' Perceptions and Use of Public Transport During a Freeway Closure: Effects of Temporary Structural Change on Cooperation in a Real-Life Social Dilemma, *Environment and Behavior*, **33**, 796–808.

Fujii, S. and R. Kitamura (2003). What Does a One-Month Free Bus Ticket Do to Habitual Drivers? An Experimental Analysis of Habit and Attitude Change, *Transportation*, **30**, 81–95.

Fujii, S. and A. Taniguchi (2005). Reducing Family Car Use by Providing Travel Advice or Requesting Behavioral Plans: An Experimental Analysis of Travel Feedback Programs, *Transportation Research D*, **10**, 385–393.

Gärling, T. (2004). The Feasible Infeasibility of Activity Scheduling. In: *Human Behaviour in Traffic Networks* (M. Schreckenberg and R. Selten, eds), Springer, Berlin, 231–250.

Gärling, T., S. Fujii, and O. Boe (2001). Empirical Tests of a Model of Determinants of Script-Based Driving Choice, *Transportation Research F*, **4**, 89–102.

Gärling, T., R. Gillholm, and A. Gärling (1998). Reintroducing Attitude Theory in Travel Behavior Research: The Validity of an Interactive Interview Procedure to Predict Car Use, *Transportation*, **25**, 129–146.

Gärling, T., R. Gillholm, J. Romanus, and M. Selart (1997). Interdependent Activity and Travel Choices: Behavioral Principles of Integration of Choice Outcomes. In: *Activity-Based Approaches to Travel Analysis* (D. Ettema and H.P.J. Timmermans, eds), Pergamon, Oxford, 135–150.

Garvill, J., A. Marell, and A. Nordlund (2003). Effects of Awareness on Choice of Travel Mode, *Transportation*, **30**, 63–79.

Gollwitzer, P.M. (1993). Goal Achievement: The Role of Intentions, *European Review of Social Psychology*, **4**, 141–185.

Gollwitzer, P.M. (1996). The Volitional Benefits of Planning. In: *The Psychology of Action: Linking Cognition and Motivation to Behavior* (P.M. Gollwitzer and J.A. Bargh, eds), Guilford Press, New York, 287–312.

Gollwitzer, P.M. and V. Brandstätter (1997). Implementation Intentions and Effective Goal Pursuit, *Journal of Personality and Social Psychology*, **73**, 186–199.

Golob, T.F., R. Kitamura, and L. Long (1997). *Panels for Transportation Planning, Methods and Applications*, Kluwer, Boston, 349–374.

Jones, P. (2003). Encouraging Behavioural Change through Marketing and Management: What Can Be Achieved? Paper presented at 10th International Conference on Travel Behaviour Research, Switzerland, Lucerne, August.

Kitamura, R. (1990). Panel Analysis in Transportation Planning: An Overview, *Transportation Research A*, **24**, 401–415.

Klayman, J. and Y.-W. Ha (1987). Confirmation, Disconfirmation, and Information in Hypothesis Testing, *Psychological Review*, **94**, 211–228.

Oulette, J.A. and W. Wood (1998). Habit and Intention in Everyday Life: The Multiple Processes by Which Past Behavior Predicts Future Behavior, *Psychological Bulletin*, **124**, 54–74.

Payne, J.W., J.R. Bettman, and J. Johnson (1993). *The Adaptive Decision Maker*, Cambridge University Press, New York.

Ronis, D.L., J.F. Yates, and J.P. Kirscht (1989). Attitudes, Decisions, And Habits as Determinants of Repeated Behavior. In: *Attitude Structure and Function* (A.R. Pratkanis, S.J. Breckler, and A.G. Greenwald, eds), Erlbaum, Hillsdale, NJ, 213–239.

Rose, G. and E. Ampt, E. (2001). Travel Blending: An Australian Travel Awareness Initiative, *Transportation Research D*, **6**, 95–110.

Schank, R.C. and R.P. Abelson (1977). *Scripts, Goals, Plans, and Understanding*, Erlbaum, Hillsdale, NJ.

Schwartz, S.H. (1977). Normative Influences on Altruism. In: *Advances in Experimental Psychology* (L. Berkowitz, ed.), Vol. 10, Academic Press, New York, 222–280.

Sheppard, B.H., J. Hartwick, and P.R. Warshaw (1988). The Theory of Reasoned Action: A Meta-Analysis of Past Research with Recommendations for Modifications and Future Research, *Journal of Consumer Research*, **15**, 325–343.

Steg, L. (2003). Factors Influencing the Acceptability and Effectiveness of Transport Pricing. In: *Acceptability of Transport Pricing Strategies* (J. Schade and B. Schlag, eds), Elsevier, Oxford, 187–202.

Svenson, O. (1990). Some Propositions for the Classification of Decision Situations. In: *Contemporary Issues in Decision Making* (K. Borcherding, O.I. Larichev, and D.M. Messick, eds), Elsevier, Amsterdam, 17–31.

Taniguchi, A. and S. Fujii (2004). A process Model of Voluntary Travel Behavior Modification and Effects of the Travel Feedback Program (TFP). In: *Proceedings of the 3rd International Conference on Traffic & Transport Psychology*, Nottingham, UK, September.

Taniguchi, A. and S. Fujii (2007). Promoting Public Transport Using Marketing Techniques in Mobility Management and Verifying Quantitative Effects: A Case Study of a Community Bus in Obihiro City, Japan, *Transportation*, **34** (1), 37–49.

Triandis, H. (1977). *Interpersonal Behavior*, Brooks/Cole, Monterrey, CA.

Verplanken, B. and H. Aarts (1999). Habit, Attitude and Planned Behavior: Is Habit an Empty Construct or an Interesting Case of Goal-Directed Automatic? *European Review of Social Psychology*, **10**, 101–134.

Verplanken, B., H. Aarts, and A. Van Knippenberg (1997). Habit, Information Acquisition, and the Process of Making Travel Mode Choices, *European Journal of Social Psychology*, **27**, 539–560.

Verplanken, B., H. Aarts, A. Van Knippenberg, and A. Moonen (1998). Habit, Information Acquisition, and the Process of Making Travel Mode Choices, *British Journal of Social Psychology*, **37**, 111–128.

Verplanken, B., H. Aarts, A. Van Knippenberg, and C. Van Knippenberg (1994). Attitude Versus General Habit: Antecedents of Travel Mode Choice, *Journal of Applied Social Psychology*, **24**, 285–300.

Verplanken, B. and S. Faess (1999). Good Intentions, Bad Habits, and Effects of Forming Implementation Intentions on Healty Eating, *European Journal of Social Psychology*, **29**, 591–604.

Verplanken, B. and S. Orbell (2003). Reflections on Past Behavior: A Self-Report Index of Habit Strength, *Journal of Applied Social Psychology*, **33**, 1313–1330.

Vlek, C. and J. Michon (1992). Why We Should and How We Could Decrease the Use of Motor Vehicles in the Future, *IATSS Resarch*, **15**, 82–93.

Threats from Car Traffic to the Quality of Urban Life
Tommy Gärling and Linda Steg (Editors)

14

Ecological Norm Orientation and Private Car Use

Ellen Matthies, Workgroup of Cognition and Environmental Psychology, Ruhr-University Bochum, Bochum, Germany
Anke Blöbaum, Workgroup of Cognition and Environmental Psychology, Ruhr-University Bochum, Bochum, Germany

ABSTRACT

Norms to cut back on driving in order to protect the local or global environment are particularly distinctive in some Western European countries. But the existence of strong ecological norms does not inevitably lead to the appropriate behaviour. On the basis of an integrative model, the chapter gives an overview of studies that have examined the influence of personal ecological norms on travel mode choice. It is shown under what conditions ecological norms may have an influence on travel mode choice. Suggestions are also made of norm-focused intervention measures.

INTRODUCTION

In Western industrial countries, the protection of the environment and the conservation of natural resources have been a topic of political and social interest for more than 30 years. Undoubtedly, environmental protection has not only developed to an important target of politics, but environmental protection has also become a social norm in everyday life. For example, in 1996 more than 50% of the Germans fully agreed with the proposition that all citizens should be ready to change their current lifestyle to protect the environment (Preisendörfer, 1999). This shows that environmental protective behaviour has become a widespread injunctive norm, or moral standard (for empirical evidence, see also Kaiser and Shimoda, 1999; Biebeler, 2000).

From the perspective of environmental protection, private car use has a particular significance, because traffic contributes strongly to the emission of CO_2, which is an important greenhouse gas and one of the main driving forces of climate change (see Chapter 2 by

Van Wee in this volume). Accordingly, in 2000 the OECD (Austrian Federal Ministry for Agriculture, Forestry, Environment and Water Management & OECD, 2000) developed a set of the so called "hard" as well as "soft" policies to achieve a reduction of private car use. Worldwide national initiatives are trying to reduce traffic through the implementation of soft measures and thereby refer to values such as environment and health (as examples, see the British programme "SustainableTravel" or the Australian programme "TravelSmart"). Especially in the case of soft measures, the initiatives implicitly rely on environmental concern. This evokes the question of how distinct ecological norms are in general and what role they play for private car use.

Since the mid-1990s, in many countries when the commitment to environmental protection had its peak, the engagement has decreased. This also applies with respect to car traffic and the readiness for the reduction of private car use. In 1996, 39% of all Germans stated that they would try to reduce their car use whenever possible to protect the environment, only 29% denied this statement. In 1998, the relations shifted so that only 26% confirmed a distinct willingness, whereas 43% refused to reduce their car use (Preisendörfer, 1999). In a survey conducted in the Netherlands, Steg and Vlek (1997) found that approximately 30% of the participants were willing to reduce their car use. Interestingly, the proportion of those who are "outraged at the unnecessary car use of others" seem to be higher than the proportion of those supporting a personal cut back (39% in Germany in 1998). This leads to the assumption that the social norm for reducing car use is more developed than the personal willingness, perhaps because the own scope of action is perceived to be limited (see also Steg and Vlek, 1997). When we refer to ecological norm orientation in the domain of car use, we thus assume that depending on country and region approximately 30% of the population hold corresponding norms to reduce their car use. However, from the fact that ecological norms are (still) widespread in parts of the population, it cannot be automatically inferred that these norms are actually being translated into corresponding behaviour. This chapter deals with exactly this topic. First, we address the question of how far ecological motives generally exert an influence on everyday environmentally friendly behaviour. We will then introduce theoretical models that explain the effect of an ecological norm orientation on behaviour, and in the following we suggest an integrated model that focuses on personal norms and their relevance for car use. An overview of relevant studies will also be given. On the basis of the presented model and findings we will then discuss how norm-focused strategies can support the reduction of car use. In concluding, we outline directions of future research.

IMPACT OF ENVIRONMENTAL CONCERN AND ECOLOGICAL NORMS

Environmental Concern

Since more than 30 years, psychologists have been dealing with the question whether environmental concern plays an important role in changing or reducing environmentally damaging behaviours. A first step in work on this topic was the development of measurements of environmental concern. Two traditions have evolved within the last 30 years. One is rooted in the early work of Maloney and Ward (1973). They developed the first multi-dimensional scale to measure environmental concern, the "ecological attitude scale." According to Maloney and Ward (1973), environmental concern was conceived of as an attitude comprising four components: affect (emotional reactions to environmental problems), verbal commitment (willingness to engage in protective actions), actual commitment (self-reported protective behaviours), and knowledge about environmental problems. This research tradition has been carried on by various psychologists, for instance, Weigel and Weigel (1978), and Schahn and Holzer (1990). The other research tradition states that environmental concern is a general value orientation or worldview. The founders of this perspective are Dunlap and Van Liere (1978, 1984). They proposed that environmental concern represents a new way of thinking about the relationship of nature and humans referred to as the "New Environmental Paradigm" (NEP). Their scale seems to be internationally the most widespread instrument to measure environmental concern (the original version from 1978 was reversed by Dunlap *et al.*, 2000, as the "New Ecological Paradigm Scale"). These two approaches towards environmental concern are linked to different assumptions about how environmental concern affects behaviour. In the first tradition, it is assumed that readiness to act in an environmentally protective way is already a component of environmental concern (Schahn and Holzer, 1990). In contrast, the second tradition suggests that ecological values do not directly influence environmentally significant everyday behaviours, and assumes the existence of variables that mediate between general values and specific behaviours.

The value belief norm (VBN) theory (Stern, 2000), a widely accepted model of the relationship between values and ecological behaviour, gives an example of a current perspective on the relation between general values, ecological norms, and environmentally protective behaviours. In this model, personal ecological norms directly influence pro-environmental behaviours (here called "environmentally significant behaviours"). However, personal norms are not directly affected by pro-environmental values, but beliefs on environmental problems and on the relevance of one's own behaviour mediate between values and personal norms.

In its assumptions about the mediating processes between a general environmental concern (an ecological worldview) and environmental protective behaviours, the VBN theory draws on a norm activation model developed by Schwartz (1970, 1977). This model was originally designed to explain pro-social behaviour, for instance, blood donation, helpfulness in everyday interactions, or helping in emergency situations. The central construct in Schwartz's theory is the feeling of personal moral obligation to take specific actions (Schwartz, 1977, p. 227), a feeling that is generated by an activation of the individual's norms.

Since the 1970s, many studies of environmentally relevant behaviours have applied parts of this model and have tried to explain the transition of environmental concern into specific behaviours. Investigated behavioural domains include littering (Heberlein, 1972), yard burning (Van Liere and Dunlap, 1978), energy conservation (Black *et al.*, 1985), recycling (Guagnano *et al.*, 1995; Hopper and Nielsen, 1991), nature protection (Blamey, 1998), and since recently also travel mode choice (Hunecke *et al.*, 2001; Nordlund and Garvill, 2003).

In the early studies, the constructs awareness of consequences (knowledge about the adverse environmental consequences of the specific behaviour) and ascription of responsibility (belief that one's own behaviour is relevant for the problem or the solution of the problem) turned out to be important factors linking ecological values to concrete actions, and are now seen as important activators or antecedents of personal norms (Stern, 2000; Hunecke *et al.*, 2001; Nordlund and Garvill, 2003).

Besides norm activating variables, Schwartz suggested that several other variables and processes influence altruistic behaviours, many of which have not been considered in the cited studies. For example, Schwartz (1977) proposed a possible neutralization of the feelings of obligation by certain defence mechanisms (reassessment and redefinition processes). These recursive processes have only rarely been included in studies of pro-environmental behaviours (but see Schahn and Bertsch, 2003) and have not been integrated in the VBN theory. In a subsequent section, we will focus on the potential of this perspective in more detail. In later publications (see Schwartz and Howard, 1981), Schwartz widened his view on altruistic behaviours and explicitly stated that besides a moral motivation (i.e., besides the activated personal norm) also other motives may influence the moral decision. He particularly mentioned social motives, or situational social norms. Because in this contribution norms are of central interest, the following section deals with the differentiation of personal and social norms in detail.

Personal and Social Norms

Schwartz defined personal norms as the feeling of personal moral obligation based on the individual's personal values (Schwartz, 1977, p. 227). Accordingly, in studies on pro-environmental behaviour personal norms are measured with items like "I feel a strong personal obligation to . . ." (Harland *et al.*, 1999) or "According to my personal value system I feel obliged to . . ." (Matthies *et al.*, 2006). Because Schwartz assumed that not acting in accordance with one's personal norms leads to feelings of guilt, we also find items like "I have a bad conscience, when I think of . . ." (Kaiser and Shimoda, 1999; Hunecke *et al.*, 2001).

While the concept of a personal norm is tightly connected with norm activation theory, the concept of a *social norm* affecting behaviour can be found in various action models. For example, in the theory of reasoned action (TRA) or in the theory of planned behavior (TPB), social influences are considered in the form of a subjective norm. Subjective norms are expectations of significant others that influence the behavioural intention in parallel to behavioural attitudes (Ajzen and Fishbein, 1980). There are numerous studies that applied the TRA or the TPB to explain pro-environmental behaviours (Goldenhar, 1991; Bamberg, 1996), and subjective norms were found to have a relevant influence on behaviours like signing a pro-environmental petition (Taylor and Todd, 1997), participating in recycling (Goldenhar, 1991; Lüdemann, 1997) or travel mode choice (Bamberg, 1996).

To examine how personal and social norms influence pro-environmental behaviour, we have to look at studies that integrate both kinds. On the basis of norm activation theory, we find two perspectives on the interplay between social and personal norms. On the one hand, social norms are seen as antecedents of personal norms (Hopper and Nielsen, 1991; Bratt, 1999). On the other, subjective norms are considered to have an influence additionally to personal norms (Klöckner *et al.*, 2003; Matthies, 2003; Blöbaum *et al.*, 2007). This perspective is corroborated by some studies that integrate variables of the TPB and norm activation theory, finding personal norms to have an additional influence in addition to the TPB constructs, or subjective norms to have an additional influence that constructs in the norm activation theory (Axelrod and Lehmann, 1993; Harland *et al.*, 1999; Hunecke *et al.*, 1999, 2001; Kaiser and Shimoda, 1999; Bamberg *et al.*, 2006; for a critical discussion, see Bamberg and Schmidt, 2003). However, subjective norms are not always in accordance with the promotion of ecological behaviour. Especially in the domain of travel mode choice, there seem to exist different social influences that are not ecologically orientated (e.g., high status of big and not very fuel-saving cars; see Steg *et al.*, 2001; see also Chapter 12 by Gatersleben in this volume).

REDUCTION OF INDIVIDUAL CAR USE

Before introducing an integrative model for the prediction of pro-environmental travel mode choice, we will have a closer look at the behaviours that actually represent a relevant contribution to the reduction of adverse effects of private car use. Looking at recommendations given by the OECD (Austrian Federal Ministry for Agriculture, Forestry, Environment and Water Management & OECD, 2000), or at national governmental programmes to reduce car use (e.g., the Australian "TravelSmart" programme, or the British programme "SustainableTravel"), there seems to be a variety of individual behaviours related to sustainable travel. All programmes suggest, walking, cycling, and using public transport instead of using the own car (see Figure 1). Another widespread suggestion is to reduce car trips and car ownership by participating in car pooling or car sharing. Some programmes suggest to drive in a fuel-efficient way (e.g., the Austrian initiative Eco-Driving Europe), or recommend the purchase of a fuel-efficient or electric car (Austrian Federal Ministry for Agriculture, Forestry, Environment and Water Management & OECD, 2000). Finally, all governmental programmes suggest measures of making (car) travel unnecessary, for instance, through teleworking.

PROGRAMME	TRAVELSMART (AU)	SUSTAINABLE TRAVEL (UK)	EST-ENVIRONMENTALLY SUSTAINABLE TRANSPORT (OECD)
SUGGESTED BEHAVIOURS	▪ Public Transport ▪ Walking ▪ Cycling ▪ Car pooling ▪ Telecommuting	▪ Public Transport: information & marketing ▪ Car clubs ▪ Car sharing ▪ Teleworking ▪ Teleconferencing ▪ Home shopping	▪ Public transport ▪ Public cars: integration of individual & public transport ▪ Greater use of non-motorized means of transport ▪ Greater use of telecommunication ▪ Buying other cars: - fuel efficient engines - electric engines - fuel cells
REFS.	TravelSmart (2003)	Cairns *et al.* (2004)	Austrian Federal Ministry for Agriculture, Forestry, Environment and Water Management & OECD (2000)

Figure 1: Behaviour Changes Suggested by Programmes to Promote Sustainable Travel

According to Kaiser and colleagues (Kaiser and Keller, 2001; Kaiser *et al.*, 2001), pro-environmental behaviours have varying difficulties that are reflected in the general popularity or frequency of these behaviours. There is evidence that also the various recommended behaviours associated with a general reduction of car use are extremely different in their popularity. Some behaviours, like driving in a fuel-efficient way, seem to be rather popular (less than 20% of respondents negated the item), others (e.g., car pooling) seem to be rather unpopular (see Table 1). The use of pro-environmental forms of transportation (the bike or public transport) instead of a car seems to have a medium level of popularity. According to Kaiser and Keller (2001), the frequency of the respective behaviours may vary from sample to sample due to situational constraints. The authors found significant differences in the frequency of using a bike/public transport instead of a car for rural and urban areas.

Within the last 20 years, many of the suggested behaviours that aim at the reduction of negative effects of car use have been investigated from a psychological perspective.

Table 1: Engagement in Different Private Car-Use Behaviours (According to Kaiser et al., 2001)

Behaviour	*Difficulty*[a]
I drive in such a way as to keep my fuel consumption as low as possible	83.9
I keep the engine running while waiting in front of a railway-crossing or in a traffic jam (−)[b]	83.2
I ride a bicycle or take public transportation to work or school	58.0
I drive my automobile in and into the city (−)	57.2
At red traffic lights, I keep the engine running (−)	56.8
I drive to where I want to start my hikes (−)	56.1
In nearby areas (within 30 km), I use public transportation or ride a bike	40.8
I own a fuel efficient automobile (less than 7 litres per 100 km)	23.1
I refrain from owning an automobile	23.0
I drive on freeways at speed under 100 kph	22.0
When I need an automobile, I rent one	21.7
I am a member of a car pool	4.3

[a]Percentage of people reporting the behaviour.
[b]Negative items are to be read "I refrain from . . .".

Already in the early 1980s, Jacobs *et al.* (1982) examined the measures through which people could be motivated to participate in car pooling. However, in their applied behaviour analysis approach, no assessment was made of internal variables like environmental concern or personal norms and their impact on behaviour.

Car sharing is a rather new topic of research that has met a strong interest from psychologists in the last years (Katzev *et al.*, 2000; Orski, 2002; Harms, 2003; Harms and Truffer, 2005), but studies have mainly concentrated on assessing the resulting travel behaviour. In a single study that thoroughly revealed motives (Harms and Truffer, 2005), it was found that environmental concern may play a role for some car-sharing users, but that the main reason to participate in car sharing seemed to be a change in personal life context (e.g., loss of car after a divorce).

Concerning the purchase of a less ecological damaging car, Klocke (2002) carried out a long-term study with adults and adolescents in Germany. He found that the intention to buy an eco-car was highly correlated with environmental concern. For adults (but not for the young car users owning their first car), he found that the former intentions to buy an eco-car was correlated with the ecological features of the car they actually own (e.g., petrol consumption).

The main body of psychological studies and theory development concerning behaviours that imply a reduction of private car use has focused on travel mode choice. Amongst these studies, we found several that investigated the impact of personal norms on travel behaviour. Other studies did not focus on specific behaviours, but tried to explain readiness to reduce car use at a more general level.

A MODIFIED NORM ACTIVATION THEORY AS AN INTEGRATIVE MODEL

There is no doubt that a moral motivation has an impact on various pro-environmental behaviours, including behaviours that aim at the reduction of car use (Nordlund and Garvill, 2003) or, more specific, travel mode choice (Hunecke *et al.*, 2001). To use this knowledge – for example for the development of "soft measures" to reduce car use – it is crucial to know more about the interplay of personal norms and other than moral factors that also influence car-use behaviour. To integrate empirical findings and to provide a basis for the development of measures, we suggest a modified norm activation theory as an integrative model (see Figure 2).

The model is based on the normative decision-making model developed by Schwartz and Howard (1981). While the early norm activation model (Schwartz, 1977) focused primarily on the conditions under which personal norms were activated, the former model has specified additional influencing factors (e.g., subjective norms or monetary costs of the behaviour) which may become relevant in the subsequent motivation and evaluation phase of the process of a moral decision. With these extensions the model is able to link up to empirical findings on the various factors influencing travel mode choice. To be able to consider blocking effects of developing habitual travel mode choice (for an overview, see Chapter 13 by Fujii and Gärling in this volume), we also include habit in the model. In the following, we will explain

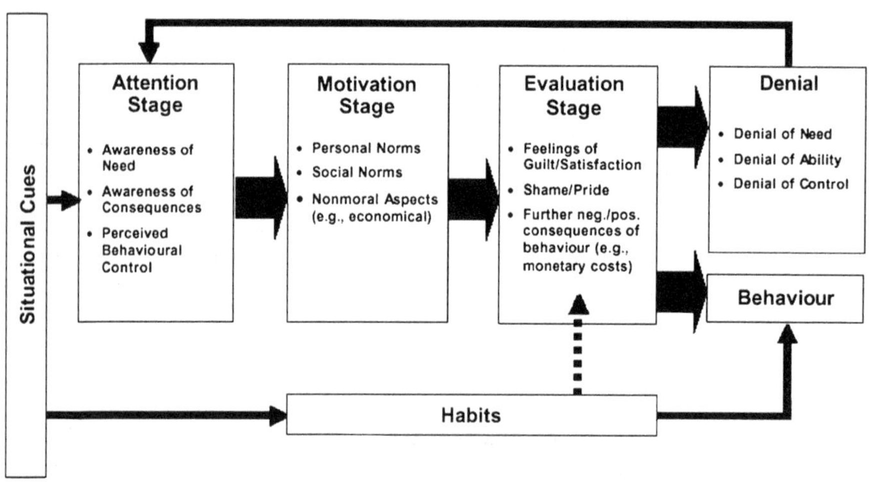

Figure 2: Modified Norm Activation Model as Suggested by Matthies (2003; Klöckner and Matthies, 2004)

the model assumptions in detail and will present empirical findings concerning the different stages of the model.

Attention Phase

In its assumption of norm activation, the theory of moral decision-making does not differ from early norm activation models. As prerequisites for the generation of a feeling of obligation to behave in accordance with one's personal norms, Schwartz and Howard (1981) mention *interpretations of the situation*: in an attention stage, the individual has to become aware that a valued object is threatened and in need of help (awareness of need). Moreover, he or she has to recognise some further aspects of the person–situation–interaction, for instance, that the individual behaviour contributes to the problem (awareness of consequences), and that the individual is able to contribute to change (perceived behavioural control). It has been proved in many empirical studies that these variables are relevant for the activation of norms with respect to environmental behaviour (Black *et al.*, 1985; Hopper and Nielsen, 1991; Guagnano *et al.*, 1995) as has already been mentioned above. Current investigations confirm the assumptions about norm activation and also for the reduction of car use and travel mode choice.

An early study was carried out by Steg and Vlek (1997). They interviewed 539 car users and recorded some of the above-mentioned variables of the interpretation of the situation (awareness of several problems following from car use, and feeling of responsibility to contribute to the solution of the problems) in order to explain car use and the willingness to accept policy measures. Results showed that respondents with higher problem awareness actually used their cars less often, felt guiltier when unnecessarily using a car, and agreed more strongly to the proposition that the government should take active measures to reduce car use.

Garvill (1999) used a combination of Schwartz's (1977) norm activation model and a social dilemma approach (Van Vugt *et al.*, 1995) to investigate the determining factors of willingness to reduce car use and readiness to support regulations to reduce car use. He interviewed 1,562 car users. As potential fundamental factors relevant for willingness, he analysed environmental concern (with respect to egoistic, anthropocentric, and biocentric values), problem awareness, and general value orientation (collective versus individual); as specific factors he recorded various cognitive factors relevant for norm activation (e.g., perceived efficacy, awareness of consequences). Results confirmed Garvill's hypotheses about the mediating effect of the interpretation of the problem and assumptions on the individual's role in it. It was shown that the effect of values on the willingness to reduce car use is mediated by factors concerning the interpretation of the situation and the current feeling of moral obligation. In a subsequent study, Nordlund and Garvill (2003) combined the Schwartz model with the VBN theory (Stern, 2000). Findings confirmed again the hypothesis that the personal norm to reduce car use is central in the causal chain of effects from general and environmental values, general and specific problem awareness on willingness to reduce car use.

Hunecke *et al.* (2001) recorded the actual travel mode choice behaviour of 160 randomly chosen potential subway users in a German city. As determining factors for the personal norm to reduce car use, they identified problem awareness and awareness of the consequences of one's own behaviour.

The assumption that problem perception and ascription of responsibility act as mediators between ecological values and specific feelings of obligation thus seems to be sufficiently confirmed also for the domain of car use.

Motivation and Evaluation Phase

In their later, elaborated model, Schwartz and Howard (1981) specified additional factors influencing moral decisions. Along with the moral obligation also social (expectations of others) and other non-moral motivations (e.g., economical motivation) are seen as relevant motivations emerging in the motivation stage. In the subsequent evaluation stage, likely competing costs (e.g., loss of comfort, costs of violating peer group norms) and outcomes (satisfaction of acting in accordance with one's norms; positive aspects of using sustainable transport modes) are assessed. It is not before this balancing process that a decision is made for or against the morally motivated behaviour.

Several empirical studies on travel choice behaviour underline that besides personal norms many factors outside the person also influence the behaviour decision. In 1999, Harland *et al.* (1999) published a first study that investigated the relevance of personal norms in a model context that considered other than moral factors. Several studies followed that evaluated the specific impact of personal norms within general action models, particularly in the context of the theory of planned behaviour.

Harland *et al.* (1999) interviewed 305 Dutch citizens recording four environmental relevant behaviours (use of unbleached paper, use of energy-saving light bulbs, turn off faucet while brushing teeth, and use of other transport forms than car). They found that the introduction of personal norm as an additional predictor to TPB variables increase the amount of explained variance significantly for all four behaviours. However, the additional influence was smallest for travel mode choice.

In their study of travel mode choice, Hunecke *et al.* (2001) applied a modified norm activation model that integrated variables from the Schwartz model (Schwartz, 1977) as well as variables from the TPB (i.e., perceived behavioural control and subjective norm), and additional external factors (distance to subway station and cost of trip). They showed that the personal ecological norm, subjective norm, and external factors (free ticket) were all relevant predictors of the actual behaviour. In this study, the personal norm explained more variance than the subjective norm, and even more than the gift of a free ticket.

Bamberg and Schmidt (2003) reported contrasting results. They explained travel mode choice behaviour (regular trips to the university) with variables of three different action models, that is the TPB, the norm activation model, and the theory of interpersonal behaviour (Triandis, 1977). They found that Schwartz's model (which indeed had the smallest set of predictors) was least successful in explaining car use and that there was no additional impact of personal norms in either of the alternative models. However, they reported a subgroup specific effect of personal norm (see Bamberg, 1999). In the subgroup of students with high ecological concern, the personal norm had a significant additional negative effect on the intention to use the car.

Taken altogether, findings confirm the assumptions of our model that besides personal norms other extra-personal motives play a significant role in the reduction of car use, particularly social norms as well as perceived or actual behavioural costs.

Denial: Reassessment and Redefinition of the Situation

The reaction to a decision against the morally motivated behaviour is an aspect that has already been recognized in the context of norm activation theory but has hardly been empirically investigated. Schwartz and Howard (1981) postulated that once a norm has been activated, it also has to be neutralized through processes of reassessment and redefinition of the situation. Schahn and Bertsch (2003) showed in that such justification tendencies for un-ecological behaviour generally exist. In the context of travel mode choice, we know of only one study in which such processes have been discussed. In their study on willingness to reduce car use, Steg and Vlek (1997) held group discussions with 336 car users on the use of different means of transportation. It was revealed that the problem awareness of the participants decreased after the discussion groups and dropped under the level of problem awareness of a comparison group. The authors reported that after the group discussion, those respondents who originally had relatively high problem awareness reduced their awareness, and that respondents with relatively low problem awareness did hardly change their judgements. The authors explained the reduction in problem awareness in the group of individuals with an initially high concern through a process of dissonance reduction. Thus, they assumed that in the discussion it became salient for the participants that there was a gap between their problem awareness and their actual behaviour causing them to reduce this cognitive dissonance (Festinger, 1957) by adjusting their problem awareness to their behaviour. In fact, this is exactly what Schwartz and Howard (1981) assumed for a situation in which people are not able to behave in accordance with their norms: a redefinition of the situation in order to neutralize the obligation to act.

The Blocking Effects of Car-Use Habits

Taking up findings on the blocking aspects of routine behaviours (Verplanken *et al.*, 1997; for an overview see Chapter 13 by Fujii and Gärling in this volume), we investigated the role of habits in the process of norm activation. In the outlined action stages, we identified several points at which habits can theoretically intervene. In the beginning, habits (in the sense of associations between specific situations and behaviours) can prevent norm activation by directly blocking cognitive processes. Also, habits can indirectly influence the decision through behavioural costs (habitual behaviour is easier to perform than new behaviour). Finally, habits can moderate the relationship between activated motives and behaviour so that, for example, norms are only translated into behaviour when no habits stand in the way. For this assumption there already exist empirical proofs. Klöckner and colleagues (Klöckner *et al.*, 2003; Klöckner and Matthies, 2004) found a moderating effect of habits on the relation between personal norms and actual behaviour in two separate studies on travel mode choice. In both studies, people with strong car-use habits were less likely to put their personal norms into action (and choose public transport instead of the car) than people with weak car habits. Klöckner and colleagues concluded that habits cut short the whole norm activation process by blocking the norm activation in the beginning. The consideration of a moderating effect of habits also provides a possible explanation for the weak explanatory power of the Schwartz model in the study of Bamberg and Schmidt (2003), in which students were inquired about their regular and thus highly habitual travel to the university.

ECOLOGICAL NORM ORIENTATION AS A STARTING POINT TO CHANGE CAR-USE BEHAVIOUR

The suggested model as presented here does not only provide an integration of different empirical findings on travel mode choice and the role of norms, it also offers perspectives for theory-driven intervention planning (see Matthies, 2005). By taking into account personal and social norms, it proposes starting points for different types of norm-focused intervention strategies, for instance, for the block leader approach, for normative feedback, or for the strategy of commitment. At the same time, the notion of other motives can be taken up for the implementation of traditional and successful psychological intervention techniques like manipulation of the situation or reward (for a review on intervention strategies in the domain of pro-environmental behaviours, see Dwyer *et al.*, 1993). Owing to its process character, the model facilitates an adequate balancing of intervention strategies and an understanding of their interplay. Looking at the activation and motivation stage, we can derive from the model that information strategies are necessary for the activation of personal norms but that a moral motivation alone will be too weak to cause a behaviour change if other strong motives are given, for example when behavioural costs are too high. The model also yields a possible risk of using norm-focused intervention strategies. If norms

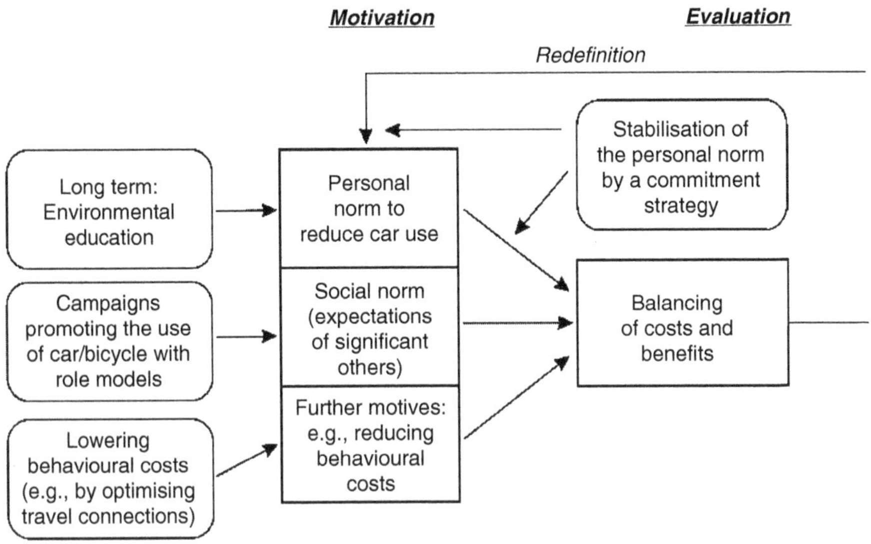

Figure 3: Measures Aiming at the Motivation and Evaluation Stage (Matthies, 2005)

are activated but cannot be translated into behaviour due to strong competing motives or blocking habits, this may lead to a denial of responsibility and redefinition processes. Therefore, the implementation of such strategies should be contemplated carefully.

Following the suggested model, there are various possibilities to promote ecological behaviours via personal norms by (1) building up ecological norms, (2) activating ecological norms, and (3) stabilising activated personal norms (see Figure 3).

Building Up Personal Ecological Norms

Building up personal norms is a long-term procedure. In the domain of pro-environmental behaviours examples are environmental education in school or social marketing of general ecological values (for an overview of social marketing strategies, see Chapter 20 by Thøgersen in this volume). According to the model, the existence of personal ecological norms is a prerequisite for the whole norm activation process, thus education and social marketing seem to be important measures, although they often are not sufficient (see Geller, 1989). As long as such educating measures are not linked to specific behavioural contexts, and thus do not activate norms, the risk of provoking counteractive redefinition processes seems to be rather low. However, when measures are being planned for the activation of specific norms, the basic conditions under which behaviour is taking place have to be taken into account.

Strategies to Activate and Stabilize Personal Norms

An obvious strategy to activate personal ecological norms to reduce car use is to present (new) information about adverse consequences of car use in general (e.g., climate change, air pollution, noise) and about possible alternatives (e.g., switch to public transport). This is what designers of soft measures in the domain of sustainable travel frequently do (see the British "Sustainable Travel" programme or the "Eco-Driving" programme). In Figure 4, it is shown how such strategies relate to the constructs of our model. However, also for these strategies it is true that they are necessary but may not be sufficient, for example, when strong competing motives are activated. To support the activated personal norm against competing motives and redefinition processes, commitment strategies can be applied.

Commitment strategies have been frequently used to improve conservation behavior (Burn and Oskamp, 1986; Katzev and Pardini, 1987/88; Wang and Katzev, 1990). Owing to their implied voluntary nature, commitment strategies address the personal norm. They emphasize the moral benefits (self-satisfaction as a result of acting in accordance with personal values), and thereby increase the costs of not-acting (which are

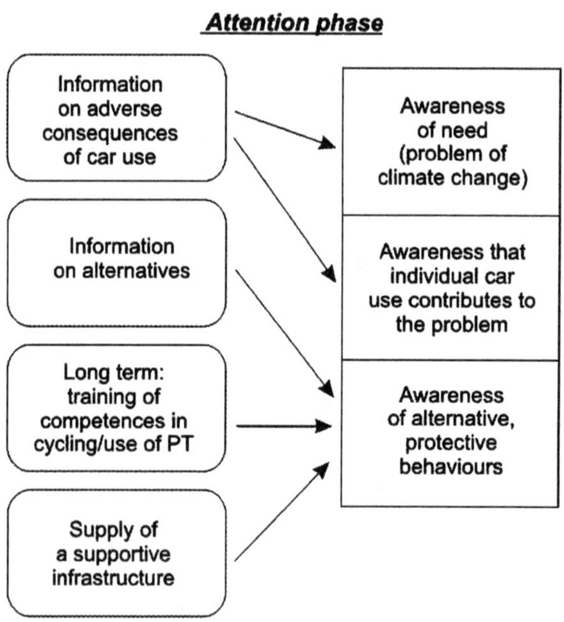

Figure 4: Measures Aiming at the Activation of Personal Norms to Reduce Car Use (Matthies, 2005)

feelings of guilt). Furthermore, it can be expected that a voluntary commitment prevents individuals from neutralizing their feeling of obligation. When signing a commitment, participants explicitly state that there is a problem, and that they are willing to contribute to a solution of the problem. Thus, a commitment should make it more difficult to deny responsibility or to redefine the situation.

Even though norm-focused strategies like commitment strategies or normative feedback have proved to be useful and effective in the long term (see Schultz, 1998), empirical studies of the effectiveness of such strategies in the field of transportation use are rare. In a study, we applied a combination of norm activating and supporting strategies (Matthies *et al.*, 2006). Car users were presented with information on the adverse consequences of their car use (norm activation) and additionally were asked to commit themselves to one or more suggested car reducing behaviours (e.g., drive in a way as to reduce petrol consumption, participate in car pooling, change to an ecological travel mode for a certain trip, etc.). However, this combination of strategies had no significant effect on the actual behaviour of habitual car users unless the commitment intervention was combined with a habit-breaking strategy.

Preconditions for the Implementation of Norm-Focused Strategies

The use of strategies that aim directly at the activation or stabilization of norms makes only sense when it can be assumed that there already exist strong personal norms in the target group that can be activated. This means that when planning to activate norms in order to reduce car use by addressing the problem of climate change, one has to be sure that people actually care about climate change. Against the background of decreasing problem awareness and ecological norm orientation this seems to be an important restriction in some contexts.

Even when personal norms are given and activated there is still the risk that personal norms are overruled by competing motives or blocked by counteracting habits. Therefore, the authors suggest to restrict the application of norm-focused techniques to target groups with (1) distinct pro-environmental norms, (2) weak car-use habits (e.g., people moving to a new city, young people who have not yet developed strong car-use habits), and (3) to situations where counteracting social norms or other motives do not exist. A further way to ensure that soft measures can be effective is to lower costs of behaviour or to break habits in advance of implementing soft measures (Matthies *et al.*, 2006).

To explore the above-mentioned preconditions for the implementation of norm-focused strategies, it seems sensible to make use of the knowledge of local experts or parts of the target group when designing norm-focused intervention measurements. Programme planners may prefer focus groups as a means of estimating the strength of existing values, norms, habits, and the other relevant variables in the target group. Also, participatory approaches can be useful, because an extensive participation of target groups and local actors can provide valuable hints of existing norms as well as on behavioural and other costs (see Matthies and Krömker, 2000; McKenzie-Mohr, 2000).

CONCLUSIONS AND FUTURE PROSPECTS

Even though the willingness to engage in environmentally friendly behaviour has decreased in the last years, ecological norm orientations are still well-developed in Western industrial countries and parts of the population are willing to contribute through their own behaviour to the problem solution. But the existence of strong personal norms to reduce private car use does not inevitably lead to the appropriate behaviour. In several empirical studies of travel mode choice (Hunecke *et al.*, 2001; Bamberg and Schmidt, 2003), several other factors besides personal norms have proved to be relevant (e.g., subjective norm, costs), compared to which personal norms are in some contexts only of minor importance (Bamberg and Schmidt, 2003). There also exist some preliminary findings about the conditions under which norms are being translated into behaviour. Especially habits appear to play a crucial role (see Klöckner *et al.*, 2003; Klöckner and Matthies, 2004).

The results of the few intervention studies that aim at activating or stabilizing ecological norms to reduce car use have been contradictory, indicating that there are limitations to the implementation of such strategies. The suggested integrative model (see Klöckner and Matthies, 2004; Matthies, 2005) can explain some of the contradictory results and provides several preconditions for the effective application of norm-focused intervention strategies. However, most of the assumptions derived from the model are not yet fully tested. Thus, there is a great demand not only for interventions, but also for theory-driven intervention studies in the domain of individual car use.

In studies of individual car use concentrating on the other car-related behaviours than the mode choice (e.g., car pooling, car sharing, or buying a new car), the role of norms has mostly been neglected. It seems however promising to consider norms as starting points for interventions in various contexts, for instance, in buying a new car (see findings of Klocke, 2002). Even though the attractiveness of ecological values and the readiness to behave according to them is currently decreasing in Western societies, this is still a valuable resource that should not thoughtlessly be given away.

ACKNOWLEDGEMENTS

Our own research reported in this chapter was funded by grant MA 1179/1-1 from the German Science Foundation (DFG) to Ellen Matthies. The authors thank Christian

Klöckner for helpful comments on an earlier draft of this paper and Regina Schulte for editing the language.

REFERENCES

Ajzen, I. and M. Fishbein (1980). *Understanding Attitudes and Predicting Social Behaviour*, Prentice-Hall, London.

Austrian Federal Ministry for Agriculture, Forestry, Environment and Water Management & OECD (2000). *EST-Synthesis Report: Environmentally Sustainable Transport: Future, Strategies and Best Practices*, Austrian Federal Ministry for Agriculture, Forestry, Environment and Water Management, Vienna, Austria.

Axelrod, L.J. and D.R. Lehmann (1993). Responding to Environmental Concern: What Factors Guide Individual Action? *Journal of Environmental Psychology*, **13**, 149–159.

Bamberg, S. (1996). *Habitualisierte Pkw-Nutzung: Integration des Konstruktes "Habit" in die Theorie des geplanten Verhaltens* [Habitualized Use of Cars: Integration of Habit into the Theory of Planned Behaviour], *Zeitschrift für Sozialpsychologie*, **27**, 295–310.

Bamberg, S. (1999). Umweltschonendes Verhalten – eine Frage der Moral oder der richtigen Anreize [Proenvironmental Behavior: A Question of Morality or Incentives]? *Zeitschrift für Sozialpsychologie*, **30** (1), 57–76.

Bamberg, S., M. Hunecke, and A. Blöbaum (2006). *Social Context, Morality, and Car Use of Public Transportation: Results from Two Field Studies*, Unpublished manuscript.

Bamberg, S. and P. Schmidt (2003). Incentives, Morality, or Habit? Predicting Student's Car Use for University Routes with the Models of Ajzen, Schwartz, and Triandis, *Environment and Behavior*, **35** (2), 264–285.

Biebeler, H. (2000). *Soziale Normen und Umweltverhalten. Theoretische und empirische Analysen zum Einfluß normativer Überzeugungen auf das persönliche Umweltverhalten [Social Norms and Environmental Behaviour. Theoretical and Empirical Analyses of the Impact of Normative Beliefs on Individual Environmental Behaviour]*, Schriftenreihe Socialia (Bd. 42), Kovac, Hamburg.

Black, J.S., P.C. Stern, and J.T. Elworth (1985). Personal and Contextual Influences on Househould Energy Adaptations, *Journal of Applied Psychology*, **70**, 3–21.

Blamey, R. (1998). The Activation of Environmental Norms. Extending Schwartz's Model, *Environment and Behavior*, **30**, 676–708.

Blöbaum, A., E. Matthies, K. Ando, and S. Ohnuma (2007). *The Impact of Moral Norms on Pro-Environmental Mobility Behavior in Japan and Germany*, Unpublished manuscript.

Bratt, C. (1999). The Impact of Norms and Assumed Consequences on Recycling Behavior, *Environment and Behavior*, **31** (5), 630–656.

Burn, S.M. and S. Oskamp (1986). Increasing Community Recycling with Persuasive Communication and Public Commitment, *Journal of Applied Social Psychology*, **16**, 29–41.

Cairns, S., L. Sloman, C. Newson, J. Anable, A. Kirkbride, and P. Goodwin (2004). *Smarter Choices – Changing the Way We Travel. Final Report of the Research Project: 'The Influence of Soft Factor Interventions on Travel Demand,'* Department for Transport, London.

Dunlap, R.E. and K.D. Van Liere (1978). The New Environmental Paradigm, *Journal of Environmental Education*, **9**, 10–19.

Dunlap, R.E. and K.D. Van Liere (1984). Commitment to the Dominant Social Paradigm and Concern for Environmental Quality, *Social Science Quarterly*, **65**, 1013–1028.

Dunlap, R.E., K.D. Van Liere, A.G. Mertig, and R. Emmet Jones (2000). Measuring Endorsement of the New Ecological Paradigm: A Revised Nep Scale, *Journal of Social Issues*, **56**, 425–442.

Dwyer, W.O., F.C. Leeming, M.K. Cobern, B.E. Porter, and J.M. Jackson (1993). Critical Review of Behavioral Interventions to Preserve the Environment: Research since 1980, *Environment and Behavior*, **25**, 275–321.

Festinger, L. (1957). *A Theory of Cognitive Dissonance*, Row, Oxford.

Garvill, J. (1999). Choice of Transportation Mode: Factors Influencing Drivers' Willingness to Reduce Personal Car Use and Support Car Regulations. In: *Resolving Social Dilemmas: Dynamic, Structural and Intergroup Aspects* (M. Foddy, M. Smithson, S. Schneider, and M. Hogg, eds), Psychology Press, New York, 263–279.

Geller, E.S. (1989). Applied Behavior Analysis and Social Marketing: An Integration for Environmental Preservation, *Journal of Social Issues*, **45** (1), 17–36.

Goldenhar, L.M. (1991). *Understanding, Predicting and Influencing Recycling Behavior: The Future Generation*, University of Michigan, Michigan.

Guagnano, G.A., P.C. Stern, and T. Dietz (1995). Influences of Attitude–Behavior Relationships: A Natural Experiment with Curbside Recycling, *Environment and Behavior*, **27**, 699–718.

Harland, P., H. Staats, and H. Wilke (1999). Explaining Proenvironmental Intention and Behavior by Personal Norms and the Theory of Planned Behavior, *Journal of Applied Social Psychology*, **29**, 2505–2528.

Harms, S. (2003). *Besitzen oder Teilen: Sozialwissenschaftliche Analyse des Car Sharings* [*Owning or Sharing? Socioscientific Analysis of Car Sharing*], Rüegger, Zürich.

Harms, S. and B. Truffer (2005). Vom Auto zum Car Sharing: Wie Kontextänderungen zu radikalen Verhaltensänderungen beitragen [From Private Car Use to Car Sharing: Context Changes as Facilitators for Radical Behavioral Changes], *Umweltpsychologie*, **9** (1), 4–27.

Heberlein, T.H. (1972). The Land Ethic Realized: Some Social Psychological Explanations for Changing Environmental Attitudes, *Journal of Social Issues*, **28** (4), 79–87.

Hopper, J.R. and J.M. Nielsen (1991). Recycling as Altruistic Behavior: Normative and Behavioral Strategies to Expand Participation in a Community Recycling Program, *Environment and Behavior*, **23**, 195–220.

Hunecke, M., A. Blöbaum, E. Matthies, and R. Höger (2001). Responsibility and Environment: Ecological Norm Orientation and External Factors in the Domain of Travel Mode Choice Behavior, *Environment and Behavior*, **33**, 830–852.

Hunecke, M., E. Matthies, A. Blöbaum, and R. Höger (1999). Die Umsetzung einer persönlichen Norm in umweltverantwortliches Handeln [Practical Application of Personal Values to Ecologically Responsible Behaviour], *Umweltpsychologie*, **3** (2), 10–22.

Jacobs, H.E., D. Fairbanks, C.E. Poche, and J.S. Bailey (1982). Multiple Incentives in Encouraging Car Pool Formation on a University Campus, *Journal of Applied Behavior Analysis*, **15**, 141–149.

Kaiser, F.G., J. Frick, and S. Stoll-Kleemann (2001). Zur Angemessenheit selbstberichteten Verhaltens: Eine Validitätsuntersuchung der Skala allgemeinen ökologischen Verhaltens [Accuracy of Self-Reports: Validation of the Skala Allgemeinen Oekologischen Verhaltens (General Ecological Behavior Scale)], *Diagnostica*, **47** (2), 88–95.

Kaiser, F.G. and C. Keller (2001). Disclosing Situational Constraints to Ecological Behavior: A Confirmatory Application of the Mixed Rasch Model, *European Journal of Psychological Assessment*, **17**, 212–221.

Kaiser, F.G. and T.A. Shimoda (1999). Responsibility as a Predictor of Ecological Behaviour, *Journal of Environmental Psychology*, **19**, 243–253.

Katzev, R., D. Brook, and M. Nice (2000). The Effects of Car Sharing on Travel Behavior: Analysis of Car Sharing Portland's First Year, *World Transport Policy and Practice*, **6**, 22–28.

Katzev, R.D. and A.U. Pardini (1987/88). The Comparative Effectiveness of Reward and Commitment Approaches in Motivating Community Recycling, *Journal of Environmental Systems*, **17**, 93–113.

Klocke, U. (2002). Bedingungen umweltrelevanter Mobilitätsentscheidungen: Umweltschutz durch staatliche Maßnahmen, bei der individuellen Verkehrsmittelwahl und beim Autokauf [Factors Influencing Environmentally Significant Mobility Decisions: Environmental Protection through Policy Measures, with Individual Travel Mode Choice and Purchase of New Cars]. In: *Mobilität im Jugend- und Erwachsenenalter* (W. Scholl and H. Sydow, eds), Waxmann, Münster, 424–469.

Klöckner, C.A. and E. Matthies (2004). How Habits Interfere with Norm Directed Behavior – A Normative Decision-Making Model for Travel Mode Choice, *Journal of Environmental Psychology*, **24**, 319–327.

Klöckner, C.A., E. Matthies, and M. Hunecke (2003). Problems of Operationalizing Habits and Integrating Habits in the Normative Decision-Making Model, *Journal of Applied Social Psychology*, **33**, 396–417.

Lüdemann, C. (1997). *Rationalität und Umweltverhalten. Die Beispiele Recycling und Verkehrsmittelwahl* [*Rationality and Environmental Behaviour: The Examples of Recycling and Travel Mode Choice*], DUV, Wiesbaden.

Maloney, M.P. and M.P. Ward (1973). Ecology: Let's Hear from the People. An Objective Scale for the Measurement of Ecological Attitudes and Knowledge, *American Psychologist*, **28**, 583–586.

Matthies, E. (2003). One to Bind Them All: How the Modified Moral Decision Making Model Can be Used for the Integration of Measures to Promote Pro-Environmental Travel Mode Choices. In: *Crossing Boundaries – The Value of Interdisciplinary Research* (T. Craig, ed.), Robert Gordon University, Aberdeen, 103–109.

Matthies, E. (2005). Wie können PsychologInnen ihr Wissen besser an die PraktikerIn bringen? Vorschlag eines neuen integrativen Einflussschemas umweltgerechten Alltagshandelns [How Can Psychologists Put across Their Knowledge to Practitioners? Suggesting a New, Integrative Influence Model of Proenvironmental Everyday Behaviour], *Umweltpsychologie*, **9** (1), 62–81.

Matthies, E., C.A. Klöckner, and C.L. Preißner (2006). Applying a Modified Moral Decision Making Model to Change Habitual Car Use – How Can Commitment be Effective? *Applied Psychology*, **55** (1), 91–106.

Matthies, E. and D. Krömker (2000). Participatory Planning – A Heuristic for Adjusting Interventions to the Context, *Journal of Environmental Psychology*, **20**, 65–74.

McKenzie-Mohr, D. (2000). Fostering Sustainable Behavior through Community-Based Social Marketing, *American Psychologist*, **55**, 531–537.

Nordlund, A.M. and J. Garvill (2003). Effects of Values, Problem Awareness, and Personal Norm on Willingness to Reduce Personal Car Use, *Journal of Environmental Psychology*, **23**, 339–347.

Orski, C.K. (2002). Car Sharing Revisited, *Innovation Briefs*, **13**, 1–2.

Preisendörfer, P. (1999). *Umwelteinstellungen and Umweltverhalten in Deutschland. Empirische Befunde and Analysen auf der Grundlage der Bevölkerungsumfragen "Umweltbewußtsein in Deutschland 1991–1998"* [*Environmental Attitudes and Environmental Behaviour in Germany. Empirical Findings on the Basis of the Public-Opinion Polls "Environmental Concern in Germany 1991–1998"*], Leske and Budrich, Opladen.

Schahn, J. and H.J. Bertsch (2003). Normdiskrepantes Verhalten im Umweltbereich Empirischer Test einer Integration des Normaktivationsmodells von Schwartz und der Neutralisationstheorie von Sykes und Matza [Neutralizing Nonnormative Ecological Behavior: Testing an Integration of Schwartz's Norm Activation Model and Sykes and Matza's Techniques of Neutralization], *Umweltpsychologie*, **7** (1), 124–148.

Schahn, J. and E. Holzer (1990). Konstruktion, Validierung und Anwendung von Skalen zur Erfassung des individuellen Umweltbewusstseins [Construction, Validation, and Application of Scales for Measuring Individual Environmental Awareness], *Zeitschrift für Differentielle und Diagnostische Psychologie*, **11**, 185–204.

Schultz, P.W. (1998). Changing Behavior with Normative Feedback Interventions: A Field Experiment on Curbside Recycling, *Basic and Applied Social Psychology*, **21**, 25–36.

Schwartz, S.H. (1970). Elicitation of Moral Obligation and Self-Sacrificing Behavior: An Experimental Study of Volunteering to be a Bone Marrow Donor, *Journal of Personality and Social Psychology*, **15**, 283–293.

Schwartz, S.H. (1977). Normative Influences on Altruism, *Advances in Experimental Social Psychology*, **10**, 221–279.

Schwartz, S.H. and J.A. Howard (1981). A Normative Decision-Making Model of Altruism. In: *Altruism and Helping Behavior: Social, Personality, and Developmental Perspectives* (J.P. Rushton and R.M. Sorrentino, eds), Erlbaum, Hillsdale, 189–211.

Steg, L., K. Geurs, and M. Ras (2001). The Effects of Motivational Factors on Car Use: A Multidisciplinary Modeling Approach, *Transportation Research A*, **35**, 789–806.

Steg, L. and C. Vlek (1997). The Role of Problem Awareness in Willingness-to-Change Car Use and in Evaluating Relevant Policy Measures. In: *Traffic and Transport Psychology. Theory and Application* (T. Rothengatter and W. Carbonell Vaya, eds), Pergamon, Amsterdam, 465–475.

Stern, P.C. (2000). Toward a Coherent Theory of Environmentally Significant Behavior, *Journal of Social Issues*, **56**, 407–424.

Taylor, S. and P. Todd (1997). Understanding the Determinants of Consumer Composting Behavior, *Journal of Applied Social Psychology*, **27**, 602–628.

TravelSmart (2003). *Training of TravelSmart Officers, Thinking about the Range of Travel Choices*. Retrieved January 25, 2007, from http://www.travelsmart.gov.au/training/thinking.html.

Triandis, H.C. (1977). *Interpersonal Behavior*, Brooks/Cole Publishing Company, California.

Van Liere, K.D. and R.E. Dunlap (1978). Moral Norms and Environmental Behavior: An Application of Schwartz's Norm-Activation Model to Yard Burning, *Journal of Applied Social Psychology*, **8**, 174–188.

Van Vugt, M., R.M. Meertens, and P.A.M. Van Lange (1995). Car Versus Public Transportation? The Role of Social Value Orientations in a Real-Life Social Dilemma, *Journal of Applied Social Psychology*, **22**, 1580–1607.

Verplanken, B., H. Aarts, and A. Van Knippenberg (1997). Habit, Information Acquisition, and the Process of Travel Mode Choices, *European Journal of Social Psychology*, **27** (5), 539–560.

Wang, T.H. and R.D. Katzev (1990). Group Commitment and Resource Conservation: Two Field Experiments on Promoting Recycling, *Journal of Applied Social Psychology*, **20**, 265–275.

Weigel, R.H. and J. Weigel (1978). Environmental Concern: The Development of a Measure, *Environment and Behavior*, **10**, 3–15.

PART 3:
POLICY MEASURES AIMED AT REDUCING PRIVATE CAR USE

Threats from Car Traffic to the Quality of Urban Life
Tommy Gärling and Linda Steg (Editors)

15

A Classification of Travel Demand Management Measures

Peter Loukopoulos, Institute for Human-Environment Systems, Swiss Federal Institute of Technology Zurich, Zürich, Switzerland

ABSTRACT

Policies that seek to influence the demand for automobile usage are referred to as travel demand management (TDM) measures. These may vary on several attributes or dimensions, which are likely to affect the key outcome variables of effectiveness in reducing travel demand, political feasibility, and public attitudes. Given that TDM measures are often implemented as a package, understanding the ways in which they are similar or differ is crucial. This chapter represents an initial attempt at doing so, while also examining the historical development of the TDM concept.

INTRODUCTION

A wide variety of policies have been proposed with the aims of alleviating the increasingly negative consequences of automobile use and, ultimately, of guiding society towards a sustainable future. Some policies have directly tackled the automobile per se, either by restricting sales and ownership or by de-marketing the car as a status symbol and convenient accessory of modern life (e.g., Wright and Egan, 2000). Such policies, however, at best affect the determinants of car use, specifically the appeal of the automobile, although ownership restrictions can be argued to not change this either. Other policies have attempted to promote technological advances, such as alternative catalysts and cleaner fuels (Borgwardt, 2001; Gärling *et al.*, 2002). Yet, such policies typically affect the consequences of car use (e.g., fewer emissions, environmentally friendly tyres). Furthermore, even in cases of large market penetration, the results of technological advances may be offset by greater driving amongst individuals who feel they can drive more given the fact that their driving is not as harmful or fuel consuming as was once the case; a phenomenon referred to as the 'rebound effect' (see Schuitema, 2003).

More importantly, emissions are only one aspect of many as regards the consequences of automobile use and, as noted for the UK context by Begg and Gray (2004), a government may meet emissions targets in the short term – even with the expansion of infrastructure – without the need to manage demand due to improvements in engine technology. Nevertheless, the demand for road use must eventually be dealt with because carbon dioxide emissions will begin to rise and because even in the essentially hypothetical case of no emissions many social consequences remain (e.g., the decreasing levels of physical activity among people), as may even congestion.

It is partly for these reasons that it is argued that demand for car use must be reduced (Hensher, 1998). That is, people's travel behaviour needs to be modified. Policies attempting to do so will be referred to as travel demand management (TDM) measures, although they are often referred to by other names with similar meanings including, but not limited to, transport system management or transportation control measures (Pendyala *et al.*, 1997), transportation demand management (Litman, 2003), and mobility management (Rye, 2002; Litman, 2003).

TRAVEL DEMAND MANAGEMENT (TDM): A DEFINITION AND HISTORICAL OVERVIEW

Humans travel more and more, longer and longer, and further and further than ever before (e.g., Cameron *et al.*, 2004). Furthermore, while the number of short trips is decreasing, the number of short trips conducted by car is on the increase (Mackett, 2001). A plethora of policies has been proposed to reverse or constrain these trends. In his review of online resources, Litman (2003) distinguishes between five categories of TDM measures: improvements in transport options; provisions of incentives to switch mode; land-use management; policy and planning reforms; and support programmes. May *et al.* (2003) distinguish six categories of intervention: land-use policies; infrastructure provision (for modes other than the private automobile); management and regulation; information provision; attitudinal and behavioural measures; and pricing. They also distinguish between 'hard measures', which aim to change behavioural opportunities and 'soft measures', which aim to change norms, motivations, and perceptions; they argue that attitudinal TDM measures empower individuals to make choices (in contrast to imposing restrictions on them). Vlek and Michon (1992) suggest that the following TDM measures are feasible ways of implementing car-use reduction policies: physical changes such as, for instance, closing out car traffic or providing alternative transportation; law regulation; economic incentives and disincentives; information, education, and prompts; socialisation and social modelling targeted at changing social norms; and institutional and organisational changes such as, for instance, flexible work hours, telecommuting, or 'flexplaces'. Louw *et al.* (1998) argued that travel by car could be influenced by policies encouraging mode switching, destination switching, changing time of travel, linking trips, substitution of trips with technology (e.g., teleworking), and substitution of trips through trip modification (e.g., a single goods delivery in lieu of a series of shoppers' trips). Marshall and Banister (2000) distinguish ten categories of TDM measures – capacity management; pricing; land-use planning; communications and technology; city/company travel policies; physical and priority measures; subsidies; access and parking restrictions; goods deliveries; and public awareness – placing heavy emphasis on the mechanisms through which these categories work: a switching mechanism (of mode, destination, or time) and a substitution mechanism (through linking trips, technology, or modification).

From the above brief overview, it is clear that TDM measures are broadly defined. Some definitions even focus on the adaptation mechanisms assumed to be triggered in response to the TDM measure. Even so, there exists one key commonality among all the various categories of TDM measures outlined above, and it is this commonality that Litman (2003, p. 245) emphasises in defining TDM as 'a general term for strategies and programmes that encourage more efficient use of transport resources (road and parking space, vehicle capacity, funding, energy, etc.)'. This definition is broad reflecting the historical progression away from an initial focus on convincing people to reduce their usage of the private car by changing modes of transport or by driving less. As Taylor and Ampt (2003) note, the term "TDM measure" now encompasses any initiative with the objective of reducing the negative impact of the car; among the benefits accruing from a reduction in car use within a community, they include less crime and children being allowed to walk or ride alone to a greater extent. This historical development of TDM can also be dealt with using Litman's (2003) definition. It should also be said that the definition of TDM reflects the broader evolutionary changes in policy measures, which have progressed from the 1960s when increasing infrastructure to alleviate traffic problems such as congestion was commonplace, to the 1970s where the emphasis was on improving management of existing infrastructure, to the 1980s and beyond when policies began to target altering human behaviour (Bovy and Salomon, 2002). An even more recent manifestation is the attempt to alter human values and change mobility culture as has occurred in, for example, some parts of Switzerland with administrations marketing a slower lifestyle and better image for public transport (City of Zurich, 2002). The policies of the 1960s are incompatible with the definition of TDM.

TDM MEASURES: ATTRIBUTES AND IMPACTS

Given that TDM is a multifaceted, umbrella term for the collection of policies seeking to minimise the negative consequences of automobile use, it is to be expected that TDM measures may vary on several attributes or dimensions (e.g., coerciveness, technical feasibility, political feasibility, cost). Figure 1 is a selection of various attributes that are assumed to have an impact on the key outcomes of public attitudes, political feasibility, and effectiveness (i.e., behavioural responses to a TDM measure).

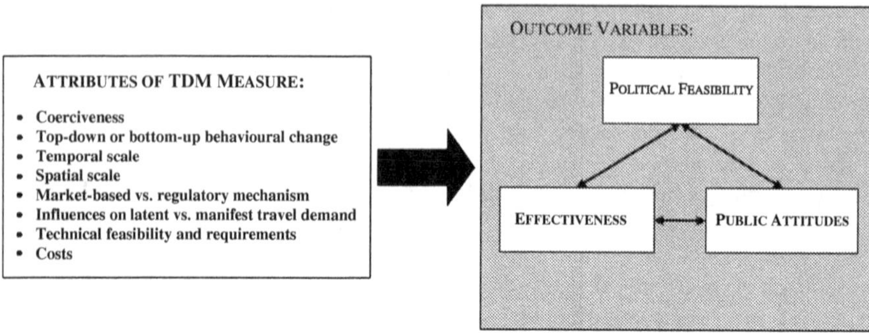

Figure 1: Schematic of the Effects of TDM – Measure Attributes on Key Outcome Variables and the Interrelationships between These Outcome Variables

Clearly, Figure 1 implies both that researchers may choose to examine any or all of effectiveness, political feasibility or public attitudes as outcome variables and that it is theoretically possible for two different TDM measures with differing attribute patterns to achieve the same level of effectiveness (in terms of trip reduction or emissions reductions or any other index of effectiveness) but to have very different levels of political feasibility or to result in different public attitudes. The effectiveness of a TDM measure, in terms of resulting behavioural changes, can be related to the measure's attribute composition. For example, it is generally the case that TDM measures covering a larger area will affect a greater number of trips (e.g., Loukopoulos *et al.*, 2005a). Similarly, the less coercive a measure (e.g., building new public transport routes), the less effective it tends to be in terms of encouraging behavioural change; research on habits is one reason as to why this is the case (e.g., Gärling *et al.*, 2001). Regarding public attitudes, however, it has generally been found that the more voluntary TDM measures allowing people to choose to change are more positively received than the more coercive measures (e.g., Thorpe *et al.*, 2000; Taylor and Ampt, 2003; Loukopoulos *et al.*, 2005b). Similarly, coercive market-based measures, such as road pricing, tend to be less popular than regulations, such as prohibition or road closures, because people do not like paying for things that were once free as it impinges on their perceived freedoms and because, from a social equity viewpoint, they disapprove of the fact that some groups are potentially disadvantaged by coercive market-based measures (Oberholzer-Gee and Weck-Hannemann, 2002; Jones, 2003).

Referring to recent research showing the importance of symbolic and affective motives for travel – travel liking or the positive utility of travel (see Mokhtarian, 2005) – TDM measures that directly or indirectly affect the non-instrumental reasons for travel may meet great resistance from large segments of the community. Furthermore, it should not be assumed that such motives are confined to leisure trips. Steg (2005) demonstrated that they also played a vital role even for the peak hour car commute of Rotterdam residents. Clearly, then, TDM measures that may in some way affect the symbolic and affective reasons of travel are likely to be less popular. However, the irony is that they may be less popular *and* less effective in that those with a greater positive utility of travel are more likely to continue driving. A distinction can thus be made between coercive policies' popularity and effectiveness as seen by reconsidering the car commute mentioned above. A road pricing TDM measure may be unpopular and less effective with respect to the commute to work than entry restrictions or prohibition in that the positive utility of travel may reduce the effectiveness of the former but not the latter TDM measure. Related to this point is the fact that TDM measures covering a greater spatial region or in operation for longer periods of time during the day are more likely to be less well received because of the greater difficulty in avoiding such TDM measures.

As regards political feasibility, many TDM measures restricting the use of the car by individuals can be difficult to implement partly because many governments consider the freedom of movement of citizens a right and consider the car an instrument of such freedom and equality (Docherty *et al.*, 2004). Additionally, more costly TDM measures are less feasible given the fiscal and budgetary difficulties facing many governments, particularly if the perceived returns on an investment in a TDM measure are seen to be relatively small. Indeed, such cost effectiveness is often an argument for many voluntary change programmes, which require little capital investment compared to, for example, the extension of existing public transport systems (Brög *et al.*, in press). Rose and Ampt (2003) further argue that, in addition to cost-benefit[1] ratios of up to 10:1, the evidence from such programmes is that their effects are relatively stable even a year after the programme's end. Furthermore, as noted by Taylor and Ampt (2003), protagonists of restrictive TDM measures may be dismissed through the democratic process and, as such, less-restrictive TDM measures are a safer option.

Yet, it is not only the case that the various TDM attributes can independently affect effectiveness, public attitudes, and political feasibility. A further implication of Figure 1 is that the various outcomes themselves are interrelated. For example, effectiveness is related to both the public attitudes generated by a TDM measure and the political feasibility of implementing the TDM measure; indeed all outcome variables are argued to be related to each other. Clearly, when attempting to minimise the negative consequences of growing car use, the more effective a measure the greater the political feasibility. However, more effective measures are also those that require greater change in behaviour and which, in turn, may be less popular amongst the voting public. This, in turn, has consequences on the political will to implement such measures. Nor should it be assumed

[1] Benefits include travel time changes for individuals or for the network, vehicle operating costs, pollution costs, accident and road trauma costs, and health and fitness (Rose and Ampt, 2003).

that the more effective a TDM measure in terms of reducing car use, the more negative the prevailing public attitudes are towards the measure. For example, because of the presumed unfairness of differential treatment of various citizen groups (e.g., the poor compared to the wealthy) as opposed to treating everyone identically, various forms of road pricing are less popular than prohibition or road closures preventing access by car to the city centre (Jones, 1995, 2003). In this case, the more effective TDM measure is preferred to the less effective TDM measure. Additionally, even in the case when there is sufficient political will to implement a highly effective TDM measure, there is the potential for negative attitudes and reactions amongst the public potentially leading to reactance, which, in turn, decreases the effectiveness of the implemented TDM measure. In brief, according to reactance theory (see Brehm and Brehm, 1981), whenever one's freedom to choose is threatened or eliminated, individuals strive to regain a feeling of autonomy, something that often results in attitudes being changed in the opposite direction to that being suggested such that individuals want the thing they are deprived of even more.

This discussion clearly demonstrates that the nature of the relationships between the various outcome variables in Figure 1 is neither simple nor straightforward. More dedicated research is needed that emphasises the importance of understanding and delineating just what it is about a TDM measure that is crucial to each of effectiveness, public attitudes, and political feasibility, and how the various attributes of TDM measures may have positive effects on one outcome variable but negative effects on another. That is, while the interrelationships between the various outcome variables are important, so, too, are the actual TDM attributes feeding into these relationships, as seen in Figure 1.

The following sub-sections examine the proposed attributes of TDM measures in greater detail. These represent initial, conceptual distinctions that highlight the need for future research testing hypotheses concerning, for example, whether and how these attributes singly or jointly influence the key outcome variables. Such research is vital both theoretically and practically, especially as it is often the case that TDM measures are implemented together with other TDM measures as part of a package. As such, it is important to know precisely how these various TDM measures are similar and different from one another. This is the key benefit of understanding how various TDM measure vary on the attributes detailed below, the focus of the present chapter.

Coerciveness

TDM measures vary in terms of whether the behavioural change is discretionary and within the control of automobile users or whether the change is forced upon automobile users in some way. For example, public transport improvements or information campaigns are non-coercive TDM measures as the decision to reduce car use is left to the individual. On the other hand, TDM measures such as road closures and prohibition within city centres are highly coercive, as individuals have no choice but to reduce car use within the designated areas. The coerciveness of other TDM measures such as congestion charging, increases in parking fees and parking restrictions, may depend on factors such as individual and household wealth.

Jones (2003), Schuitema (2003), Steg and Vlek (1997), Stradling *et al.* (2000), and Thorpe *et al.* (2000) distinguish between push and pull measures. Push measures discourage car use by making it (or car ownership) less attractive. Coercive measures tend to be classified as push measures. Pull measures encourage the use of alternative modes to the car by making such modes more attractive. As such, the less-coercive measures, such as cheaper public transport or new bike lanes or even car pooling subsidies, tend to be classified as pull measures. There is a close correspondence between push and pull measures and between TDM measures encouraging attitude change and those forcing behavioural change. For example, the idea underpinning the Individualised Marketing Programme (IndiMark, see Brög *et al.*, in press) is that much opposition to public transport is due to lack of information and motivation. By bringing such information to the individual and by showing when public transport can be of benefit to the individual, the expectation is that attitudes change such that individuals become more willing to reduce the use of the car for certain trips and purposes. Forced behaviour change, on the other hand, disregards individual attitudes and imposes restrictions.

As Vlek and Michon (1992) and Gärling *et al.* (2002) point out, forced changes may have negative side effects outweighing the expected benefits, such as costs or sacrifices that households will not accept as well as potential negative health impacts arising from increases in time pressure and stress. Additionally, the less-coercive measures (attitude change or pull measures) may be based on untenable assumptions about how much households are willing and able to change their car use, particularly since car use habits frequently interfere with information search and processing related to new public transport services. There appears to be a consensus that a package of measures needs to be introduced consisting of coercive measures that break a habit (by, for example, making car use no longer possible or making it prohibitively expensive) and non-coercive measures (such as increased public transport services or new routes) encouraging the use of other modes (Meyer, 1999; Gärling *et al.*, 2002).

Behaviour Change Resulting from Top-Down or Bottom-Up Processes

A recently made distinction between various TDM measures has been described by Taylor and Ampt (2003). They distinguish between traditional, top-down approaches, which tell people what to do, and bottom-up approaches, which allow people to choose to change their travel behaviour. The term given to such bottom-up approaches for the reduction of automobile use is voluntary travel-behaviour-change approaches (Taylor and Ampt, 2003); voluntary in the sense that nothing is changed in the transport system and that people are provided with better information concerning their transport options (Stopher *et al.*, 2005). Although the distinction between voluntary travel-behaviour-change approaches and pull measures is not clear-cut[2], the goal of the former is to empower people to change travel behaviour as opposed to expecting or forcing a response to external stimuli or pressures (coercive or otherwise). A key principle of voluntary travel-behaviour-change approaches is that individuals should define their own goals, in accordance with their own needs and existing lifestyle. That is why change must be initiated as part of a bottom-up process, with households deciding whether or not they wish to participate in a voluntary behaviour change programme (Taylor and Ampt, 2003).

Empowerment implies providing options and permitting individuals or households to choose to change. As such, one can consider land-use planning TDM measures as being a bottom-up process. Consider the effect of neighbourhood or city design on physical activity. While the possibility of engaging in physical activity for leisure purposes is available in any built environment, there is a clear difference between car-oriented and pedestrian-oriented environments in terms of providing choices for physical activity associated with daily life (e.g., commuting or shopping): only the latter provides such possibilities (King *et al.*, 2002).

Time Scale

A somewhat forgotten aspect is the variation in time scale required for both the implementation of TDM measures and for responses to various TDM measures. Goodwin (1998) noted that while many effects can be realised immediately, the cumulative effects of policies may not be felt for up to 50 years or more. For example, prioritising public transport and new fares policies have impacts on demand within the first year of implementation, with longer-term elasticities being twice as great after 5–10 years. Perhaps, the most obvious example of a TDM measure requiring long-term planning and with long-term consequences is that of land-use planning. Goodwin (1998) claims that the short-term effects of land-use planning to reduce travel distances are small, with the larger cumulative effects being felt only after 20 years. Related to land-use planning is pedestrianisation, which Goodwin (1998) claims has an immediate effect on traffic and with effects on pedestrians and retailers being felt after 5 years. This is consistent with the work of Gemzoe (2001) who examined the 30-year process of pedestrianisation and bicycle promotion in central Copenhagen. He argues that that the whole concept and use of public space in the city has changed, as has traffic culture as evidenced by the fact that more people travel to work by cycle than by car.

---

[2] For example, land-use planning (e.g., mixed zoning) can be considered to be a long term, pull measure even though it may arguably be excluded from the definition of voluntary behaviour change if one considers land-use planning to involve a change to the transport system. The key issue, in a sense, is where the transport system ends and land-use system begins.

The temporal nature of a TDM measure can also refer to its operational specifications. Congestion pricing, for example, typically operates during peak periods or during the day but not at evenings or at the weekend. Prohibition measures can also operate in similar fashion (Cambridgeshire County Council, 2005), although most road closures tend to be permanent. TDM measures can also affect the temporal nature of the activity per se: work-hour management strategies attempt to affect vehicle trip demand by reducing that demand or shifting it to less-congested time periods (e.g., flexible work hours, staggered work hours, modified work schedules such as a four-day week, and telecommuting services) (Golob, 2001). Such strategies have been shown to be effective, but it is also known that any savings often generate new, longer trips for non-commuting purposes such as leisure activities or maintenance activities, many of which were previously linked via chaining to the commuting trip (Black, 2001).

Spatial Scale

TDM measures' scope of influence may vary from the local to the national. Road pricing or congestion charging, for example, is a local initiative aimed at easing traffic flows in urban areas and at improving local air pollution levels, as well as at improving the liveability of urban areas (Foo, 1997, 2000; Goh, 2002; Banister, 2003). Road closures and pedestrianisation measures are also local initiatives. An example of a TDM measure with a large spatial zone of influence is kilometre charges for road transport involving the payment of a certain charge for each kilometre driven by the vehicle user (Ubbels *et al.*, 2002). A further example is that of public transport discounts for certain groups, typically those outside the employment market (e.g., pensioners, unemployed).

TDM measures may be initiated nationally but have local, even company-specific, impacts. An example is legislation requiring employers to implement strategies to reduce transportation impacts of employees, suppliers, visitors, and customers (Rye, 2002; Enoch and Potter, 2003). Such strategies vary from employer to employer but may include car-pooling schemes, coordination of specific transport routes with local public transport providers, and parking restrictions. Methods aimed at specifically encouraging the commercial sector to minimise the travel of its employees has been a growing TDM area (Enoch and Potter, 2003). An example of an employer-based TDM measure is proximate commuting where large decentralised employers (e.g., a bank with many geographically dispersed branches) reassign willing employees to job locations closer to the worker's residence so as to reduce commuting (Rodríguez, 2002).

Finally, an alternative conception of the spatial reach of TDM measures is whether or not they target the origin or destination of trips. For example, it is possible in a monocentric city to make car use less attractive by means of traffic calming and access restrictions in both the city centre (i.e., a common destination) and residential areas (i.e., the typical origin). Such a policy has been implemented in the city of Enschede, the Netherlands (Louw and Maat, 1999).

Market-Based Versus Regulatory Mechanisms

TDM measures also vary in terms of whether they can be classified as market-based (i.e., based on economic principles and pricing mechanisms) or regulatory in nature (i.e., based on legislation, standards, and legal principles). Examples of the latter include road closures, maximum parking ratios, enforced speed limits, and mandatory employer trip-reduction programmes (Meyer, 1999). Violations of such TDM measures are met with some form of punishment, fine or disapproval and the assumption is that such regulations or laws are internalised under the threat of punishment (Vlek and Michon, 1992). These policy measures are also referred to as command-and-control measures insofar as an authority assumes responsibility for the management of a transport system and controls it so that it, in principle, functions effectively. In contrast, examples of TDM measures based on market-based mechanisms include road and congestion pricing (Foo, 1997, 2000; Goh, 2002; Banister, 2003), kilometre charges (Ubbels *et al.*, 2002), fuel excises and parking charges (Meyer, 1999), and public transport discounts and travel vouchers (Root, 2001).

The principles underlying market-based mechanisms can be found in classical economics: People's behaviour is based on the principle of supply-and-demand and explicit cost-benefit analysis, such that if the price of a product (i.e., transportation) increases, the demand for this product will decrease, and vice versa (Schuitema, 2003). Using congestion pricing as an example, the idea is to raise costs so that the congestion externality is internalised (Emmerink *et al.*, 1995). In fact, this principle is clearly seen in Singapore where prices are increased or decreased dependent on the level of road utilisation (Foo, 1997, 2000; Goh, 2002). Furthermore, market-based mechanisms have become increasingly popular in recent years, particularly amongst politicians who appear to have embraced a new competition paradigm emphasising less governmental control. However, the increasing popularity amongst politicians has been matched by an increasing scepticism amongst researchers with regards to the claimed first-best nature of congestion pricing and market-based TDM measures in general. For a TDM measure to be viewed as a first-best instrument for tackling the problems of automobile usage, certain requirements must be fulfilled, two of which are that (i) individual behaviour must be rational, based on utility maximisation and (ii) full information is available on all costs involved (Emmerink *et al.*, 1995). However, research demonstrating the habitual nature of automobile use and the consequences for predecisional information search (e.g., Gärling *et al.*, 2001) renders these two assumptions suspect, even though Emmerink *et al.* (1995) have demonstrated that congestion pricing, for example, can be considered to be a second-best strategy with some advantages over other second-best strategies (but also some problems). Furthermore, independent lines of research have revealed low-price elasticities to be associated with various pricing policies (at least in the short term, although some higher elasticities have been obtained in the long term) (Hensher and King, 2001; Sipes and Mendelsohn, 2001).

In any case, the popularity of some form of market-based measure continues to grow amongst politicians who also see the potential of such measures to yield additional revenues, either for other environmentally friendly transport modes or for other services such as health and education (Odeck and Bråthen, 2002). It is also known that hypothecating revenues from market-based measures is one method to increase the acceptability and feasibility of implementation for any given TDM measure (e.g., Oberholzer-Gee and Weck-Hannemann, 2002).

Influences on Latent and Manifest Travel Demand

Variations in TDM measures in terms of the nature of the demand they manage is seldom taken into consideration. Yet, there are clear and important distinctions between various types of demand. Latent demand can be defined as the demand for services or resources that goes unsatisfied for various reasons (e.g., slow travel times or congestion, in the case of travel by car). Road construction has historically been driven by a 'predict-and-provide' approach (SACTRA, 1994), where the argument was that latent demand should be satisfied because better and more roads were required as a matter of individual freedom (the right to use the automobile) and economic competitiveness (the need for efficient road links for business). However, as cogently reviewed by Noland and Lem (2002), increases in urban capacity do not yield faster or more efficient travel times but rather, counterintuitively, make congestion worse. The reasons for this are summarised by Downs (1992) as being due to the fact that free road space produced from marginal reductions in commuting time (or from expansion of road capacity) is consumed quickly by (i) those travelling just outside of the peak time period who would shift back in; (ii) those driving on less optimal routes who would take advantage of lowered congestion on the most popular freeways; and (iii) those on slower public transit modes who would prefer driving if there were any more space on the road. Outside of commuting, demand induced by increased road capacity could include new vehicle trips by people who would not have otherwise made the trip or trips resulting from drivers who select an alternative destination (i.e., shoppers who prefer a new out-of-town retail development over the city centre). Noland and Lem (2002) even point out that new transportation capacity induces long-run changes in land-use development patterns, while Hansen and Huang (1997) estimate that the five-year elasticity of vehicle travel with respect to highway lane miles is 0.6–0.7 at the county level and 0.9 at the metropolitan level in California. The implications are that most of the trips using a new road in California are trips that would not have occurred had the road not been built. That is, the 'demand' generally used to justify new roads will not exist if the road is not built (Hansen and Huang, 1997; Noland and Lem, 2002).

In contrast to influencing latent travel demand, many TDM measures influence manifest travel demand (i.e., observable behaviour). Road or congestion pricing or kilometre-based charges are attempts at changing the observed driving behaviour of many people by increasing the cost of driving (Foo, 1997, 2000; Goh, 2002; Ubbels *et al.*, 2002; Banister, 2003) or, in the case of public transport rebates and discounts, by decreasing the costs of alternative modes (Root, 2001). Road closures or prohibition make it impossible for a behaviour or demand to manifest itself in certain areas or at certain times (Cambridgeshire County Council, 2005). The initial waves of TDM measures focused on better management of existing resources (Bovy and Salomon, 2002) and, as such, emphasised influencing manifest travel demand.

Yet, many recent TDM measures have also begun to specifically influence latent travel demand. One example is the attempt to alter human values and change mobility culture so that a less mobile society is not seen as a negative (Vlek and Michon, 1992; see also City of Zurich, 2002). Another example is the increasing use of information technology systems (e.g., Internet shopping, telecommuting) (Golob, 2001). Yet another example of such a TDM measure is land-use planning. Research has demonstrated that intensities and mixtures of land use significantly influence decisions to drive alone, car pool, or use public transport (Cervero, 2002). The assumptions made by proponents of such measures are that land-use patterns influence the time cost of travel and that the variations in time cost due to land use are of sufficient size to induce changes in travel behaviour (e.g., Boarnet and Crane, 2001). In summary, therefore, whereas the construction of road infrastructure assisted the satiation of latent travel demand by allowing it to be manifested in actual car use so that individuals could drive to their activity, land-use policies promoting, for example, mixed zoning satiate latent travel demand by bringing the activity to the individual.

Technical Feasibility and Costs

The technical feasibility and requirements of TDM measures often play a key role in the achievement of sustainable transportation. Having said this, however, technological development in transportation is notoriously slow due to the scale and costs of projects, the long periods required for research and development, and the long-life expectancy of infrastructure and mobile equipment (Shiftan *et al.*, 2003). Additionally, technical feasibility and requirements often play a larger role in the success of the more coercive TDM measures as opposed to TDM measures espousing voluntary behaviour change. The greatest concern has been, put simply, whether or not the technology will work, whether the technology will reliably detect vehicles so that drivers do not escape without paying or so that they are not wrongly charged, and whether high levels of non-compliance will overwhelm the system causing it to lose credibility (Jones, 2003). Evidence from successfully implemented road pricing schemes is testimony to the possibility that the technological requirements can be met; lack of appropriate technology is not likely to be a constraint in the near future (Blythe, 2005).

Technical feasibility and requirements have been most extensively examined with respect to road pricing schemes. Foo (1997) highlighted the technical features of the then Area Licensing Scheme in Singapore and noted that pre-payment is necessary to avoid the creation of congestion at entry points. Also important is the clear demarcation of a payment area with relatively few entry points for ease of enforcement, proper training of enforcement officers, and the provision of alternative bypass or escape routes so that motorists are not accidentally forced into a payment area. If anything, it is possible to argue that the technical feasibility of the majority of TDM measures that are likely to be implemented is within reach and, therefore, not an important influence on effectiveness. Indeed, the failure to take full advantage of the available technology can severely limit the long-term effectiveness of a TDM measure, bringing its entire viability into question. Nowhere is this more evident than in London where congestion charging has failed to deliver the anticipated revenue for reinvestment in transport infrastructure, a shortfall due to (i) the greater than expected reductions in traffic and (ii) the higher operational costs associated with the manually intensive process of call-centre registrations of licence plates (due to the use of video/photo technology to trace vehicles in and/or entering the charging zone) and follow-up activities (i.e., those related to enforcement and penalty notices processes) (Murray-Clarke, 2004). A fully electronic scheme (as in Singapore or in several Norwegian cities), taking full advantage of the available technology, would not only drastically minimise such operational costs but also potentially increase effectiveness both by making it more difficult to cheat the system and by allowing scarce resources to be wisely devoted elsewhere (e.g., enforcement).

As is no doubt evident from the above discussion, the implementation and operational costs of a TDM measure are of considerable importance. However, it is difficult to see how the costs of a TDM measure are necessarily related to effectiveness – both costly and less costly TDM measure may meet with success or failure. The operational costs described previously with respect to the London congestion-charging scheme are a potential example, although these are argued to be more related to the issue of technical feasibility. Nevertheless, a TDM measure's costs are important for political feasibility and possibly for public attitudes (May *et al.*, 2003).

CONCLUSIONS

This chapter has presented a discussion on the ways in which TDM measures may vary on several attributes that are potentially important for understanding the ways in which such measures impact the transportation system and its users. The presented conceptual framework has two key benefits. First, it can be used as a theoretical starting point for future empirical research to determine exactly how TDM measures vary from each other and just how it is that the various attributes influence key outcome variables, such as effectiveness, public attitudes, and political feasibility. Secondly, there is a practical interest in that TDM measures are often implemented as part of a policy package and a conceptual framework – and, hopefully, as alluded to above, an empirically tested framework – that can be used to classify TDM measures ought to prove invaluable to planners and politicians when designing such a policy package.

ACKNOWLEDGEMENTS

Financial support of its preparation and some of the author's research reported in this chapter was obtained through grant #2002-00434 from the Swedish Agency for Innovation Systems and grant #25.9/2001-1763 from the Swedish Research Council for Environment, Agricultural Sciences, and Spatial Planning. I thank the editors for comments.

REFERENCES

Banister, D. (2003). Critical Pragmatism and Congestion Charging in London, *International Social Science Journal*, 55, 249–264.

Begg, D. and D. Gray (2004). Transport Policy and Vehicle Emission Objectives in the UK: Is the Marriage between Transport and Environmental Policy Over? *Environmental Science and Policy*, 7, 155–163.

Black, W.R. (2001). An Unpopular Essay on Transportation, *Journal of Transport Geography*, 9, 1–11.

Blythe, P.T. (2005). Congestion Charging: Technical Options for the Delivery of Future UK Policy, *Transportation Research A*, 39, 571–587.

Boarnet, M. and R. Crane (2001). The Influence of Land Use on Travel Behavior: Specification and Estimation Strategies, *Transportation Research A*, 35, 823–845.

Borgwardt, R.H. (2001). Platinum, Fuel Cells, and Future US Road Transport, *Transportation Research D*, 6, 199–207.

Bovy, P.H.L. and I. Salomon (2002). Congestion in Europe: Measurements, Patterns and Policies. In: *Travel Behaviour: Spatial Patterns, Congestion and Modelling* (E. Stern and I. Salomon, eds), Edward Elgar, Cheltenham, UK, 143–179.

Brehm, S. and J.W. Brehm (1981). *Psychological Reactance: A Theory of Freedom and Control*, Academic Press, New York.

Brög, W., E. Erl, and N. Mense (in press). Individualised Marketing: Changing Travel Behaviour for a Better Environment. In: *Road Pricing, the Economy, and the Environment* (C. Jensen-Butler, B. Madsen, O. Anker Nielsen, and B. Sloth, eds), Elsevier, Amsterdam.

Cambridgeshire County Council. (2005). Vehicle Access to Cambridge City Centre, http://www3.cambridgeshire.gov.uk/transport/around/city_access. Retrieved 24 January, 2005.

Cameron, I., T.J. Lyons, and J.R. Kenworthy (2004). Trends in Vehicle Kilometres of Travel in World Cities, 1960–1990: Underlying Drivers and Policy Responses, *Transport Policy*, **11**, 287–298.

Cervero, R. (2002). Built Environments and Mode Choice: Toward a Normative Framework, *Transportation Research D*, 7, 265–284.

City of Zurich. (2002). *Mobilität ist Kult(ur). Die neue Mobilitätsstrategie – Wie wir miteinander weiterkommen [Mobility is Cult(ure). The New Mobility Strategy – How We Can Move Forward Together]*, Civil Engineering Office, City of Zurich, Zurich.

Docherty, I., J. Shaw, and M. Gather (2004). State Intervention in Contemporary Transport, *Journal of Transport Geography*, **12**, 257–264.

Downs, A. (1992). *Stuck in Traffic*, Brookings Institution Press, Washington, DC.

Emmerink, R.H.M., P. Nijkamp, and P. Rietveld (1995). Is Congestion Pricing a First-Best Strategy in Transport Policy? A Critical Review of Arguments, *Environment and Planning B*, **22**, 581–602.

Enoch, M. and S. Potter (2003). Encouraging the Commercial Sector to Help Employees to Change Their Travel Behaviour, *Transport Policy*, **10**, 51–58.

Foo, T.S. (1997). An Effective Demand Management Instrument in Urban Transport: The Area Licensing Scheme in Singapore, *Cities*, **14**, 155–164.

Foo, T.S. (2000). An Advanced Demand Management Instrument in Urban Transport, *Cities*, **17**, 33–45.

Gärling, T., D. Eek, P. Loukopoulos, S. Fujii, O. Johansson-Stenman, R. Kitamura, R. Pendyala, and B. Vilhelmson (2002). A Conceptual Analysis of the Impact of Travel Demand Management on Private Car Use, *Transport Policy*, **9**, 59–70.

Gärling, T., S. Fujii, and O. Boe (2001). Empirical Tests of a Model of Determinants of Script-Based Driving Choice, *Transportation Research F*, **4**, 89–102.

Gärling, T., A. Gärling, and P. Loukopoulos (2002). Forecasting Psychological Consequences of Car Use Reduction: A Challenge to an Environmental Psychology of Transportation, *Applied Psychology: An International Review*, **51**, 90–106.

Gemzoe, L. (2001). Copenhagen on Foot: Thirty Years of Planning and Development, *World Transport Policy and Practice*, 7 (4), 19–27.

Goh, M. (2002). Congestion Management and Electronic Road Pricing in Singapore, *Journal of Transport Geography*, **10**, 29–38.

Golob, T.F. (2001). Travelbehaviour.com: Activity Approaches to Modeling the Effects of Information Technology on Personal Travel Behaviour. In: *Travel Behavior Research: The Leading Edge* (D. Hensher, ed.), Pergamon, Amsterdam, 145–183.

Goodwin, P. (1998). The End of Equilibrium. In: *Theoretical Foundations of Travel Choice Modeling* (T. Gärling, T. Laitila, and K. Westin, eds), Elsevier, Amsterdam, 103–132.

Hansen, M. and Y. Huang (1997). Road Supply and Traffic in California Urban Areas, *Transportation Research A*, **31**, 205–218.

Hensher, D.A. (1998). The Imbalance between Car and Public Transport Use in Urban Australia: Why Does It Exist? *Transport Policy*, **5**, 193–204.

Hensher, D.A. and J. King (2001). Parking Demand and Responsiveness to Supply, Pricing and Location in the Sydney Central Business District, *Transportation Research A*, **35**, 177–196.

Jones, P. (1995). Road Pricing: The Public Viewpoint. In: *Road Pricing: Theory, Empirical Assessment and Policy* (B. Johansson and L.-G. Mattsson, eds), Kluwer, Dordrecht, The Netherlands, 159–179.

Jones, P. (2003). Acceptability of Transport Pricing Strategies: Meeting the Challenge. In: *Acceptability of Transport Pricing Strategies* (J. Schade and B. Schlag, eds), Elsevier, Amsterdam, 27–62.

King, A.C., D. Stokols, E. Talen, G.S. Brassington, and R. Killingsworth (2002). Theoretical Approaches to the Promotion of Physical Activity, Forging a Transdisciplinary Paradigm, *American Journal of Preventive Medicine*, **23** (2S), 15–25.

Litman, T. (2003). The Online TDM Encyclopedia: Mobility Management Information Gateway, *Transport Policy*, **10**, 245–249.

Loukopoulos, P., T. Gärling, and B. Vilhelmson (2005a). Mapping the Potential Consequences of Car-Use Reduction in Urban Areas, *Journal of Transport Geography*, **13**, 135–150.

Loukopoulos, P., C. Jakobsson, T. Gärling, C.M. Schneider, and S. Fujii (2005b). Public Attitudes towards Policy Measures for Reducing Private Car Use: Evidence from a Study in Sweden, *Environmental Science and Policy*, **8**, 57–66.

Louw, E. and K. Maat (1999). Enschede: Measures in a Package, *Built Environment*, **25**, 118–128.

Louw, E., K. Maat, and S. Mathers (1998). Strategies and Measures to Reduce Travel by Car in European Cities, Paper presented at the 8th World Conference on Transport Research, Antwerp, Belgium.

Mackett, R.L. (2001). Policies to Attract Drivers out of Their Cars for Short Trips, *Transport Policy*, **8**, 295–306.

Marshall, S. and D. Banister (2000). Travel Reduction Strategies: Intentions and Outcomes, *Transportation Research A*, **34**, 321–338.

May, A.D., A.F. Jopson, and B. Matthews (2003). Research Challenges in Urban Transport Policy, *Transport Policy*, **10**, 157–164.

Meyer, M.D. (1999). Demand Management as an Element of Transportation Policy: Using Carrots and Sticks to Influence Travel Behavior, *Transportation Research A*, **33**, 575–599.

Mokhtarian, P.L. (2005). Travel as a Desired End, not Just a Means, *Transportation Research A*, **39**, 93–96.

Murray-Clarke, M. (2004). The London Congestion Charging Scheme, Paper presented at the 11th World Congress on Intelligent Transportation Systems and Services, Nagoya, Japan, October.

Noland, R.B. and L.L. Lem (2002). A Review of the Evidence for Induced Travel and Changes in Transportation and Environmental Policy in the US and the UK, *Transportation Research D*, 7, 1–26.

Oberholzer-Gee, F. and H. Weck-Hannemann (2002). Pricing Road Use: Politico-Economic and Fairness Considerations, *Transportation Research D*, 7, 357–371.

Odeck, J. and S. Bråthen (2002). Toll Financing in Norway: The Success, the Failures and Perspectives for the Future, *Transport Policy*, 9, 253–260.

Pendyala, R.M., R. Kitamura, C. Chen, and E.I. Pas (1997). An Activity-Based Micro-Simulation Analysis of Transportation Control Measures, *Transport Policy*, 4, 183–192.

Rodríguez, D.A. (2002). Examining Individuals' Desire for Shorter Commute: The Case of Proximate Commuting, *Environment and Planning B*, 29, 867–881.

Root, A. (2001). Can Travel Vouchers Encourage More Sustainable Travel? *Transport Policy*, 8, 107–114.

Rose, G. and E. Ampt (2003). Travel Behavior Change through Individual Engagement. In: *Handbook of Transport and the Environment* (D. Hensher and K. Button, eds), Elsevier, Amsterdam, 739–755.

Rye, T. (2002). Travel Plans: Do They Work? *Transport Policy*, 9, 287–298.

SACTRA (Standing Advisory Committee on Trunk Road Assessment). (1994). *Trunk Roads and the Generation of Traffic*, The Stationery Office, London.

Schuitema, G. (2003). Pricing Policies in Transport. In: *Human Decision Making and Environmental Perception: Understanding and Assisting Human Decision Making in Real-Life Settings* (L. Hendrickx, W. Jager, and L. Steg, eds), University of Groningen, Groningen, The Netherlands, 203–222.

Shiftan, Y., S. Kaplan, and S. Hakkert (2003). Scenario Building as a Tool for Planning a Sustainable Transportation System, *Transportation Research D*, 8, 323–342.

Sipes, K.N. and R. Mendelsohn (2001). The Effectiveness of Gasoline Taxation to Manage Air Pollution, *Ecological Economics*, 36, 299–309.

Steg, L. (2005). Car Use: Lust and Must. Instrumental, Symbolic and Affective Motives for Car Use, *Transportation Research A*, 39, 147–162.

Steg, L. and C. Vlek (1997). The Role of Problem Awareness in Willingness-to-Change Car Use and in Evaluating Relevant Policy Measures. In: *Traffic and Transport Psychology. Theory and Application* (T. Rothengatter and W. Carbonell Vaya, eds), Pergamon, Amsterdam, 465–475.

Stopher, P., R. Alsnih, and E.S. Ampt (2005). Evaluating Voluntary Travel Behaviour Change Strategies – A Case Study, Paper presented at the 84th Annual Meeting of the Transportation Research Board, Washington, DC.

Stradling, S.G., M.L. Meadows, and S. Beatty (2000). Helping Drivers out of Their Cars. Integrating Transport Policy and Social Psychology for Sustainable Change, *Transport Policy*, 7, 207–215.

Taylor, M.A.P. and E.S. Ampt (2003). Travelling Smarter Down Under: Policies for Voluntary Travel Behaviour Change in Australia, *Transport Policy*, 10, 165–177.

Thorpe, N., P. Hills, and S. Jaensirisak (2000). Public Attitudes to TDM Measures: A Comparative Study, *Transport Policy*, 7, 243–257.

Ubbels, B., P. Rietveld, and P. Peeters (2002). Environmental Effects of a Kilometre Charge in Road Transport: An Investigation for the Netherlands, *Transportation Research D*, 7, 255–264.

Vlek, C. and J.A. Michon (1992). Why We Should and How We Could Decrease the Use of Motor Vehicles in the Near Future, *IATSS Research*, 15, 82–93.

Wright, C. and J. Egan (2000). De-Marketing the Car, *Transport Policy*, 7, 287–294.

16

SUSTAINABLE URBAN FORM: TRANSPORT INFRASTRUCTURE AND TRANSPORT POLICIES

Peter Newman, Institute for Sustainability and Technology Policy, Murdoch University, Murdoch, WA, Australia
Jeff Kenworthy, Institute for Sustainability and Technology Policy, Murdoch University, Murdoch, WA, Australia

ABSTRACT

Urban transport is shaped by physical structure, economic factors and cultural factors. The chapter focuses on how physical planning can change travel. Data are provided on how cities are structured so that there is no increase in travel times. Other factors become important only when the options for travel fit within the restrictions of the travel time budget. In order to provide policy options to reduce traffic, more sustainable transport options must be provided that can be competitive with the car in time. Data are provided to support this and also on the minimum densities in centres required to make viable sustainable transport options.

INTRODUCTION

Sustainability is more than a good idea – it is *the* idea of our age. It is powerful in that it can provide policy direction. It is relevant because it can work at every level from household, community, firm and local government to city, region and globe. The State Sustainable Strategy developed for Western Australia (Government of Western Australia, 2003) defines sustainability for settlements as:

> "Reducing the ecological footprint of settlements (resource inputs and waste outputs) whilst simultaneously improving their quality of life (health, housing, accessibility, community. . .)."

This is based on the Extended Metabolism Model (Newman and Kenworthy, 1999). The definition provides a simple way to approach the complexities of cities and sustainability. The approach requires transport to be considered as a central means of creating a sustainable future for cities as it shapes the ecological footprint and the quality of life in cities as shown below.

TRANSPORT AND CITY FORM

Cities are shaped by transport priorities and this is increasingly understood to be due to the uniform travel time budget of around 1 hour on average per person per day for all trip purposes (Marchetti, 1994). This concept of a constant travel time budget for all daily travel has been repeatedly supported in the literature (e.g., Schafer and Victor, 1997). The Marchetti constant means that people in the workforce, on average, travel no more than half-an-hour to work and half-an-hour home again and have little time therefore left in their daily travel budget for other travel. The average of around 30 minutes for the journey-to-work has been found to apply to a significant sample of international cities (the average for 33 cities in North America, Australia, Europe and Asia was 29 minutes, with a range from 20 minutes to 44 minutes (Kenworthy and Laube, 1999). Historical studies have also shown that people in cities have chosen to spend around 30 minutes travelling to work (Manning, 1978; Pederson, 1980; Zahavi and Ryan, 1980; Zahavi, 1982). In particular, data on UK cities have shown that the amount of time spent on journeys-to-work has remained stable for 600 years (SACTRA, 1994).

Travel times by people in cities for both work trips and all trip purposes together follow a normal distribution, meaning that there are many people who spend significantly less than one hour and many that spend a good deal more, but the literature generally supports the idea of an average daily travel budget for all residents of any city of no more than about 1.1 hours (Schafer and Victor, 1997).

Due to the respective speed capabilities of different transport modes, this constant travel time budget has meant historically that:

- *Walking Cities* were (and are) dense, mixed use areas no more than 5 km across. These were the major urban form for 8000 years.
- *Transit Cities* from 1850 to 1950 were based on trams and trains, which meant they could spread 20–30 km with dense clusters of corridors following the rail lines and stations.
- *Automobile Cities* from the 1950s on could spread 50 km in all directions and at low density.

Thus it can be seen that the urban form of cities through the centuries has been more or less determined or constrained by the available transport technologies and infrastructure, or the dominance of a particular technology, for whatever reasons, where there has been a choice of different modes available. Urban form and transport infrastructure therefore form a close nexus, and most cities today have some walking city, transit city and automobile city characteristics in different areas. Many cities, however, are reaching the limits of their urban form based on these transport systems. This is due to local and global sustainability issues, which help us to see what the key sustainability policies for cities are.

As already shown, in the 33 cities from across the globe where journey-to-work trip times are available (Kenworthy and Laube, 1999), the Marchetti constant largely holds in terms of an average of one hour (i.e., 58 minutes). In those cities where travel time has begun to go over that figure, transport issues are big on the political agenda (e.g., New York, Sydney, London, Paris, Singapore, Hong Kong, Jakarta, Seoul, Bangkok, Manila). The biological or psychological basis of the Marchetti constant seems to be a need for a more reflective or restorative period between home and work, but it cannot go for too long before people become very frustrated due to the need to be occupied in more productive pursuits, rather than just "wasting" time between activities. Economic imperatives are also likely to be important in explaining this constant.

TRANSPORT AND LAND USE PATTERNS

To examine sustainability in cities, it is necessary to understand some of the broad land use and transport patterns in the world's cities. The data provided come from the ISTP Global Cities Study that is being constantly expanded and now contains around 100 cities on all continents (Kenworthy *et al.*, 1999; Newman and Kenworthy, 1999; Kenworthy and Laube, 2001; Laird *et al.*, 2001). These basic data have been extensively compared and explained in numerous publications and it is not necessary to repeat this in detail here (see, e.g., Kenworthy and Laube, 1996, 1999, 2005; Kenworthy, 2002).

In all of this work "cities" have been defined for the most part as the functional urban regions (e.g., in the case of New York it is the Tri-State Metropolitan Region of 20 million people, not just the City of New York at the centre of the region). In a few cases availability of data for all the items being sought did not permit this type of broader city definition, so a smaller area had to be accepted (e.g., Frankfurt). Within the chosen city area, we are careful to define the urbanised or built-up portion by collecting the actual urbanised land area (which excludes larger areas of undeveloped land, regional open spaces, forests, natural areas, etc). This is particularly important to ensure that correct densities that are relevant to transport discussions are used (see also footnote 1). This comment applies equally to whole cities and the analyses undertaken in this chapter on smaller areas within cities. Therefore, all the observations we make between per capita travel and urban form are legitimately matched to each other, that is, they refer to travel only by inhabitants of particular areas for which the urban form of that area is well defined.

Some of the key factors in the international comparisons of relevance to this discussion are:

- There are huge variations in car use per capita in cities around the world, which bear no consistent relationship to income (city GDP per capita). Car use is taken here to mean annual vehicle kilometres of travel (VKT) or passenger kilometres of travel (i.e., VKT multiplied by the average 24 hour, 7 day per week car occupancy for all trip purposes) by inhabitants of the defined city area. Travel on the city's roads by cars originating outside the defined area is not included in car use per capita data for the cities. Car ownership has a closer link to wealth, but not car use. For example, wealthy Asian cities such as Hong Kong, Singapore, Tokyo are similar in car use to developing Asian cities such as Bangkok, Manila, Surabaya, Jakarta, Seoul and Kuala Lumpur, but are about ten times wealthier.
- Public transport variations are even greater in extent, with Asian and European cities (especially Eastern European) significantly higher than all other cities.
- Walking and cycling constitute between 25% and 65% of total daily trips in all cities in our studies except the automobile cities of North America and Australia where they average only 8% to 16%.
- Density variations are enormous with Asian cities highest (around 150 persons per ha); European transit cities are medium density (around 50 persons per ha) and Australian, New Zealand and American cities are uniformly low in density (around 15 persons per ha).
- Transport infrastructure varies enormously with the North American and Australian cities having very high road and freeway provision. Other cities tend to have far greater networks of premium transit infrastructure in the form of reserved rights-of-way for transit than they do in the case of premium road infrastructure in the form of freeways. Automobile cities also have far higher levels of parking in their central areas than all other cities.
- Significantly, as far as policy is concerned, there is a clear difference between the average speed of traffic and the average speed of transit. The automobile-dependent cities, which are often dominated by bus-based transit systems, have traffic speeds that are significantly faster than transit. European and wealthier Asian cities have much faster transit with speeds either exceeding that of traffic or much closer to it. However, many developing cities such as in parts of Asia, Latin America and the Middle East have slower transit than traffic, even though the traffic itself is very slow (e.g., Bangkok has average traffic speeds of 15 kph but transit speed average of just 10 kph).

These data help to explain automobile dependence. In Australian, Canadian and American cities, many people have no choice but to use a car, due to land use patterns. Sustainable mobility options are too difficult to provide other than by inconvenient, irregular bus services or by long bike or walk trips. Either way, the one-hour travel budget is threatened unless a car is used, or where a competitive train service is available (mainly for central city bound trips). Many people in such low-density cities do not have the option of a speed-competitive rail due to inadequate system coverage and thus have little choice other than to use a car.

In European and wealthy Asian cities, much more sustainable patterns of transport occur because the city is better structured and the infrastructure for sustainable modes is more available (Kenworthy and Laube, 2005). A train can be faster for a trip along a corridor, and local trips (linked to the train or not) are just as easy by walking and biking where land use is more compact and mixed in character as it is in European and Asian cities. As shown below, there is evidence that some of these characteristics are being lost in these cities as they develop new car-dependent suburbs on their fringes.

In most developing cities, especially those in transition economies, such as in many parts of Asia, and especially in China, a new kind of car dependence has emerged (Kenworthy and Hu, 2002; Kenworthy and Townsend, 2002). Despite being structured for non-motorised modes and transit (with dense corridors of development), the main priority in recent times has been to facilitate cars through the construction of large roads and extensive parking facilities. As a result, most of these cities have only poor-quality bus services operating in heavy congestion and there is little, if any, emphasis on biking or walking. Indeed, non-motorised modes are being forced out in many of these cities either through direct policies to exclude them from streets, or through the sheer hostility and danger of these cities' public environments from intense traffic. Thus, people are coerced into cars and motor cycles just to make up a little time and to avoid otherwise dysfunctional travel time budgets. The city therefore spirals down into traffic chaos or "traffic saturation" because the city's traditional walking and transit oriented urban form cannot cope with the space demands of private transport modes. This kind of car dependence is just as real as land use induced car dependence as people have few options other than a car or motor cycle within their Marchetti travel time budgets. However, as shown below, this type of rapidly accelerating car dependence in developing cities may be easier to solve, at least from a physical planning perspective.

THE PROBLEMS OF AUTOMOBILE DEPENDENCE

Serious economic, environmental and social problems are linked to automobile dependence. Economically, car dependence is not good for cities with the most auto-dependent high-income cities spending some 12%–14% of their metropolitan GDP on passenger transport, while cities in Europe and high-income parts of Asia, oriented to transit and non-motorised modes, spend only 7%–8%. This is understandable in terms of the sheer capital and operating costs of cars relative to much more economical transit systems (in both operating and infrastructure investment terms) and highly cost-effective walking and cycling.

The comparative urban data also show that transport deaths are primarily related to road use with US cities having some 13 deaths per 100,000 people, while Western European and wealthy Asian cities have 7–8 deaths per 100,000 people. This is a large external cost, as is air pollution and other environmental impacts associated with high energy use in transport. Apart from the environmental implications of high energy use, oil supplies and cost will also become major sustainability issues in the next few decades as global production peaks and declines (Campbell, 2003). This will be a major constraint on the future sustainability of car-dependent suburbs.

Health has become a new dimension for assessing car dependence (Frumkin *et al.*, 2004). The importance of the link to obesity in car-dependent cities is that if the whole travel time budget is taken up by driving, then little time for walking or riding a bike will be found on top of this. Building cities where walking and cycling are part of the travel time budget is a major reason for creating less car dependent cities.

Socially, the loss of community and the subsequent dysfunctional aspects of cities, such as social isolation, certain aspects of crime, degraded public environments and so on, can be related to excessive car dependence. Always having to be in a car means streets are mostly empty of people and children; the elderly and socially disadvantaged are unable to have any independent mobility. All this takes away from the community building processes of local place (Putnam, 2000). Counter to this, it has been theorised as early as the 1960s that car-based cities with their greatly expanded networking opportunities across metropolitan regions would provide "community without propinquity" and the model city put forward for this by Melvin Webber was Los Angeles (Webber, 1963, 1964). Since then, such theories have had some of the shine removed, especially in the scathing criticisms that have emerged on Los Angeles, not as a place of regional communities and networks, but as "Fortress LA" and the "Ecology of Fear" (Davis, 1990).

In summary, restraining sprawl or the land use-induced car dependence of cities has many rationales, as has the restriction of car dependence due to traffic saturation effects in public environments in many more intensively developed cities in lower income countries.

OVERCOMING AUTOMOBILE DEPENDENCE

In the western world there is considerable policy interest in how to overcome automobile dependence. This is coming from a shift in values associated with the desire for more sustainable outcomes, a younger demographic, which is much more urban than suburban, and the other dysfunctional aspects of urban sprawl that go beyond the Marchetti time budget (see Newman, 2003). Time constraints mean that people will tend to arrange their location and their mode of travel accordingly. The basis for overcoming automobile dependence then becomes whether people can access the amenities of a city without needing to use a car.

Thus there is a New Urbanism trend in urban design and a continuing move to reduce the funds directed into highway building and re-direct them into transit and other sustainable mobility projects (e.g., the American ISTEA legislation and its subsequent re-authorisations which seek to provide more funds for transit). The changes are slow but the evolutionary steps are happening. New infrastructure for rail systems is finally being funded and re-urbanisation rather than sprawl is becoming more favoured. Case studies are appearing in many parts of the world such as Perth, Australia (Newman, 2001). The importance of walking and cycling for personal health and local communities is being recognised as well (Tolley, 2003; Frumkin *et al.*, 2005). Car dependence is still the dominant reality in very many cities, but it is an urban ideology that is now exposed and found seriously wanting.

The vision for transforming sprawling car-dependent cities into sustainable cities of the future involves preventing further sprawl at the fringes and instead directing growth inwards to re-urbanise around sub-centres, especially rail stations. This Transit Oriented Development (TOD) has strong rationales in economic, social and environmental factors, but will not occur effectively unless the developments can be based around quality transit systems where the speed of transit is higher than the speed of traffic along major corridors; and where local transport options based on walking and cycling are more attractive than car trips. Thus there are infrastructure requirements as well as urban design requirements to ensure that sustainable modes are favoured. Wherever this has occurred, the resulting "urban village" has had extremely favourable market responses, as many people prefer the lifestyle options that are available in such places. Creating such options in sufficient numbers will determine the sustainability of future Australian and American cities.

In European cities, the trend in recent decades has been to simultaneously move inwards and outwards: to re-urbanise the older city centres and inner areas whilst at the same time growing at the urban fringes in suburbs not unlike those in Australia and America. The move inwards appears to have slowed as the centres are largely full and cannot be substantially redeveloped without impacting on the heritage of the cities. The outwards movement has begun hitting limits similar to those found in the new world cities, especially over land constraints and traffic generated by these car-dependent urban forms (Boeckl, 2003). Thus, the European Eco City Program for example is trying to establish models for how the new suburbs can be re-urbanised with centres that have transit at their heart and also feature other eco-city technologies (www.ecocityprojects.net). This is the same model as outlined above and is rooted in the need for a more focussed urban form that in particular can create viable transit systems for these car-dependent areas and ensure that future developments are much more transit oriented and walkable.

In the rapidly growing cities of Asia and throughout the developing world, there is a rush to greater motorisation. Partly, this is the pressure of globalisation, but limits to its growth are very obvious as the sheer space restraints on these cities makes it very difficult to imagine that anything like the low-density urban form and car use patterns of the United States and Australia could occur. Thus, although car ownership may have to be accommodated, as in all cities, the necessity for high car use and dependence will need to be addressed.

A similar analysis to the above has been made by Barter (2000) and Barter *et al.* (2003) on Asian cities and Kenworthy and Hu (2002) and Kenworthy and Townsend (2002), specifically on Chinese cities. The analysis suggests some policy directions for these Asian cities based on the data that show these cities do not yet have the same sprawling land use problem as many western cities (though some fringe car-based land use is occurring and will continue unless urban development is led by and anchored to new fixed route transit systems, rather than facilitated by freeways and tollways). Essentially, the dense, mixed-use centres and corridors that currently comprise the major part of Asian cities and many other lower income cities are still potentially good walking and transit city forms. They just rarely do justice to them in terms of providing infrastructure for transit, walking and cycling, and these dense areas are increasingly burdened by the traffic generated by more car-dependent areas further out.

Thus, policy analysis suggests the need for an evolutionary process of moving cities progressively towards less automobile dependence through changes in infrastructure priorities. The following steps are suggested for the different Asian city types and similar cities:

- Traffic-saturated walking/cycling cities like many in China need to provide bus-priority on major boulevards, ensure that the generous cycle lanes on existing main roads are not removed for road widening, and retain priority for pedestrians and cyclists on smaller streets. Unless this is done car traffic growth will always force itself upon urban policy makers. Once car traffic is facilitated to flow faster than local buses or local walk/cycle trips, then there will be an explosion in car use, pressure will mount for lower density land uses to accommodate cars, and the cities will spiral into the kind of traffic chaos symbolised by Bangkok. Ultimately such cities will need to move to superior rail-based transit systems if the lure of cars is to be countered as incomes rise.
- Traffic-saturated bus cities such as Bangkok, Jakarta and Manila need to provide rail rapid transit systems that are much faster than traffic. If they do not do this they will continue to be plagued by traffic; there is now no way back but to create high-capacity, fast, electric rail systems that can meet the enormous passenger demand of these dense cities and can easily out-perform in speed, reliability and comfort, the buses stuck in corridors of traffic. Once car dependent traffic chaos has been created, there is no other option than expensive mass transit in corridors. This becomes more and more compelling as incomes and expectations rise in such cities. But as shown in earlier discussions, these initially more expensive

rail systems provide economic rewards to the cities through significantly lower overall costs of passenger transport, which can make the cities more economically competitive and attractive for investment.

- Transit-oriented cities like Singapore, Tokyo and Hong Kong need to continue to build closely around their fast transit systems and not allow traffic speeds to improve above the transit system. Every freeway built will move the city closer to enabling car dependence to gain a foothold. There is little doubt that the "success" of these cities in a broad global context is at least partly underpinned by the fact that they still have highly functional and reliable transport systems that make them more liveable and attractive. In all three of these cities, but most strongly in Singapore, the functionality of their transport systems and their reduced car dependence are also underpinned by pricing polices which make car ownership and use a very expensive proposition (Newman and Kenworthy, 1999).

Similar analysis has been done on a more extensive number of Asian cities and other rapidly developing cities in Africa, the Middle East and Latin America (Kenworthy and Townsend, 2007).

In all cities, other policies will be needed to supplement these fundamental infrastructure and land use policies. These policies include transport demand management, parking restrictions, congestion taxes and travel behaviour modification and direct marketing programmes. However, while the car remains a quicker option, while transport investment priorities favour new roads, and there is no evolutionary urban structure building around rapid transit, walking and cycling, little happens to promote more sustainable transport.

WHAT KIND OF CENTRES ARE NEEDED?

The link between urban intensity and automobile dependence has been repeatedly confirmed by numerous authors and studies (Newman and Kenworthy, 1989, 1999; Holtzclaw, 1990, 1994; Naess, 1993a, 1993b; Cervero, 1995, 1998; Kenworthy and Laube, 1999). Taking the international comparative data referred to earlier, when measures of urban density (population per urban hectare or jobs per urban hectare or both together – called activity intensity) are correlated with either car use or passenger transport energy use per capita, extremely strong relationships are found with R^2 values between 0.77 and 0.82.

Figure 1 shows the relationship between per capita car travel and activity intensity in 58 higher income metropolitan areas around the world, revealing an extremely tight relationship where some 82% of the variance in car use internationally can be explained by variations in activity intensity (Kenworthy and Laube, 2001). When activity intensity is

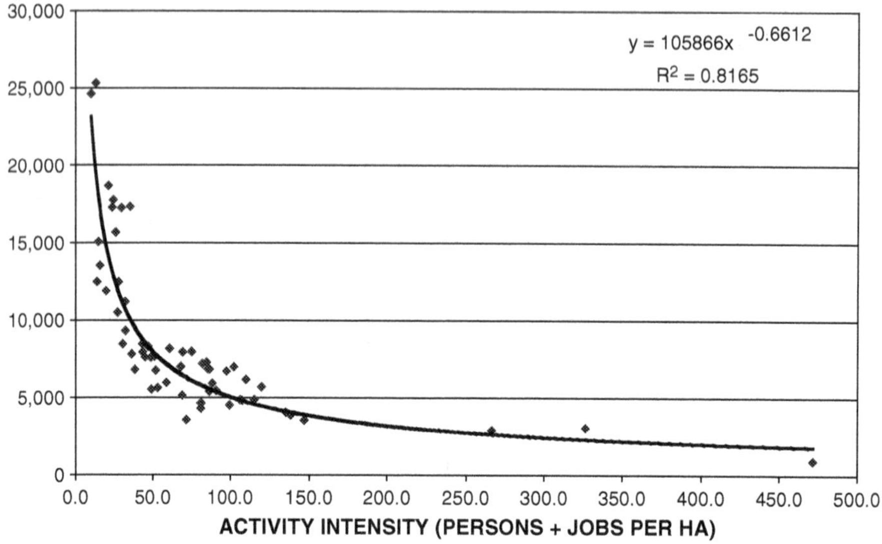

Figure 1: Activity Intensity Versus Passenger Car Use in 58 Higher Income Cities, 1995

used on the horizontal axis, the point on the graph where rapid acceleration in car dependence appears to occur seems to be around 35 people and jobs per ha; the error band is around 30 to 40 per ha[1].

The similar-shaped relationships can be found within cities. Transport energy studies by the authors on Sydney and Melbourne local government areas over a 22-year period show a sharp reduction in per capita transport energy use (and by implication car use) as density reaches certain cut-off levels. Studies from 1980 for Melbourne and 1981 for Sydney reveal R^2 values of 0.64 in each case when activity intensity is correlated with per capita passenger transport energy use. The data have been recently re-collected for 2002 for these two cities and show an even stronger relationship between transport energy use per capita and land use intensity with an R^2 value of 0.74 for Melbourne (Figure 2) and 0.70 for Sydney. The same clear influence of urban land use intensity thus emerges, despite the fact that in this 20-year period, the inner, higher density local areas

[1]Note that the density figures referred to here are genuine "urban density", in that they use total urbanised land as the denominator (residential, commercial, industrial land, local parks and open spaces, plus roads and any other urban land uses). The denominator excludes large areas of undeveloped land such urban zoned yet to be developed land, regional scale open spaces, agriculture and forestry land etc. The relationships shown in this paper will not work with other measures of density such as residential density because these are irrelevant to transport issues.

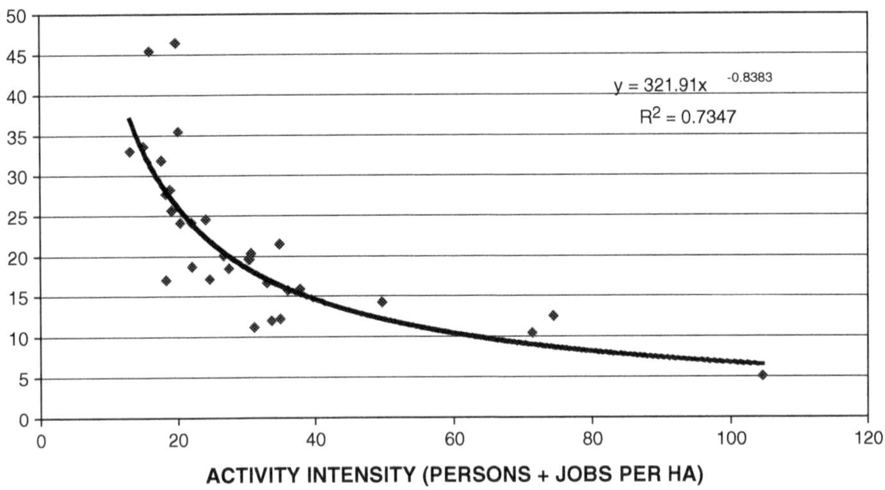

Figure 2: Per Capita Passenger Transport Energy Use Versus Activity Intensity in Melbourne LGAs, 2002

have become significantly wealthier than the outer areas. Wealth seems to be much less of an issue in determining transport activity than urban design.

The interesting threshold point at around 35 people and jobs per ha is not something that has been unnoticed by others who have collected urban data. Calthorpe (1993) came to a similar figure based on his impressions of what makes a viable centre. Holtzclaw *et al.* (2002) have more recently compiled detailed data on San Francisco, LA and Chicago with the same pattern of sharp increases in car use below this kind of density. Data have been examined for cities as different as Paris (INRETS, 1995), New York (Newman and Kenworthy, 1989) and San Francisco (Holtzclaw, 1990). Data for the latter two are shown in Table 1. Naess (1993 a, 1993b) developed similar data for Scandinavian cities. Jeffrey Zupan, quoted in Owen (2004), states the following about the idea of a critical threshold density:

> "The basic point is that you need density to support public transit. In all cities, not just in New York, once you get above a certain density two things happen. First, you get less travel by mechanical means, which is another way of saying you get more people walking and biking; and second, you get a decrease in trips by auto and an increase in trips by transit. That threshold tends to be around seven dwellings per acre. Once you cross that line, a bus company can put buses out there, because they know they're going to have enough passengers to support a reasonable frequency of service".

Table 1: Urban Density Versus Gasoline Use for San Francisco and New Regions

City and Area	*Gasoline Use (MJ Per Capita)*	*Urban Density (Persons Per Ha)*
San Francisco (1990)		
Central San Francisco	17,449	128
Inner city (city of San Francisco)	33,337	57
Middle suburb with strong sub-centre (Rockridge/Berkeley)	45,548	25
Middle suburb with no sub-centre (Walnut Creek)	49,641	10
Outer suburb (Danville-San Ramon)	67,090	5
New York (1980)		
Central city (New York County, including Manhattan)	11,860	251
Inner city (city of New York)	20,120	107
Whole city (Tri-State Metropolitan Region)	44,033	20
Outer suburbs	59,590	13

Seven dwellings per acre at a reasonable dwelling occupancy is equivalent to around 35–40 persons per ha.

All these studies reveal similar-shaped relationships to those seen in Figures 1 and 2 between travel patterns (as reflected by either per capita car use or passenger transport energy use) and density of urban activity. The question therefore arises as to what may be behind this[2]. Questions of wealth do not appear to be driving this phenomenon, as there is an inverse relationship between urban intensity and household income in Australian cities and a variety of wealth patterns across all the cities quoted above. As the data on Melbourne below (Table 2) indicate, it is the poorer households who are driving more, using public transport less and walking less. There are obviously other factors than the intensity of activity affecting transport, otherwise there would be an even stronger relationship within cities between activity intensity and transport patterns; such factors include the network of services provided, income, prices, cultural factors etc., but all of these can also be linked back to the intensity of activity in various ways. Thus, although many discussions have tried to explain transport in non-land use terms (e.g., Brindle, 1996; Mindali *et al.*, 2004), the data suggest that the physical layout and form of a city does have a fundamental impact on movement patterns and provides the most reliable, strong and "scientifically repeatable" explanation available of travel patterns within any city in the world and between different cities. The role of this chapter is to

[2] Brindle (1994) suggested that our attempts to link transport and density were flawed by a statistical problem involving per capita relationships on both axes. Evill (1995) explained how the graphical relationship was quite correct. If the per capita factor is removed from both then you just relate transport for the whole city to the area of the city. This is no difference when it comes to policy on how to reduce transport energy (reduce the area for the city, i.e., increase its density), but it is harder to follow. The emphasis on ecological footprint of cities today perhaps makes it easier to see by just presenting transport energy and area of the city, but in terms of planning for people and jobs it is more policy relevant to see the threshold density figure of 35 per ha and how cities or parts of cities become much more or less auto-dependent as density varies.

Table 2: Differences in Wealth and Travel Patterns from the Urban Core to the Fringe in Melbourne

	Core	*Inner*	*Middle*	*Fringe*
Percentage of households earning >$70,000 pa	12	11	10	6
Car use (trips/day/cap)	2.12	2.52	2.86	3.92
Public transport (trips/day/cap)	0.66	0.46	0.29	0.21
Walk/bike (trips/day/cap)	2.62	1.61	1.08	0.81

Source: Kenworthy and Newman (2001).

take the next step and explain how the relationship between transport and activity intensity works.

Analyses by us to explain car dependence show that culture, climate, politics, income, prices, education and other influences, although at work to different degrees in different settings, are not able to provide the consistent insights provided by the physical characteristics of the built form of cities (Newman and Kenworthy, 1989; 1999). Urban design, in particular urban form as reflected by population and job densities, emerged as the most significant determinant of the travel patterns we find both between and within cities around the world. We are therefore seeking to pursue a more fundamental explanation in terms of the physical design of cities and in particular the design and intensity of activity in centres.

Some policy-relevant physical design parameters can be calculated by using the activity intensity data suggested above as the threshold or critical value for creating less car dependent urban centres. A pedestrian catchment area or "Ped Shed", based on a 10-minute walk, creates an area of approximately 220–550 hectares for walking speeds of 5–8 km/h. Thus for an area of around 300 hectare developed at 35 people and jobs per ha, there is a threshold requirement of approximately 10,000 residents and jobs within this 10-minute walking area. The range would be from about 8,000 to 19,000 based on the 5–8 km/h speeds. Some centres will have a lot more jobs than others, but the important physical planning guideline is to have a combined minimum activity intensity of residents and jobs necessary for a reasonable local centre and a public transport service to support it. Other authors support these kinds of numbers for viable local centres and public transport services (Frank and Pivo, 1994; Ewing, 1996; Pushkarev and Zupan, 1997; Cervero *et al.*, 2004). The number of residents or jobs can be increased to the full 10,000, or any combination of these, as residents and jobs are similar in terms of transport demand. Either way, the number suggests a threshold below which services become non-competitive without relying primarily on car access to extend the catchment area.

Many new car-dependent suburbs have densities more like 12 per ha and hence have only one-third of the population and jobs required for a viable centre. When a centre is built for such suburbs, it tends to just have shops with job densities little higher than the surrounding population densities. Hence the Ped Shed never reaches the kind of intensity, which enables a walkable environment to be created. Many New Urbanist developments are primarily emphasising changes to improve the legibility and permeability of street networks, with less attention to the density of activity. As important as such changes are to the physical layout of streets, we should not be surprised when the resulting centres are not able to attract viable commercial arrangements and have only weak public transport. However, centres can be built in stages with much lower numbers to begin with, provided the goal is to reach a density of at least 35 per ha through enabling infill at higher intensities.

If a 30-minute Ped Shed for a Town Centre is used then the area of the catchment extends to between 2000 and 5000 hectares. Thus 35 people and jobs per ha provide approximately 100,000 residents and jobs within this 30-minute walking area (based on 3000 hectares). The range again is from around 70,000 to 175,000 people and jobs. This number could provide for a viable Town Centre based on standard servicing levels. Fewer numbers than this means services in a Town Centre are non-viable and it becomes necessary to increase the centre's catchment through widespread dependence on driving from much farther afield. This also means that the human design qualities of the centre are compromised because of the need for excessive amounts of parking. Of course, many driving trips within a walking Ped Shed still occur. However, if sufficient amenities and services are provided then only short car trips are needed, which is still part of making the centre less car dependent. "Footloose jobs", particularly those related to the global economy, can theoretically go anywhere in a city and can make out the difference between a viable centre and a non-viable centre. However, there is considerable evidence that such jobs are located in dense centres of activity due to the need for networking and quick "face to face" meetings between professionals. High-amenity, walking scale environments are better able to attract such jobs because they offer the kind of environmental quality, liveability and diversity that these professionals are seeking.

POLITICS OF SUSTAINABLE TRANSPORT

Developing more sustainable transport and land use patterns has been made difficult over recent decades, as car-dependent suburbs have been facilitated by the construction of fast roads. Few examples can be provided where cities have been led by professionals towards more sustainable policies, as outlined above (see Newman and Kenworthy, 1999, for case studies of those cities that have demonstrated some success). Mostly, such changes have been demanded by the public and have been achieved through political intervention (e.g., Newman, 2001).

The reason for this seems to be a professional reticence to push sustainable modes and in particular a deep seated anti-rail sentiment. The idea that "anything a train can do, a bus can do better and cheaper" has remained dominant; this has meant that in cities where bus systems are stuck in traffic, the ideology of car dependence has been meekly acceded to. This is seen in organisations such as the International Energy Agency (IEA) and World Bank (Schipper *et al.*, 2000, World Bank, 1996). Such analyses accept second best for transit systems and, in essence, car traffic is seen as inevitable and requiring higher priority. Car dependence has been the major outcome. This is highly problematic as it is simply not sustainable from either a local or global perspective.

Most citizens who experience car dependence can understand the phenomenon, since they directly feel and bear its economic, social and environmental consequences. They want other options provided for them. As cities continue to evolve, the politics of sustainable transport will demand both more liveable and less car dependent options for the future.

The growth of cities globally will continue. Wealth is likely to increase across all cities and in the past this has meant greater mobility based on greater car use. Just as has occurred with electricity growth being decoupled from wealth, it is entirely possible to imagine mobility being decoupled from wealth. The international comparative data discussed earlier in this chapter plus other trends suggest that this is already beginning to occur (Newman and Kenworthy, 1999; Newman, 2003; Kenworthy and Laube, 2005). The key to this move towards sustainability is better provision of access to transit that is faster than cars along corridors, and better provision for walking and cycling in local areas, associated with a supportive land use structure of intensive centres with minimum land use activity intensity of 35 people and jobs per ha. This is due to a fundamental need to ensure that the more sustainable transport modes have a competitive speed advantage for long trips (transit) and for short trips (bike/walk) within centres. Such change is evolutionary but it will always require political leadership.

CONCLUSIONS

Sustainability is a powerful and relevant concept for managing the cities of the world as we move rapidly towards an urbanised global economy. Sustainable transport is not just a case of developing better technology. We must also develop ways of overcoming car dependence. This has been explained in terms of the Marchetti constant on travel time budgets and the average speed of modes in cities. Data have been presented that show the constraints of the Marchetti budget on cities around the world. They mean that car dependence is land use induced in Australian and North American cities and increasingly in the outer areas of European cities, but primarily traffic induced in the rapidly motorising cities of Asia and other rapidly developing lower income countries. With few options for sustainable transport modes if Marchetti budgets are to be maintained, the main planning agenda for cities in the 21st century is to build reduced automobile dependence through better infrastructure and urban design. Other economic and social programmes will help, but if the infrastructure to enable people to travel by sustainable transport modes within the Marchetti budget limits is not provided, then it will be very hard to make much change happen. Sustainable transport will require ways of accommodating car ownership but not car dependence. It will require that viable, convenient and speed-competitive sustainable transport modes be provided in such a way that people will choose them for most trip purposes. Cars will become a mode of last resort. Professional advice on how to advance sustainability will need constant political intervention as the modernist assumptions of automobile dependence, supplemented by slow buses, is not able to adequately answer the demands of an increasingly sophisticated public with more disposable income looking for better and more sustainable options.

REFERENCES

Barter, P.A. (2000). Urban Transport in Asia: Problems and Prospects for High Density Cities, *Asia Pacific Development Monitor*, **2** (1), 33–66.

Barter, P.A., J.R. Kenworthy and F. Laube (2003). Lessons from Asia on Sustainable Urban Transport. In: *Making Urban Transport Sustainable* (N. Low and B. Gleeson, eds), Palgrave-Macmillan, Basingstoke, UK, 252–270.

Boeckl, M. (ed.) (2003). *Stadtumbau-Urban Conversion: Recent International Examples*, Edition architektur.aktuell 04, Springer, Wien.

Brindle, R.E. (1994). Lies, Damned Lies and "Automobile Dependence" – Some Hyperbolic Reflections. In: *Australasian Transport Research Forum 94*, 117–131.

Brindle, R.E. (1996). Transport and Urban Form: The Not So Vital Link. In: *Transport and Liveable Cities*, District 8/ITE Australia Section Inaugural Regional Conference, Melbourne.

Calthorpe, P. (1993). *The Next American Metropolis: Ecology, Community and the American Dream*, Harvard University Press, Boston.

Campbell, C.J. (2003). The Peak of Oil: An Economic and Political Turning Point for the World. In: *Making Urban Transport Sustainable* (N. Low and B. Gleeson eds), Palgrave-Macmillan, Basingstoke, UK, 42–66.

Cervero, R. (1995). Sustainable New Towns: Stockholm's Rail Served Satellites, *Cities*, **12** (1), 41–51.

Cervero, R. (1998). *The Transit Metropolis*, Island Press, Washington, DC.

Cervero, R., G.B. Arrington, J. Smith-Heimer, and R. Dunphy (2004). *Transit Oriented Development in America: Experiences, Challenges and Prospects*, Transportation Research Board, National Research Council, Washington, DC.

Davis, M. (1990) *City of Quartz: Excavating the Future in Los Angeles*, Vintage, London.

Evill, B. (1995). Population, Urban Density, and Fuel use: Eliminating Spurious Correlation, *Urban Policy and Research*, **13** (1), 29–36.

Ewing, R. (1996) Transit Oriented Development in the Sun Belt, *Transportation Research Record*, **1552**, 145–153.

Frank, L. and G. Pivo (1994). *Relationships between Land Use and Travel Behaviour in the Puget Sound Region*, Department of Transportation, WA-RD 351.1, Washington State.

Frumkin, H., L. Frankand, and R. Jackson (2004). *Urban Sprawl and Public Health: Designing, Planning and Building for healthy Communities*, Island Press, Washington, DC.

Government of Western Australia (2003). *Hope for the Future: A Vision for Quality of Life in Western Australia*, The State Sustainability Strategy, Department of the Premier and Cabinet, Perth, Western Australia.

Holtzclaw, J. (1990). *Explaining Urban Density and Transit Impacts on Auto Use*, Natural Resources Defense Council and The Sierra Club, San Francisco.

Holtzclaw, J. (1994). *Using Residential Patterns and Transit to Decrease Auto Dependence and Costs*, Natural Resources Defense Council, San Francisco.

Holtzclaw, J., R. Clear, H. Dittmar, D. Goldstein, and P. Haas (2002). Location Efficiency: Neighbourhood and Socio-economic Characteristics Determine Auto Ownership and Use – Studies in Chicago, Los Angeles and San Francisco, *Transportation Planning and Technology*, **25** (1), 1–27.

INRETS (1995). *Budgets Energie Environnement des Déplacements (BEED) en Ile-de-France – Analyse de la Dépense Energétique et des Emissions Polluantes Liées à la Mobilité des Franciliens.* Rapport de Convention ADEME-INRETS n°690-9306-RB, Institut National de Recherche sur les Transports et leur Sécurité, Arceuil, Paris.

Kenworthy, J.R. (2002). Traffic 2042 – a More Global Perspective, *Transport Policy*, **9**, 11–15.

Kenworthy, J.R. and G. Hu (2002). Transport and Urban Form in Chinese Cities: An International and Comparative Policy Perspective with Implications for Sustainable Urban Transport in China, *DISP*, **151**, 4–14.

Kenworthy, J.R., and F.B. Laube (1996). Automobile Dependence in Cities: An International Comparison of Urban Transport and Land Use Patterns with Implications for Sustainability, *Environmental Impact Assessment Review, Special Issue: Managing Urban Sustainability*, **16** (4–6), 279–308.

Kenworthy, J.R. and F.B. Laube (1999). Patterns of Automobile Dependence in Cities: An International Overview of Key Physical and Economic Dimensions with Some Implications for Urban Policy, *Transportation Research A*, **33**, 691–723.

Kenworthy, J.R. and F.B. Laube (2001). *The Millennium Cities Database for Sustainable Transport*, International Union of Public Transport (UITP), and Institute for Sustainability and Technology Policy (ISTP) – CD ROM database, Brussels and Perth.

Kenworthy J.R. and Laube F.B. (2005) An International Comparative Perspective on Sustainable Transport in European Cities, *European Spatial Research and Policy*, **12** (1), 11–50.

Kenworthy J.R., F.B. Laube, P. Newman, P. Barter, T. Raad, C. Poboon, and B. Guia (1999). *An International Sourcebook of Automobile Dependence in Cities, 1960–1990*, University Press of Colorado, Boulder.

Kenworthy, J. and P. Newman (2001). *Melbourne in an International Comparison of Urban Transport Systems*, A Report to the Department of Infrastructure, Melbourne as Part of the Melbourne Strategy, Institute for Sustainability and Technology Policy, Perth.

Kenworthy, J.R. and C. Townsend (2002). An International Comparative Perspective on Motorisation in Urban China: Problems and Prospects, *IATSS Research*, **26** (2), 99–109.

Kenworthy, J.R. and C. Townsend (2007). A Comparative Perspective on Urban Transport and Emerging Environmental Problems in Middle-Income Cities. In: *Scaling Urban Environmental Challenges, From Local to Global and Back* (G. McGranahan and P.J Marcotullio, eds), Project, Earthscan Publications, London.

Laird, P., P. Newman, M. Bachels, and , J.R. Kenworthy (2001). *Back on Track: Rethinking Australian and New Zealand Transport Policy*, University of NSW Press, Sydney.

Manning, I. (1978). *The Journey to Work*, Allen and Unwin, Sydney.

Marchetti, C. (1994). Anthropological Invariants in Travel Behaviour, *Technical Forecasting and Social Change*, **47** (1), 75–78.

Mindali, O., A. Raveh, and I. Saloman (2004). Urban Density and Energy Consumption: A New look at Old Statistics, *Transportation Research A*, **38**, 143–162.

Naess, P. (1993a). *Energy Use for Transport in 22 Nordic Towns*, NIBR Report No 2, Norwegian Institute for Urban and Regional Research, Oslo.

Naess, P. (1993b). Transportation Energy in Swedish Towns and Regions, *Scandinavian Housing and Planning Research*, **10**, 187–206.

Newman, P.W.G. (2001). Railways and Reurbanisation in Perth. In: *Case Studies in Planning Success* (J. Williams and R. Stimson eds), Elsevier, New York, 295–310.

Newman, P.W.G. (2003). Global Cities, Transport, Energy and the Future: Will Ecosocialisation Reverse the Historic Trends? In: *Making Urban Transport Sustainable* (N. Low and B. Gleeson, eds), Palgrave-Macmillan, Basingstoke, UK, 25–41.

Newman, P.W.G. and J.R. Kenworthy (1989). *Cities and Automobile Dependence: An International Sourcebook*, Gower, Aldershot, UK.

Newman, P.W.G. and J.R. Kenworthy (1999). *Sustainability and Cities: Overcoming Automobile Dependence*, Island Press, Washington, DC.

Owen, D. (2004). Green Manhattan: Why New York is the Greenest City in the US, *The New Yorker*, October 18, 111–123.

Pederson, E.O. (1980) *Transportation in cities*, Pergamon, New York.

Pushkarev, B. and J. Zupan (1997). *Public Transportation and Land Use policy*, Indiana Press, Bloomington and London.

Putnam, R. (2000). *Bowling Alone: The Collapse and Revival of American Community*, Simon and Schuster, New York.

SACTRA (1994). *Trunk Roads and the Generation of Traffic*, Department of Transport, London.

Schafer, A. and D. Victor (1997) The Past and Future of Global Mobility, *Scientific American*, 277 (4), 36–39.

Schipper, L., C. Marie-Lilliuand, and R. Gorham (2000). *Flexing the Link between Transport and Greenhouse Gas Emissions: A Path for the World Bank*, International Energy Agency, Paris.

Tolley, R. (ed.) (2003). *Sustainable Transport: Planning for Walking and Cycling in Urban Environments*, Woodhead, Cambridge, UK.

Webber, M. (1963). Order in Diversity: Community without Propinquity. In: *Cities and Space: The Future Use of Urban Land* (L. Wingo ed.), John Hopkins Press, Baltimore.

Webber, M. (1964). The Urban Place and the Non-place Urban Realm. In: *Explorations in Urban Structures*, University of Pennsylvania Press, Philadelphia, PA, 79–153.

World Bank (1996). *Sustainable Transport: Priorities for Policy Reform*, World Bank, Washington, DC.

Zahavi, Y. (1982) Discussion of J. Supernal, Travel-time Budget: A Critique. *Transportation Research Record*, 879, 25–28.

Zahavi, Y. and J.M. Ryan (1980). Stability of Travel Components over Time, *Transportation Research Record*, 750, 19–26.

Tommy Gärling and Linda Steg (Editors)

17

EFFECTIVENESS, PUBLIC ACCEPTANCE, AND POLITICAL FEASIBILITY OF COERCIVE MEASURES FOR REDUCING CAR TRAFFIC

Tommy Gärling, Department of Psychology, Göteborg University, Göteborg, Sweden
Peter Loukopoulos, Institute for Human-Environment Systems, Swiss Federal Institute of Technology Zurich, Zürich, Switzerland

ABSTRACT

We draw on a conceptual framework briefly identifying the main reasons why non-coercive measures for changing private car use in urban areas may fail to be effective and why coercive measures are effective. On the basis of attributes describing travel demand management measures (see Chapter 15 in this volume), we conceptualize coerciveness as a continuum from non-coercive to coercive that has both objective and subjective definitions. We note that public and political acceptability is a critical issue related to the implementation of coercive measures and discuss possible reasons for this.

INTRODUCTION

There are many conceivable policy measures that may reduce the adverse effects of car use. Some measures may not require a reduction in car use (e.g., increasing capacity of road infrastructure, improving car technology or limiting speed), if they succeed in sufficiently reducing the environmental impact per car. However, a general assessment of the current state is that measures reducing the demand for car use must be implemented in many urban areas (e.g., OECD, 1996). Furthermore, it is necessary to change car use with respect to when and where people drive, particularly on major commuter arteries during peak hours and in downtown areas. Since the proposed measures focus on changing or reducing demand for car use, they are generally referred to as *travel demand management* (TDM) measures (Kitamura *et al.*, 1997).

Several TDM measures have been proposed and some have been implemented with the aim of reducing or changing car use. Loukopoulos (Chapter 15 in this volume) reviews definitions and historical development of TDM measures. He also examines eight underlying attributes of systems of classification of TDM measures (coerciveness, top-down vs. bottom-up process, time scale, spatial scale, market-based vs. regulatory, influences on latent or manifest demand, technical feasibility, and costs) that have been shown or are assumed to be related to effectiveness (strength of effect on car use), public acceptance, and political feasibility. In this chapter, we focus on the attribute of *coerciveness.* Table 1 lists the most important TDM measures. Prohibition of car traffic (frequently physically as well as legally) is the most obvious coercive measure, while information and education are the most obvious non-coercive measures. Other coercive measures include parking control and speed limits. Economic measures (e.g., road or congestion charging) may fall in between coercive and non-coercive measures.

We assume that coercive TDM measures are effective. Why then do politicians not endorse such measures? Does the current state of affairs (i.e., the adverse effects of car use) not call for this? In other chapters of this book (e.g., Chapter 2 by Van Wee), the argument is made that this may soon be the case. Therefore, it is important to illuminate what seem to be impediments to the implementation of coercive TDM measures.

In the next section, we will first give some reasons for why non-coercive measures may fail to have any substantial effects on car use. Then we discuss why coercive measures

Table 1: Travel Demand Management Measures

Provision of physical alternatives and physical changes	Improving alternatives for car use (e.g., public transport, infrastructure for walking, and biking) Park and ride schemes Avoiding major new road infrastructure Land-use planning encouraging shorter travel distances
Legal measures	Prohibition of car traffic in city centers Parking control Decreasing speed limits
Economic measures	Taxation of cars and fuel Road or congestion pricing Kilometer charge
Informational and educational measures	Individualized marketing Public information campaigns about the negative effects of driving Social modeling where prominent public figures use alternative travel modes

Source: Adapted from Steg (2003a).

may be effective. In a following section, building on the attributes outlined by Loukopoulos in Chapter 15 of this volume, we note problems with the definition of coerciveness, as well as the fact that it varies with a number of other attributes for any given TDM measure. Finally, in the fourth section we identify factors preventing the implementation of coercive measures.

EFFECTIVENESS OF NON-COERCIVE TDM MEASURES

Why non-coercive TDM measures are effective or not in changing people's car use may be explained with the help of the conceptual framework proposed by Gärling *et al.* (2002a) (see Figure 1). This framework posits that car users choose among travel options, which are defined as bundles of attributes describing trip chains (including purposes, departure and arrival times, travel times, and monetary costs). Drawing on self-regulation theory in social psychology (Carver and Scheier, 1998), it is assumed that car users compare the current situation to a reference value or goal. If they experience a discrepancy, some action is undertaken to reduce the discrepancy. Following this, it is claimed that a reduction or change in car use is jointly determined by (1) changes in attributes characterizing trip chains and (2) reduction or change goals set by car users.

If a TDM measure is implemented, a car-use reduction or change goal may be set when car users experience a worsening of travel options, such as increased monetary costs (e.g., increased gasoline tax) or travel times (i.e., congestion). After having set a car-use reduction or change goal, car users form an implementation plan consisting of sets of predetermined choices contingent on specified conditions. In making such a plan, they

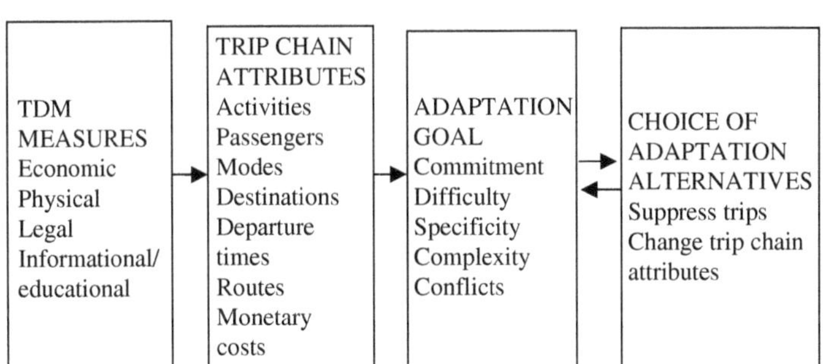

Source: Adapted from Gärling *et al.* 2002a

Figure 1: Conceptual Framework

may choose from among a wide range of options, such as staying at home, thereby suppressing trips for engagement in out-of-home activities, perhaps using electronic communication instead of driving, car pooling, traveling to closer destinations, combining trips, changing route or using other travel modes. They may also consider longer-term strategic changes, such as moving to another residence, changing work place, or changing work hours (e.g., compressing the work week).

Based on the proposed conceptual framework, we claim that there are three main reasons as to why non-coercive TDM measures may not be effective: (1) TDM measures may fail to make car use less attractive; (2) TDM measures may fail to activate goals to change car use; and (3) TDM measures may fail to facilitate the implementation of goals to change car use. Evidence consistent with these claims comes, for example, from analyses of travel diary data reported by Loukopoulos *et al.* (2005a), observing that the number affected trips varied greatly depending on the temporal and spatial specifications of TDM measures. That is, one cannot assume that a TDM measure makes car use less attractive in a given area without first examining existing travel patterns. Furthermore, Loukopoulos *et al.* (2004) demonstrated that individualized marketing may lead to significantly smaller car-use reduction goals being set than either road pricing or prohibition. The same study also revealed that, for example, people believed that their shopping activities would not be affected because such activities would be conducted outside the typical hours of operation of road pricing; that is, car use reduction goals were not activated in this instance. Finally, more recently, Loukopoulos *et al.* (2006) found evidence suggesting that the adaptation to TDM measures took the form of a change hierarchy, which proceeded according to a general cost-minimization principle, with the less costly adaptation alternatives being selected first. The exact nature of the hierarchy varied from trip purpose to trip purpose, suggesting that less costly alternatives for one trip purpose (e.g., public transport for work trips) are less costly for another (e.g., public transport for shopping trips). The implication is that non-coercive TDM measures will not be effective if the resulting alternatives are too costly, since these alternatives are unlikely to be chosen.

EFFECTIVENESS OF COERCIVE TDM MEASURES

A coercive measure such as prohibition does not need to rely on car users setting car-use reduction or change goals. An issue is instead, whether it can be applied such that it changes people's travel options to a sufficient and an appropriate degree. Another issue concerns, whether there are alternatives that fulfil the manifest travel demand that the car fulfilled or not. Both these issues have been discussed below.

Only a limited number of published studies seem to have assessed the effects of coercive TDM measures on different households' travel options. Still, it may be concluded from an extensive study by Timmermans *et al.* (2003) that the results of such assessments would most likely generalize to different urban areas, partly because their spatial structure is similar, and partly because human activities motivating travel are similar for different spatial structures. Loukopoulos *et al.* (2005a) showed that closing out car traffic from downtown may not affect many car trips. In fact, depending on the timescale (time of day in which restrictions were in place: weekday morning peak hours, weekday daytime hours or 24 hours a day, seven days a week) and the spatial scale of the prohibited zone, the percentage of affected car trips varied from about 2% to close to 50%. This suggests that prohibition changes the travel options for only a small proportion of the population. Thus, effectiveness is not guaranteed unless time- and spatial scales are chosen appropriately. In the particular case investigated by Loukopoulos *et al.* (downtown Göteborg, Sweden), it is still likely that prohibition achieved the goals the municipality had set for their traffic and environmental policy.

As an additional example, Santos and Rojey (2004) demonstrated that the effects of road pricing might be progressive, regressive or neutral depending on where it is implemented, where people live, where they work, and existing modal splits. A progressive scheme would imply that those with higher incomes or wealth pay a higher proportion of their assets than those with lower incomes. Regressive schemes, in contrast, imply that those with lower income pay a higher proportion of their incomes as a result of the implantation of the road-pricing scheme. A neutral scheme is neither progressive nor regressive. Thus, it may be concluded that road pricing is coercive only for some car users. Of course, the fee level is decisive in determining how many and who are affected.

The second issue is whether or not alternatives are available. Today, fast and effective transportation is essential for the welfare of industrialized societies. A coercive measure, such as prohibition directly targets car use. However, if it is assumed that few trips can be suppressed, improvements to alternative modes of transport (e.g., better public transport, more cycle paths) are essential. Research has, however, demonstrated that it is not often the case that such improvements in alternative travel options succeed. There are several reasons for this. Two of them have to do with the habitual nature of car use, which interferes with the information acquisition process regarding alternative modes (see Fujii and Gärling, Chapter 13 in this volume), and with the fact that the alternative modes are not perceived to have been made more attractive, such that the car remains the mode with superior travel options even for those who do not drive (see Steg, 2003b; Jakobsson, Chapter 11 in this volume). Still, in the case of coercive TDM measures, the problem is to assess what constitutes satisfactory quality of public transport to those who use it (see, e.g., Friman *et al.*, 1998). The results of studies of travel mode choice are thus less relevant.

Alternatives to the car may still never be fully acceptable substitutes. The question therefore needs to be raised, concerning what negative side effects are to be expected. Possible side effects were discussed by Gärling *et al.* (2002b). Although research is relatively silent on the issue, it does not seem far-fetched to believe that immediate direct effects include less flexibility and increased time pressure. It is an open question whether this only results in a greater need for personal planning or to more devastating indirect effects, such as increased stress levels or impaired family function. Of course, such negative effects must be weighed against increased quality of urban environments.

COERCIVENESS

It may be assumed that coerciveness of TDM measures forms a continuum from coercive to non-coercive along which TDM measures vary. As an example, prohibition is likely to be more coercive than road pricing, which is likely to be more coercive than individualized marketing. In real-world applications, the situation is not so simple as other TDM measure attributes, which co-vary and interact with coerciveness, are likely to play a role (e.g., spatial and temporal scale; see Loukopoulos, Chapter 15 in this volume), as may individual and psychological characteristics (e.g., household income; see Steg and Schuitema, Chapter 19 in this volume). Still another qualification is that perceptions may play a role for how coercive a TDM measure is.

Setting aside these caveats for the time being, it is argued that prohibition (where one cannot drive in a certain area) and individualized marketing (where one has total freedom to choose if and when the car is to be used) form the endpoints of a *coerciveness continuum*, with road pricing falling somewhere in between. Indeed, previous research has found evidence consistent with this continuum. For example, Loukopoulos *et al.* (2005b) examined beliefs concerning the effects of prohibition, road pricing, and individualized marketing on urban environmental quality (i.e., air pollution, noise pollution, green area preservation), car accessibility (i.e., traffic flow, accessibility in downtown areas), accessibility for modes other than the car, and travel costs. Prohibition, presumed to be most coercive, was seen by the participants to yield the greatest improvement in urban environmental quality and in accessibility for other modes, and it was seen to yield the greatest decrease in car accessibility. Individualized marketing had the smallest impacts, with road pricing falling in between. Travel costs were believed by participants to increase most for road pricing.

Participants in Loukopoulos *et al.* (2005b) were provided with three real-world descriptions of the TDM measures and were asked to indicate their attitudes toward these measures if they were implemented in Göteborg, Sweden. This is an important point. The assumed continuum of coerciveness, while relatively easy to define, assumes that other attributes of TDM measures are held constant, for example, for any given spatial scale, prohibition is more coercive than road pricing.

This leads to the question of how to define coerciveness and whether or not objective or subjective definitions should be stressed. While in many cases, changes in objective coerciveness may be easy to argue for (e.g., an increase in the spatial scale of prohibition holding other attributes constant, or an increase in the costs of a congestion charging scheme), in other cases this is not so simple, particularly when comparing across different types of TDM measures: Is a congestion charging scheme which increases its spatial extent, but decreases the requisite toll more or less coercive than before? Or is prohibition in a small area more coercive than congestion charging in a large area with a small toll? Does increasing a toll by a small amount result in an increase in objective coerciveness, but not subjective coerciveness?

The presence of such ambiguity suggests that subjective definitions of coercion are more likely to be of importance when examining the effectiveness of a TDM measure in reducing car use, or for that matter when examining impacts on public attitudes and acceptance, as well as political feasibility. As an example, research has found that policy design features (e.g., whether or not revenues from road pricing are earmarked for other transport-system improvements) and individual and household factors (e.g., problem awareness, trust in governments, income levels) influence the acceptability of TDM measures (see Jones, 1991; Steg and Schuitema, Chapter 19 in this volume, for a review of these factors for the specific case of transport pricing). It is conceivable that many of these policy and demographic characteristics also influence the perceived coerciveness of a TDM measure. Further research on the exact nature of coerciveness and its determinants are needed, as it is the coerciveness of a TDM measure that is explicitly related to whether or not a TDM measure noticeably makes travel options by car worse, such that a car-use reduction goal is activated and implemented.

IMPEDIMENTS TO IMPLEMENTATION OF COERCIVE TDM MEASURES

We argue that public support is an important precondition for implementing TDM measures, especially when they are perceived to be coercive. Thus, coercive TDM measures may be effective but not acceptable to the public. If public opposition is strong, politicians are also likely to be hesitant to implement such measures. We believe that they otherwise would be attractive to politicians, since they are effective in achieving a car-use reduction. These relationships are illustrated in Figure 2.

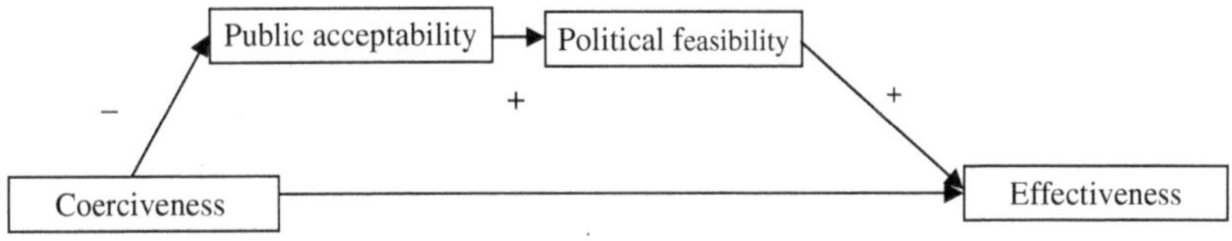

Figure 2: Assumed Mediation of Public Acceptability and Political Feasibility in the Relationship between Coerciveness and Effectiveness of TDM Measures

We thus identify coerciveness as an important attribute of TDM measures that affect the possibility of implementing such measures. Still, as will be discussed in the following subsections, other factors also affect public acceptability and political feasibility.

Public Acceptability

Car users are opposed to TDM measures if they are perceived to be ineffective and limit their freedom to drive; in particular, if they believe that they cannot reduce car use without large sacrifices (Bamberg and Rölle, 2004; Jakobsson *et al.*, 2000). There may also be opposition, because the TDM measure is perceived as unfair if some suffer more than others. Yet, prohibition of car use is perceived as fairer (and also more effective) than road pricing, presumably because the latter is believed to allow wealthy people to buy their way out (Jones, 1995, 2003). Another possibility is that one believes — sometimes mistakenly, as reviewed earlier with reference to work by Santos and Rojey (2004) — that predominantly poor people will suffer. As shown by social justice research (e.g., Deutsch, 1985), even though minimizing the differences in outcomes between groups is the most important goal (referred to as equality), when this goal implies that unprivileged groups will suffer, need will dominate equality as a fairness principle.

Other characteristics of a partially coercive TDM measure, such as road pricing include fee level and the allocation of revenues. It has been shown that revenues that benefit the individual car user, for instance, by decreasing road or fuel taxes, are more acceptable as compared to revenues that benefit society as a whole, such as general public funds (Steg and Schuitema, Chapter 19 in this volume). Thus, a possible way of increasing the acceptability of economic sanctions is to have revenues compensate for the infringement on freedom and unfairness.

Political Feasibility

Coercive TDM measures may not be politically feasible because they are not acceptable to the public. Yet, there are also other reasons. In political decision-making, solutions must be found for conflicts between different goals, either because different political parties (ideologies) prioritize different goals or because a TDM measure in itself leads to conflicting goals. In Europe, reducing car traffic is attractive to environmental or green parties, but not to social democratic or conservative parties, since they view it as a threat to economic development and, in the longer-term, the welfare of society (Johansson *et al.*, 2003). An additional reason for the conservative opposition is a strong belief in deregulation and free-market solutions. A similar goal conflict exists in political decision-making between the municipal and national levels. Although politicians at both levels are sensitive to public opposition, they may be so more at the local and municipal level

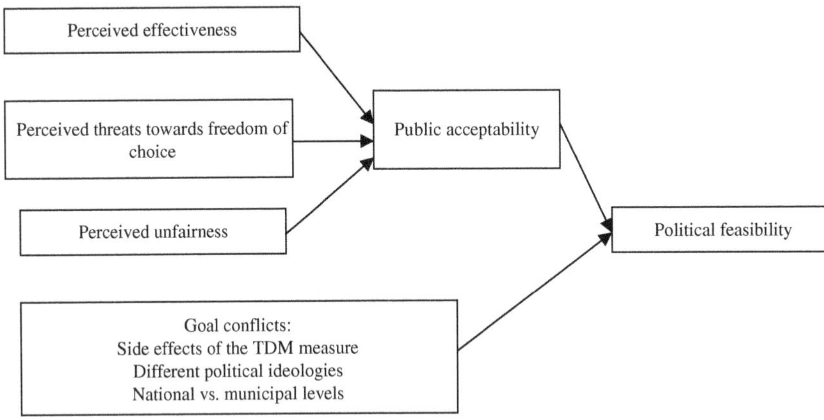

Figure 3: Factors Assumed to Affect Political Feasibility

because of closer ties between politicians and their voters. It may also be the case that public opposition is stronger in a municipality where implementation of the TDM measure is considered.

A conflict also emanates from the fact that TDM measures have both intended and unintended effects. For instance, road pricing may alleviate congestion if prices are set sufficiently high. At the same time, this may reduce the profits of businesses if their gains from faster transportation due to less congestion fail to offset the increased costs. In a similar vein, poor households may no longer afford leisure travel that they consider highly desirable, at least not without cutting expenditures for, perhaps, other essential products and services. Prohibition may have even stronger side effects unless exceptions are made for essential road use.

The assumed relationships are represented in Figure 3. It is assumed that political feasibility increases with public acceptability, but is also affected by the existence of conflicts between different political goals. Given public acceptability of a coercive TDM measure, implementation of the measure is likely only if such goal conflicts can be resolved, so that the goal of car-use reduction is not compromised.

CONCLUSIONS AND IMPLICATIONS

In this chapter, we have argued and provided some evidence for that voluntary, non-coercive measures are not likely to succeed in curtailing current car use in urban areas. Still, if current knowledge about the effectiveness of non-coercive measures is applied to a greater

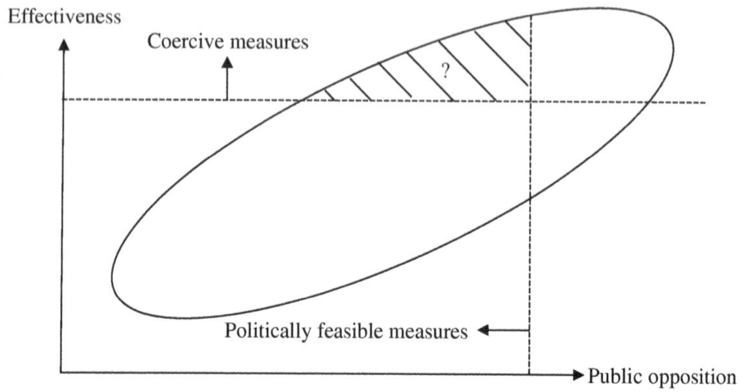

Figure 4: Possible Existence of Coercive TDM Measures that are Effective and Politically Feasible

extent, such measures may be made more effective (see Thøgersen, Chapter 20 in this volume). Furthermore, from the existing knowledge one would also expect that events such as, for instance, a temporary freeway closure might have lasting effects if acceptable alternatives are available (Fujii and Gärling, 2005). The underlying theoretical principle is to change old "bad" habits to new "good" habits (see Chapter 13 by Fujii and Gärling in this volume). Since effective coercive TDM measures appear to meet with public opposition and are not politically feasible, the current problems appear impossible to be solved. Is there any hope? As illustrated in Figure 4, this raises the question of whether any effective coercive TDM measures exist that are acceptable to the public and are they politically feasible? In our view, this is perhaps the most important question that research in this area should address.

In the shortterm any large changes in current transport systems are not feasible (Goodwin, 1998). Land-use patterns and infrastructure investments are forceful counteracting factors. A reduction of car use in the order of 10%–25% would probably be possible and would have beneficial effects (Gärling *et al.*, 1998). It is likely that this degree of reduction would be accomplished by coercive measures. If *car use* is targeted, rather than car owners, there may be less public opposition and coercive measures may become more politically feasible (Wright and Egan, 2000). Prohibitive rationing schemes have been tried in the past (Lyons and Chatterjee, 2002) and may be tried again. It may be essential that the implementation of coercive measures is accompanied by launching public information campaigns targeting attitude change, improving alternatives, increasing the monetary costs of driving, and installing advanced information technology (i.e., intelligent transportation systems, see Gärling *et al.*, 2004).

ACKNOWLEDGEMENTS

The authors' own research was financially supported by grants from the Swedish Agency for Innovation Systems (#2002–00434) and the Swedish Research Council for Environment, Agricultural Sciences, and Spatial Planning (#25.9/2001–1763).

REFERENCES

Bamberg, S. and D. Rölle (2004). Determinants of People's Determinants of Pricing Measures. In: *Acceptability of Transport Pricing Strategies* (J. Schade and B. Schlag eds), Elsevier, Amsterdam, 235–248.

Carver, C.S. and M.F. Scheier (1998). *On the Self-Regulation of Behavior*, Cambridge University Press, Cambridge.

Deutsch, M. (1985). *Distributive Justice: A Social Psychological Perspective*, Yale University Press, New Haven, CT.

Friman, M., B. Edvardsson, and T. Gärling, (1998). Perceived Service Quality Attributes in Public Transport: Inferences from Complaints and Negative Critical Incidents, *Journal of Public Transportation*, **2**, 67–89.

Fujii, S. and T. Gärling, (2005). Temporary Structural Change: A Strategy to Break Car-Use Habit and Promote Public Transport. In: *Traffic and Transport Psychology* (G. Underwood, ed), Elsevier, Amsterdam, 585–592.

Gärling, T., D. Eek, P. Loukopoulos, S. Fujii, O. Johansson-Stenman, R. Kitamura, R. Pendyala, and B. Vilhelmson, (2002a). A Conceptual Analysis of the Impact of Travel Demand Management on Private Car Use, *Transport Policy*, **9**, 59–70.

Gärling, T., A. Gärling, and P. Loukopoulos, (2002b). Forecasting Psychological Consequences of Car Use Reduction: A Challenge to an Environmental Psychology of Transportation, *Applied Psychology: An International Review*, **51**, 90–106.

Gärling, T., R. Gillholm, and A. Gärling, (1998). Reintroducing Attitude Theory in Travel Behavior Research: The Validity of an Interactive Interview Procedure to Predict Car Use, *Transportation*, **25**, 147–167.

Gärling, T., C. Jakobsson, P. Loukopoulos, and S. Fujii, (2004). Roles of Information Technology for Households' Adaptation of Private Car Use to Travel Demand Management Measures, *Journal of Intelligent Transportation Systems*, **8**, 186–194.

Goodwin, P. (1998). The End of Equilibrium. In: *Theoretical Foundations of Travel Choice Modeling* (T. Gärling, T. Laitila, and K. Westin eds), Elsevier, Amsterdam, 103–132.

Greene, D.L. and M. Wegener, (1997). Sustainable Transport, *Journal of Transport Geography*, **5**, 177–190.

Jakobsson, C., S. Fujii, and T. Gärling, (2000). Determinants of Private Car Users' Acceptance of Road Pricing, *Transport Policy*, **7**, 153–158.

Johansson, L.O., M. Gustafsson, G. Falkemark, T. Gärling, and O. Johansson-Stenman, (2003). Goal Conflicts in Political Decision Making: A Survey of Municipality Politicians' Views of Road Pricing, *Environment and Planning C: Government and Policy*, **21**, 615–624.

Jones, P. (1991). Gaining Public Support for Road Pricing Through a Package Approach, *Traffic Engineering and Control*, **32**, 194–196.

Jones, P. (1995). Road Pricing: The Public Viewpoint. In: *Road Pricing: Theory, Empirical Assessment and Policy* (B. Johansson and L.-G. Mattsson, eds), Kluwer, Dordrecht, The Netherlands, 159–179.

Jones, P. (2003). Acceptability of Transport Pricing Strategies: Meeting the Challenge. In: *Acceptability of Transport Pricing Strategies* (J. Schade and B. Schlag eds), Elsevier, Amsterdam, 27–62.

Kitamura, R., S. Fujii, and E.I. Pas, (1997). Time-Use Data, Analysis and Modelling: Toward the Next Generation of Transportation Planning Methodologies, *Transport Policy*, **4**, 225 – 235.

Loukopoulos, P., C. Jakobsson, T. Gärling, S. Meland, and S. Fujii, (2006). Understanding the Process of Adaptation to Car-Use Reduction Goals, *Transportation Research F*, **9**, 115–127.

Loukopoulos, P., C. Jakobsson, T. Gärling, C.M. Schneider, and S. Fujii, (2004). Car User Responses to Travel Demand Management Measures: Goal Setting and Choice of Adaptation Alternatives, *Transportation Research* D, **9**, 263–280.

Loukopoulos, P., T. Gärling, and B. Vilhelmson, (2005a). Mapping the Potential Consequences of Car-Use Reduction in Urban Areas, *Journal of Transport Geography*, **13**, 135–150.

Loukopoulos, P., C. Jakobsson, T. Gärling, C.M. Schneider, and S. Fujii, (2005b). Public Attitudes Towards Policy Measures for Reducing Private Car Use: Evidence From a Study in Sweden, *Environmental Science and Policy*, **8**, 57–66.

Lyons, G. and K. Chatterjee. (eds) (2002), *Transport Lessons From the Fuel Tax Protests of 2000*, Ashgate, Aldershot, UK.

OECD, (1996). *Towards Sustainable Transportation*, OECD Publications, Paris.

Santos, G. and L. Rojey, 2004. Distributional Impacts of Road Pricing: The Truth Behind the Myth, *Transportation*, **31**, 21–42.

Steg, L. (2003a). Factors Influencing the Acceptability and Effectiveness of Transport Pricing. In: *Acceptability of Transport Pricing Strategies* (J. Schade and B. Schlag, eds), Oxford, Elsevier, 187–202.

Steg, L. (2003b). Can Public Transport Compete With The Private Car? *IATSS Research*, **27**, 27–35.

Timmermans, H., P. van der Waerden, M. Alves, J. Polak, S. Ellis, A.S. Harvey, S. Kurose and R. Zandee, 2003. Spatial Context and The Complexity of Daily Travel Patterns: An International Comparison, *Journal of Transport Geography*, **11**, 37–46.

U.S. Department of Energy (2002). *Annual Energy Review 2000*, Energy Information Administrator, Washington, D.C.

Wright, C. and J. Egan (2000). De-Marketing The Car, *Transport Policy*, **7**, 287–294.

Tommy Gärling and Linda Steg (Editors)

18

THE ECONOMIC THEORY OF TRANSPORT PRICING

Barry Ubbels, NEA Transport Research and Training (member of Panteia), Rijswijk, The Netherlands
Erik Verhoef, Department of Spatial Economics, Free University of Amsterdam, Amsterdam, The Netherlands

ABSTRACT

This chapter gives an overview of the economics of transport pricing and provides some basic understanding of key issues important for prices in transport markets. The pricing policy adopted depends upon objectives. The objective of economic efficiency may be most important to society; prices must then be equal to marginal social costs. In most cases, however, actual transport prices deviate from marginal costs due to specific transport market conditions (e.g. externalities) and the presence of constraints (e.g. practical and legal). The regulator has then obviously to resort to 'second-best' pricing: setting prices that are available optimally under the constraints applying.

INTRODUCTION

Transport forms an important part of everyone's life. The spread of production, trade and ideas, and the economic ascendancy of mankind all depend upon movement. Adequate transport is something that we tend to take for granted in the industrialised world and if it is not available public concerns soon emerge. The reasons for this are not difficult to find. Comprehensive transport provision is perceived as an important input into the efficient functioning of modern industry and commerce. It also affords individuals and households the benefits of mobility.

Transport has some characteristics that makes it different from other goods. Possibly the most important characteristic of transport is that it is often not really demanded in its own right (Button, 1993). People wish, in general, to travel so that some benefit can be obtained at the final destination. Similarly, users of freight transport perceive transport as a cost in their overall production function and seek to minimise it wherever possible.

While the demand for transport has particular, if not unique, features, also certain aspects of supply are entirely peculiar to transport. More specifically part of the plant is mobile – almost by definition – and is entirely different in its characteristics to the fixed plant (for example, roads, airports, etc.). The fixed component is usually extremely long-lived and expensive to replace. Further, few pieces of transport infrastructure have alternative uses.

Demand and supply work together to determine the market price in competitive markets. The price of a good or a service is what must be given in exchange for the good or service (Stiglitz and Driffill, 2000). When the forces of supply and demand operate freely, price measures scarcity. In addition, in the competitive model, the equilibrium price of an object will normally equal its cost of production (including the amount needed to pay a firm's owner to stay in business rather than seek some other form of employment). Elementary economics tells us that in the long-run price will then be equated with the marginal (and average) costs of each supplier. But the transport market is different. Simple market economic theory cannot directly be applied to transport for a variety of reasons. Since journeys are unique in space and time monopoly is likely to arise in varying degrees, especially when technological change offers an advantage to a particular mode or where economies of scale affect one mode more than another. This situation also affects the pricing of transport services. Transport prices do not simply result from the law of supply and demand. This contribution addresses theoretical transport pricing principles and some complexities that arise due to market conditions generally observed in the transport sector. The aim is to provide a basic exposition of key issues important for prices in transport markets. Although general pricing issues in transport will be discussed, particular attention will be paid to pricing of passenger car transport. This means that we do not discuss pricing issues particularly relevant to other modes (such as the Mohring effect in public transport pricing).

This chapter comprises several sections. We start with a general discussion on pricing issues in transport. The third section will focus on efficient pricing. It also discusses the existence of externalities and the implications for (efficient) pricing. The fourth section deals with equity issues. First-best pricing is a useful benchmark, but the practical relevance seems limited due to various constraints. Therefore, the fifth section will focus on some second-best pricing issues, which are nowadays often analysed in transport economic literature. Finally, the sixth section summarises.

TRANSPORT PRICING

Pricing can be seen as a method to affect resource allocation. Pricing strategies permit specified aims to be achieved, there is no such thing as the right price independent of the aims pursued. The pricing policy adopted by any transport undertaking with some degree of market power depends upon its basic objectives. For example, an optimal price aimed at achieving profit maximisation may differ from that needed to maximise social welfare[1], or to ensure highest sales revenue. In some cases, there is no attempt to devise a price to maximise or minimise anything, but prices are rather to permit lower level objectives (for example, security, minimum market share) to be attained. Further, prices may be set to achieve certain objectives for the transport supplier in terms of his welfare. This is normally the case of private enterprise transport undertakings, while in other areas prices may be set to improve the welfare of consumers (as has been the case with publicly owned transport undertakings). This distinction is important, as many undertakings consider that the employment of the pricing mechanisms to achieve their objectives is automatically to the benefit of customers.

It is clear that pricing objectives differ depending on the provision of transport services (public or private) and market conditions. The following pricing objectives can be distinguished:

- Economic efficiency[2];
- Profit maximisation;
- Cost coverage;
- Environmental sustainability;
- Equity (including redistributive objectives);
- Objectives transcending the boundaries of transport markets, including macro-economic objectives.

The objective of economic efficiency is usually important to governments as it reflects the aim to maximise welfare of all inhabitants, it will be discussed in more detail in the following section. Profitability reflects the traditional economic assumption that firms set prices as to maximise profits. Variations on this theory suggest that many undertakings adopt prices that maximise sales revenues (Baumol, 1962) when in an expansive phase, or simply price to ensure that certain satisfactory levels of profit or market domination are achieved (Simon, 1959). A third possible objective is that of cost coverage. Most publicly owned firms are not so much focused on making profits, but rather to stay in business and recoup their costs, often induced to do so for political or fiscal reasons. Protection of the environment has become an important objective for governments in recent years. Transport in general, and road transport in particular, are widely recognised

[1] Social welfare refers to the measure used to express a society's aggregate well-being. It can be defined in many ways, most of which take individuals' utilities as a building block. Applied research often uses (weighted) sums of individual welfare measures, which is also true for 'social surplus' as we will use below. There is no objective criterion for the specification of a social welfare function; i.e. economists cannot define it objectively. See, for example, Atkinson and Stiglitz (1980) for further details.

[2] Economic efficiency is concerned with the use of society's resources such that no mutually beneficial transactions remain possible.

as an important source of pollution which threatens environmental sustainability (see Chapter 2 by Van Wee, this volume). Pricing measures have been suggested or introduced to deal with these problems. It is arguable that promoting environmental objectives is consistent with the aim of securing welfare maximisation through economic efficiency, in particular when social welfare incorporates environmental social costs and benefits.

Equity objectives and the distribution of real incomes in society are important issues to a government, reflected in the pattern of taxation and public expenditures. Whilst transfer payments, such as benefits and pensions, are a major means of redistributing income, the provision of services, such as transport at subsidised prices, is often considered to be equally important (United Nations, 2001). Moreover, tax policies (or other policies) aimed at regulating transport and the various possible allocations of tax revenues, will have distributional consequences that may or may not match more generally formulated distributional targets, and may therefore motivate adjustments in currently used (distortive) taxes, which in turn implies that indirect efficiency effects may occur elsewhere in the economy. Finally, public bodies are concerned with macroeconomic policy objectives. Governments usually focus on four target variables: the level of unemployment, the rate of inflation, the balance of payments, and the rate of growth of national output (see Stiglitz and Driffill, 2000). The level of investment in, and the pricing of, transport infrastructure and transport services both affects and is affected by macroeconomic policies.

These sorts of objectives mentioned above are complex and are often not compatible (see Table 1 for some examples). Whilst there are many transport pricing objectives, economists often focus on the pursuance of economic efficiency in the transport sector alone. Prices that are socially optimal are seen as the first-best benchmark, which is in most cases politically desired.

Nevertheless, an expanding body of literature on transport pricing is emerging that considers pricing and revenue allocation in the context of a wider – general equilibrium – framework, in which (tax)distortions elsewhere in the economy and distributional

Table 1: Pricing Policy Objectives and Possible Conflicts

Pricing Policy Objectives	*Conflicts*
Economic efficiency vs profit maximisation (or cost coverage)	Efficient pricing of the use of transport capacity may lead to financial losses for the owner
Profitability vs income distribution	Pricing for profitability may lead to higher transport prices with adverse effects on poor income groups
Economic efficiency vs macroeconomic policy	Macroeconomic price restraint policies may conflict with the need to increase transport prices during periods of congestion and excess demand

Source: Adapted from United Nations (2001).

objectives as represented in social welfare functions are considered explicitly (e.g. Mayeres and Proost, 1997; Parry and Bento, 2002).

In conclusion, the efficient use of resources and the optimisation of social welfare is clearly an important objective. It may even be the most important to governments involved setting prices in mobility. In the following, the consequences of realising maximum welfare for price setting in general and transport pricing in particular will be discussed. We start by considering what is often called a 'first-best' world: apart from the (transport) price to be determined, all other markets have efficient pricing, there are no other constraints on the transport prices, and no market failures to be considered[3]. Second-best pricing issues will be discussed in the fifth section.

ECONOMIC EFFICIENCY AND EXTERNALITIES

Efficiency: Marginal Cost Pricing

The concept of economic efficiency is derived from the theory of welfare economics, and is concerned with the allocation of resources in an economy. Welfare economics takes a rather wide view of pricing, looking upon price as a method of resource allocation that maximises social welfare rather than simply the welfare of the supplier (Button, 1993). According to this view, prices should equate with marginal social cost (MSC) to obtain maximal social welfare. Sometimes, private provision of the good or service may also result in maximising the social welfare. Otherwise, regulatory policies may be applied to private companies so that their pricing policy is modified to maximise social rather than private welfare. This may take the form of price regulation, or taxing firms so that their prices become socially optimal. This will be explained below.

Deriving socially optimal prices needs an objective function (describing the target to be optimised, in this case social welfare). The most general form of this function is a social welfare function. Formally, a social welfare function $W = W(V^1 \ldots V^n)$ has as its arguments the indirect utility functions V^i of individuals i, $i = 1 \ldots N$ (Varian, 1999). These indirect utility functions indicate the maximum utility levels of the individuals at given prices, incomes, and magnitudes of externalities such as congestion and pollution. The social welfare function inevitably incorporates welfare judgements with respect to the distribution of economic resources. These value judgements will be reflected in the policy prescriptions based on the welfare function.

[3] First-best pricing is often a theoretical benchmark since real world is usually different from many distortions. Economists therefore focus often on setting prices that are available optimally, under the constraints applying, the so-called second-best prices (see the fifth section).

Several alternatives can be used as objective. These include the search for local improvements to welfare (rather than a global optimum), Pareto improvements, potential Pareto improvements, compensating and equivalent variation, and social surplus[4]. This last one is the most commonly used for applied welfare analysis. Social surplus is defined to be the sum of producers' surplus and consumers' surplus (CS). Consumers' surplus represents the benefit to consumers, as expressed by their willingness to pay, in excess of the price to be paid to obtain a particular quantity or level of output. Producers' surplus represents the revenue in excess of the (variable) cost of providing that level of output. The principal advantage of CS in applied work is that it can be calculated using aggregate demand curves. It measures the change in the area under the demand curve for a certain good as the welfare effect of a price change. Intuitively, the area under the demand curve and above the prevailing price measures the additional amount of money the consumer would have been willing to pay for the units of the goods consumed by him. It can therefore be regarded as the amount of money that has not to be spent on this good and remains available for other purposes. A drawback of this welfare measure is that aggregate CS makes the implicit welfare judgement that welfare distributional concerns should not matter, by giving equal weight to the surplus of all individuals.

Despite this latter disadvantage, the considerable practical advantages of CS make it attractive for applied work. CS is most commonly used to measure welfare changes associated with the change in the price of a good. Willig (1976) has shown that CS serves as a good approximation to the compensating and equivalent variation, which are exact measures of changes in welfare, in particular as long as any one of three conditions hold:

1. The price change is not large;
2. Expenditure on the commodity is not a large fraction of income;
3. The income elasticity of demand is not large.

Most applied studies assume that interpersonal utility comparisons can be made, and seek to maximise the sum of consumers' and producers' surpluses. Diagrammatically, this can be illuminated as in Figure 1. In the diagram, MC is the marginal cost curve or, under perfectly competitive conditions, the supply function. Marginal cost is the increase in total cost that occurs from producing one more unit of output or service. In the figure, it is assumed that marginal costs increase when output increases. However, this is not necessarily the case in the transportation industry, where firms may face large

[4]Compensating and equivalent variation and social surplus are measures for individual welfare changes, for instance when prices change. Compensating variation expresses the change in utility as the change in income necessary to restore the consumer to his original utility. Equivalent variation measures the impact of a price change by asking how much money would have to be taken away from the consumer before the price change to leave him as well off as he would be after the price change. Measures for group welfare include Pareto efficiency (or economic efficiency) which says that one situation should be preferred to another if all persons are at least as well of and at least one is strictly better off. If an allocation allows for a Pareto improvement, it is Pareto inefficient, when no improvements are possible the situation is Pareto efficient.

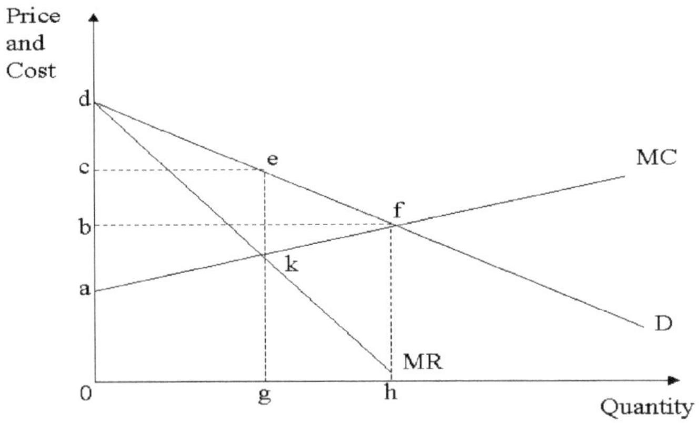

Figure 1: Welfare Maximisation in a Partial Equilibrium Setting

fixed costs. D and MR represent the inverse demand and marginal revenue curves, respectively. A profit-maximising firm would produce the output *g* at the price *c* (marginal revenues equals marginal cost). Consumers' surplus (assuming no income effects) is equal to the area *cde*, and producers' surplus is represented by the area *acek*, leading to a total social surplus of *adek*.

If this firm sets its price equal to marginal cost, output would expand to *h* and price would fall to *b*. Consumers' surplus would be given by the area *bdf*, and producers' surplus by the area *abf*, resulting in a social surplus of *adf*. When we compare this with the previously obtained aggregate, it appears that the triangle *kef* makes the difference. The sum of the two surpluses is at the maximum when price equals marginal cost. In other words, social welfare is maximised when price is equated with MSC (Pareto efficiency).

What marginal cost pricing does, in effect, is to result in transport services being provided up to the point where the benefit for the marginal unit is equated with the costs of providing that unit (Button, 1993). The traditional theory suggests that such a condition prevails in the long term when perfect competition exists, despite the fact that each firm is trying to maximise its own profits, i.e. producers' surplus. The ability to exercise any degree of monopoly power, however, permits a firm to set prices above marginal cost, so that it can achieve additional profit at the expense of reduced output and at a cost to the consumer. This higher price, charged by the profit maximising monopolist, will deny some consumers the use of a service, even though they are prepared to pay at least the MSCs involved. This fear has been explicitly high in the transport market, leading to a widespread price regulation in many sectors by numerous governments.

The so-called first-best optimum of point *f* in Figure 1, which will prevail as a market equilibrium under certain conditions, among which:

- Perfect competition;
- No distortions in other market segments;
- No externalities;
- Perfect information;
- No subsidies or indivisibilities of demand or supply.

An allocation is to be said first-best if it maximises social welfare subject to the irreducible technological constraints of production (Dreze and Stern, 1987). A first-best optimum in transport is an allocation defined by quantities of goods, including passenger and freight transport volumes, that maximises $W(\cdot)$ given the prevailing technology such as vehicle fuel consumption and emissions, and the capital stock including transport infrastructure (MC-ICAM, 2002). This definition encompasses externalities if their costs are internalised in the decisions of agents who generate them and included in their utility functions. Economic efficiency then implies that the full costs of transport services are accounted for, including social and environmental costs (no externalities).

Externalities: Transport Pricing and Congestion

The previous revealed that equality of prices and MSCs leads to an efficient use of resources in an otherwise ideal world. But the practical world is not ideal. Actual (market) prices may deviate from marginal costs for a number of reasons. The presence of imperfect information or externalities in a market may cause a deviation from the optimal solution. In this subsection, we deal with one particular type of market failure: external effects. Specific attention will be paid to one particular type of externality that is an important threat to the quality of urban life: congestion (see Part 1 of this book for other threats to quality of life from car traffic).

Externalities

Formally, externalities exist when the activities of one group (either consumers or producers) unintentionally affect the welfare of another group, without any payment or compensation being made (Button, 1993). Most attention in transport is paid to the negative (costs) externalities, although also positive externalities (benefits) have been identified (see for a discussion on this latter issue, Verhoef, 1996). It is quite clear from everyday experience, that there are costs associated with transport that are not directly borne by those generating them. Transport generates many negative externalities, including noise, accidents, pollution, and congestion. Road travellers, for example, impose noise and vibration costs on those living adjacent to highways. These costs may be significant. Recent estimates of the external costs give an indication of the significance of these costs (for more information, on measurement for instance, we refer to Chapter 7 by Domergue and Quinet in this volume). Total external costs (excluding congestion) in Western Europe (EU 15 plus Norway and Switzerland) have been estimated to amount to €530 billion for 1995, or 7.8% of the total GDP in these countries (INFRAS/IWW, 2000). Accidents are the most important cost category, and responsible for 29% of total external costs. Air pollution (25%) and climate change (23%) costs are other important cost categories. The congestion costs are separately estimated in the same study and amount approximately €33.3 billion (0.5% of total GDP). Although the congestion costs seem relatively small compared to other externalities, they may be substantial when looking at marginal external cost estimates (e.g. marginal external cost estimates from a Dutch study indicate that the external congestion costs exceed the other external cost categories by several orders of magnitude (CE, 1999).

A result from the clear presence of externalities in transport is that the early neo-classical writers studying market failures frequently illustrated their viewpoints using transport examples. Dupuit illustrated efficient pricing of public goods in 1844 (Button and Verhoef, 1998). Coase (1960) considered sparkles from a railway when studying the absence of property rights in relation to the existence of externalities. Another well-known example is that of a congested road, including optimal congestion charges (Pigou in 1920). They all showed that the market mechanism fails to allocate resources efficiently.

It should be emphasised that optimality does not imply the total elimination of congestion and pollution, but rather the achievement of optimal levels of external cost. Ideally, externalities should be contained to the point where the costs of further reductions exceed the marginal social benefits. Let us analyse the problem of externalities by examining road traffic congestion together with the use of pricing methods. Similar analyses will also hold for other transport modes and other externalities, but road traffic congestion is, in a marginal sense, a very important externality, and offers a useful basis for analysis.

Congestion

The demand for transport is not constant over time. In large cities, there are regular peaks in commuter travel while on holiday routes, both within the country and to overseas destinations, there are seasonal peaks in demand. Transport infrastructure, although flexible in the long run, has a finite capacity at any given period of time. When users of a particular facility begin to interfere with other users because the capacity of the infrastructure is limited, then congestion externalities arise. One should add that congestion does not only impose costs on the road user in terms of wasted time and fuel (also referred to as the pure congestion cost) but the stopping and starting it entails can also worsen atmospheric and other forms of pollution. This latter problem is particularly acute with local forms of pollution, because road traffic congestion tends to be focused in urban areas.

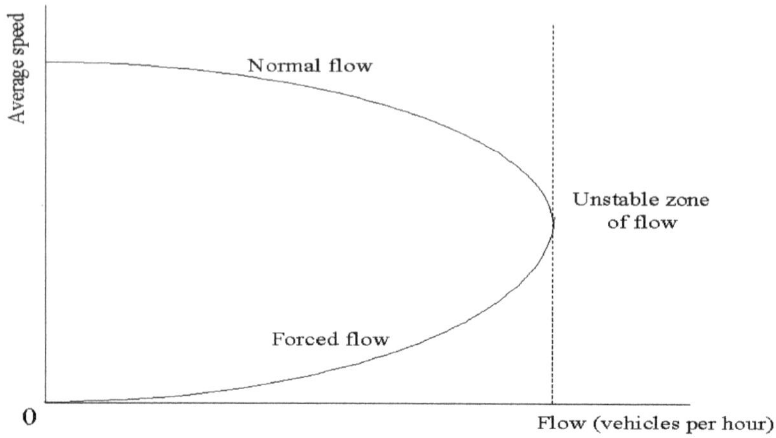

Source: Button (1993).

Figure 2: The Speed–Flow Relationship

One idea for optimising the level of congestion is to use the price mechanism to make road users more fully aware of the impedance they impose upon one another. The idea that motorists should pay for the additional congestion they create, when entering a congested road, was first suggested by Pigou (1920). Ideally, they should negotiate with the actual road users affected, but practically this is impossible, so the idea is that an organisation (e.g. the relevant road authority) should be made responsible for collecting the charges.

The economic costs of road congestion can be calculated using the engineering concept of the speed–flow relationship. If we take a straight one-way street and consider traffic flows along it over a period over time at different speed levels then the relationship between speed and flow would appear as in Figure 2. It shows that as vehicle numbers entering a road increase, average speed at first does not significantly change as does flow, measured by vehicles per hour. At low volumes of traffic, when vehicle impedance is zero, high speeds are possible, constrained only by the capability of the vehicle and legal speed limits, but as the number of vehicles trying to enter the road increases (so they interact with existing traffic) drivers will slow one another down. Up to a point, flow will continue to rise because the effect of additional vehicle number outweighs the reduction in average speed. The maximum flow is reached when increased traffic volume ceases to offset the reduced speed (the road's engineering capacity). Absence of information causes motorists continue to enter the road beyond this volume leading to further drops in speed and resulting in the speed–flow relationship turning back on itself. These levels are known as forced flows in the engineering literature; economists often use the term 'hypercongestion'.

Speed–flow curves can be used to measure the economic costs of congestion. Generalised travel costs include the motoring costs plus the travel time cost for a certain journey. Broadly, faster travel in urban areas means cheaper travel in terms of generalised cost – vehicles are used more effectively and travel times are reduced. The average generalised costs of a trip will increase as flow increases up to the maximum flow rate. It will continue to rise as hypercongestion sets in and produces a reduction in the rate of vehicle flow.

Pigou's economic analysis of road pricing and costs will be explained in the following (Figure 3). We use a simplified framework and are not concerned with matters such as pollution and safety and suppose a simple road without junctions. In these circumstances, road users are identical apart from their marginal willingness to pay for a trip, represented by the inverse demand curve D = MPB = MSB (marginal private and social benefits, respectively)[5]. Individual users entering the road will only consider the costs they personally bear (marginal private cost, MPC), but not the external congestion costs they impose on other road users. The sum of MPC and the marginal external costs gives the MSC. The difference between MSC and MPC is caused by congestion. The individual

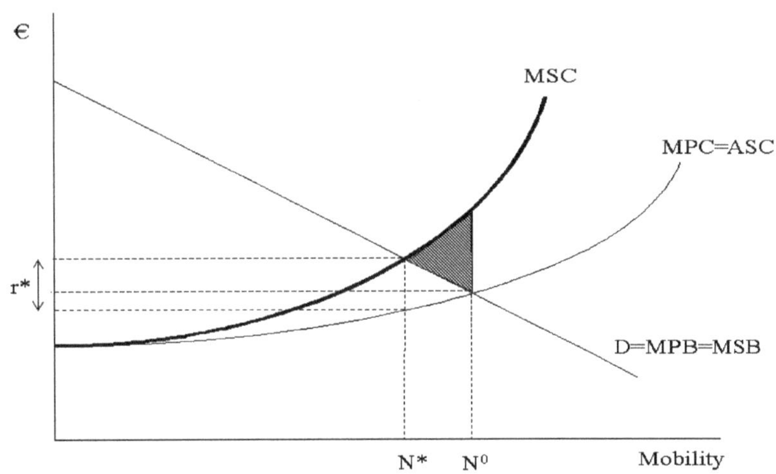

Source: Button and Verhoef (1998).

Figure 3: The Economics of Road Pricing

[5] A popular way of measuring the benefits from consumption in applied economic research is based on the idea that the benefits attached to consumption of a certain unit, expressed in monetary terms, can be equated to the maximum willingness to pay for that unit. A consequence is that the inverse demand function shows the marginal benefits of consumption at every quantity. It shows the benefits from the consumption of the last unit consumed, for an individual (MPB), or aggregated for a group (MSB).

motorist will only consider the average costs experienced by road users and take no account of the impact of their trip on other vehicles. Consequently, the MPC is equal to the average social cost (ASC).

The fact that potential trip makers tend to consider only the private costs of any trip and ignore the external or social costs means that effectively ASC is the perceived supply curve for private motorists. This means that road utilisation will be N^0, at which MPB equals MPC (the free market equilibrium outcome). This is excessive from a social point of view, since at traffic flow N^0, the MSC (including congestion cost) is (much) higher. The socially optimal level of traffic flow is N^*. To internalise the externality or external costs of congestion, it would be necessary to impose a congestion tax. The charge that accomplishes this optimum is the Pigouvian charge r^*. This equals to the difference between the MSCs and MPCs at the optimum, also known as the marginal external congestion costs. The shaded area reflects the welfare gain realised through introducing this charge. The Pigouvian tax is a charge imposed on the traveller or car driver, equal to the amount of the marginal external cost of the transport activity concerned, ensuring a Pareto-efficient allocation of resources. Theoretically, this policy strategy is preferable in a market system, but its implementation may encounter great difficulties as a result of high transaction costs.

Many parts of the transport industry experience a systematic pattern of demand fluctuations leading to congestion within a given time period, the pattern repeating itself from period to period. The duration of the fluctuations is too short to permit capacity to be varied to match them, while it is too expansive to store spare capacity to meet the requirements of the peak demand period. This has led Vickrey (1969) to assume that traffic congestion takes the form of cars queuing behind a bottleneck. This alternative model of traffic congestion endogenised individuals' departure times that generated a wealth of insights into urban rush-hour auto congestion. For example, in its most basic form, the model shows that reduction in overall demand and adjustments in arrival times at destinations are not needed to move from an equilibrium situation to an optimum in which all queuing is eliminated. Adjustment of departure times from the origin suffice. Also the generalised price of travelling does not rise in the optimum compared to the equilibrium, suggesting that travellers are not worse off due to optimal pricing before revenues are redistributed. Arnott *et al.* (1998) discuss these and other results in greater detail.

TRANSPORT PRICING: SOCIAL WELFARE AND EQUITY

From the previous discussion, it becomes clear that in a fully competitive and distortion-free economy, each price equilibrium is a Pareto optimum. This means that the market mechanism is able to guarantee efficiency for any initial distribution of resources over the population. However, this distribution might not be consistent with policy objectives and the public may perceive it as unfair. This section deals with pricing and its consequences in terms of equity. Equity is important from the viewpoint of distribution of income (or other items) and for the acceptability of pricing policies by the public. There are various types of equity concepts important in transport and different strategies to deal with distributional issues.

Equity is important in the context of acceptability of pricing. Stakeholders raise objections about pricing measures, which they perceive to be unfair. And perceived unfairness could create significant acceptability problems. Transport pricing can often be perceived to be regressive, allowing only those with enough money to access a resource (e.g. infrastructure) that was once considered free, or yielding travel time gains that are valued more highly by higher income groups. Implementation strategies may therefore focus on allowing certain sections of the community to be exempted from pricing, or compensating some groups with a lump-sum transfer. The problem of who should receive extra compensation (e.g. tax exemptions), and the wider problem of making sure price measures are both equitable and perceived to be so, are important to be included in any successful implementation strategy.

Public finance and tax literature makes the distinction between horizontal equity and vertical equity. Horizontal equity refers to the principle that those who are in identical or similar circumstances should pay identical or similar amounts in taxes (Stiglitz and Driffill, 2000). It requires that those with equal status – whether measured by ability or some other appropriate scale – should be treated the same. If income were the only measure of a person, for instance, then two persons with equal incomes would be treated as equals. Vertical equity says that people who are better off should pay more taxes (Stiglitz and Driffill, 2000). This generally requires that those with less ability be treated favourably relative to those with greater ability.

The role of these concepts in transport can be illustrated by describing the implementation of road pricing and the use of the revenues. *Horizontal equity* implicates that similar users should pay identical tolls. But the question who 'deserves' the benefit (or revenues) according to this criterion is matter of debate. It can be defined as those who actually pay the toll, or it could also include those who change their behaviour (travel pattern), thereby incurring costs in terms of inconvenience, and providing congestion reduction benefit to the toll payers. So the difficulty is that the initial users of the road have become 'unlike' after the implementation of the charge and should be compensated. The use of road charges to fund public transport is an example. Horizontal equity is further complicated by the existence of externalities from motor vehicle use, including accident risk and environmental degradation. That vehicle use imposes costs on other people itself represents horizontal inequity. If the criterion is horizontal equity and external impacts are recognised, then revenues may be used to compensate for external costs (Litman, 1996). Funding candidates may include environmental and social programs that mitigate the harms of motor vehicle use. Compensation of external costs, however, may in turn, induce inefficient behaviour by recipients of externalities, in the sense that insufficient incentive is provided for avoiding reception of the externality (Oates, 1983; Verhoef, 1994). This implies that (also) from this perspective, there may be trade-offs between efficiency and equity in the regulation of externalities.

Vertical equity is concerned with the treatment of individuals and classes that are unlike. By this principle, the distribution of costs and benefits should reflect people's needs and abilities. Progressive tax rates, and need-based services such as programs to help the poor, seniors, and disabled people, are examples of policies addressing vertical equity. Vertical equity is often measured with respect to income. This is an imperfect metric, since people with the same income may have very different needs and abilities. Road pricing is usually considered vertically inequitable because charges impose a relatively larger burden on the poor. For example, a €2 per day toll might be horizontally equitable (everybody pays the same amount), but vertically inequitable because it represents a larger portion of income for a low-income driver than for a high-income driver. This fact is, of course, tempered by the fact that lower income people drive less on average than those with higher incomes.

Another equity issue refers to spatial or *geographical equity*. This term is concerned with the treatment of individuals located in various regions or cities. Congestion pricing could be considered as unfair from this point of view as charges (depending on time and place) will differ among regions. Another illustration of spatial equity concerns in transport is the Sydney City Council, who decided that transport availability should not depend on the geographical area in which a person lives. Transport services should be equally available to people across the Sydney metropolitan region.

It should not be forgotten that it remains a political issue to decide what allocation of road pricing revenues is equitable. Whereas economic efficiency and equity can be evaluated with a certain degree of objectivity, political acceptability will reflect popular perceptions and the distribution of political power. When policymakers are not satisfied with the situation, implementation of individual specific transfers can satisfy the distributional goals. If this redistribution may be realised without distorting the market mechanism, equity and efficiency can both be reached. The theoretically proper instrument for carrying out such transfers is a lump-sum tax, i.e. a tax that does not depend on any action of the individual; there is no way that he can change the tax liability (Atkinson and Stiglitz, 1980). A lump-sum tax is therefore one that induces income effects alone, no substitution effects. If such lump-sum transfers were possible there would be no reason to worry about distributional concerns in the transportation pricing policy domain. Unfortunately, in practice, it is almost impossible to implement transfers without distorting the market mechanism.

It appeared from the previous that consumers' surplus and the social surplus weight monetary gains and costs equally for all individuals. The fact that these objective functions are so often used in transportation policy analysis does not imply that researchers deliberately want to ignore distributional issues. The main reason for using social surplus as the welfare measure is that it is easier to use, mainly because it requires only information at the market level.

It can be concluded that there will always be some form of tension between efficiency and equity of pricing in transport. An important reason for this is that average and also the marginal external costs are most probably not constant. In this case, the concept of user pays together with optimal prices leads to the situation where the user will not pay a total amount that is equal to the total of the external (non-private) costs.

SECOND-BEST PRICING ISSUES

The previously discussed pricing principles and Pigouvian taxation to correct for the congestion externality implicitly assumed a theoretical first-best world. Such first-best pricing is increasingly recognised as being of limited practical relevance, but it might serve as a useful theoretical benchmark. Various constraints and barriers may exist that prevent a regulator from charging prices that it ideally would like. Verhoef (2002) mentions the following important constraints:

- Technological and practical constraints – first-best pricing requires charges that vary continuously over time, place, route chosen, type of vehicle, driving style, etc., which might be too sophisticated and not understood by drivers or impossible to implement under available charging technologies.
- Acceptability constraints – there may be too much resistance and uncertainty (e.g. about objective and necessity of the measure) that may make it preferable to start with a few small-scale demonstration projects.
- Institutional constraints – one example is where local or regional governments cannot affect some transport charges that are set by a higher level government.
- Legal constraints – ideal prices might not be possible on the basis of legal arguments (e.g. when taxes should be predictable).
- Financial constraints – for instance, the prior definition of minimum or maximum tax revenue sums to be collected.
- Market interaction constraints – transport taxes will have many consequences for other markets, among the most important is the labour market.
- Political constraints – charges may become a political issue much more than an economic question.

Under such conditions, the regulator has to resort to second-best pricing: setting the prices that are available optimally, under the constraints applying. Examples of second-best tolling include the use of toll cordons around cities instead of tolling each road in the network and the use of step tolls instead of smoothly time-varying tolls. It is safe to state that second-best pricing will be the rule for the implementation of marginal cost-based pricing in reality.

Second-Best Pricing: An Overview of Some Results

It is clear that some aspects of the transport sector do not correspond to first-best conditions. We then enter the world of second-best where the imperfectness of one variable has consequences for the optimal value of others. The conclusions that emerge from second-best analyses are dependent on the details of the situation under consideration and it is hard to draw general conclusions. A substantial technical literature has emerged over the last decade, that addresses various types of second-best pricing schemes (i.e. what do the tax rules look like, how do they deviate from Pigouvian charges), and towards the relative efficiency of these schemes. Attention has turned in the recent literature to more realistic types of 'second-best' pricing, in which various costs or constraints deter or prevent the setting of first-best tolls. Much of this literature is reviewed in Lindsey and Verhoef (2001), whereas MC-ICAM (2002) gives insight into the kind of analysis. The following briefly presents some of the most relevant second-best problems that have been studied in transportation (congestion) pricing and draws heavily on these studies.

Networks

First-best pricing in a network assumes that each link of a road network is efficiently priced. This is often impossible due to excessive costs, the requirement of toll-free alternatives by governments and the likeliness of incremental implementation rather than at once. The question under study is then how second-best tolls should be set on toll roads given unpriced congestion on untolled roads elsewhere in the network.

This network problem is one of the most widely studied where the simplest version concerns a simple network in which there are two links connecting the same origin and destination. Verhoef *et al.* (1996) demonstrate that if one of the links is often congested, the optimal second-best toll of the other link can be negative. This study also shows that the optimal toll depends on the relative free-flow travel times and capacities of two routes, and on the price elasticity of travel demand. Welfare gains from second-best pricing are, according to this study, a small fraction of the benefits from the first-best benchmark (around 10%). Other studies have looked at ways to enhance efficiency and incorporated the possibility of dynamic (time varying) tolls and sorting of drivers according to value of travel time. This indeed yields higher relative efficiency gains.

Most network studies assume a unimodal network. In reality, a traveller has the possibility to choose between modes. The leading example is the choice between public transport and the private car. The main difference between second-best problems on networks and those for mode choice is that in the former case an assumption of perfect substitutability is often made. Although at first sight the two-mode problem appears to be relatively simple, it has proved to be difficult to solve (MC-ICAM, 2002). Results are very much restricted by the assumptions made (such as fixed capacity and a fixed toll) and may be complicated and difficult to interpret (see for a review of second-best choices in a model with two modes, for instance, Arnott and Yan, 2000).

Heterogeneity

Travellers and road vehicles differ in a number of characteristics. Vehicles vary for example in the road space they occupy and weight and acceleration capabilities. Travellers have different values of time, desired speed, and so on. First-best pricing often requires to distinguish between different vehicle types and users (because of different marginal costs). It is important to know whether first-best congestion pricing can still be implemented given these dimensions of heterogeneity, and if not, how second-best tolls are optimally determined. In this context, a distinction is often made between anonymous tolling schemes (independent of vehicle type and the identity of the driver) and non-anonymous (type-specific) tolls.

Many studies have been conducted towards the implications of the problem of heterogeneity and pricing. The topics range from heterogeneity in drivers' values of time and trip timing preferences to the heterogeneity in travel speed. For example, Verhoef and Small (1999) consider a differentiation of tolls across parallel traffic lanes by using a static model. They show that an anonymous toll may still be optimal on each lane separately, and efficient segregation of drivers is achieved without regulation other than through anonymous taxes. Optimal anonymous tolling may entail segregation of vehicle or driver types onto separate routes.

Interactions with Other Sectors

Imperfections in other sectors of the economy may have consequences for second-best optimal pricing of transport services. Conventional modelling of the transport sector typically assumes that the rest of the economy operates under first-best conditions. Although applied with the aim to simplify the modelling exercise, it is usually well off the mark. For example, the existence of distortionary taxes on other markets (especially the labour market) and income distributional concerns (which might be suboptimal) can be motivated by governmental objectives other than for transport, but have implications for pricing in transport. Intersectoral issues matter because transport pricing interacts with other markets that may be strongly distorted.

The existing literature on intersectoral issues shows the importance of them for optimal pricing: they are almost always relevant (MC-ICAM, 2002). The most relevant relation is with the labour market, which is heavily distorted: labour taxes exist mainly due to the governmental need for revenues and equity reasons, but strongly distort the labour-leisure choice. This implicates that whenever there is a reform in transport prices, the effect on the labour market distortion should be taken into account. Moreover, since distributional objectives can hardly be achieved by using distortionary labour taxes alone, the distributional concern will almost always be relevant in transport pricing.

In this context, also the spending of revenues from pricing schemes is important for the overall success of the measure. These revenues might be used to subsidise public transport, but also to reduce labour taxes, to increase government spending on other services and so on. The effects for other sectors should not be neglected when looking at the effects of pricing. Parry and Bento (2001), for instance, show that the general equilibrium effects of road pricing schemes are sensitive to the spending of revenues, and may deviate considerably from partial equilibrium outcomes. Recycling the revenues in public transport fare subsidies rather than tax cuts appears to be less efficient according to their analysis. Revenue-spending targets that are acceptable to the public may therefore not necessarily be efficient.

CONCLUDING REMARKS

This chapter gave an overview of the economics of transport pricing and provided some basic exposition of key issues important for prices in transport markets. It addressed a range of complexities that arise due to the nature of the cost structure and different market conditions generally observed in the transport sector.

Pricing is a method of resource allocation. There exists no right price as such, rather there are optimal pricing strategies to permit specified aims to be achieved. Several transport pricing objectives can be distinguished, such as economic efficiency, profitability, and cost coverage. Whilst there are many transport pricing objectives, economists will usually argue that the pursuance of economic efficiency should take precedence, in any case as long as political power does not indicate its preference about redistribution and equity. Optimising the welfare of society may even be the most important to governments involved in setting prices in mobility. The concept of economic efficiency is derived from welfare economics, and is concerned with the allocation of resources in an economy. According to this theory, prices should be equal to MSC (throughout the economy) to obtain maximum social welfare. Marginal cost pricing results in transport services being provided up to the point where the benefit for the marginal unit is equated with the costs of providing that unit. But, this efficient price will only prevail as a market equilibrium under certain conditions, i.e. all other markets have also efficient pricing, there are no constraints on transport prices and market failures are non-existing.

However, this optimal outcome need not satisfy all policy objectives. Equity is an important issue when it comes to transport pricing. Stakeholders often raise objections against pricing measures, which they perceive to be unfair. Governments may then decide to differentiate prices among different groups or to make individual specific transfers (and if these can be realised without distorting the market, equity and efficiency can both be reached).

In most cases, however, actual market prices deviate from MSCs. The transport market has some characteristics making it very unlikely that, without regulation, optimal transport prices will result. For instance, transport prices do not include external costs. These costs, such as noise and accidents, may be significant. In this chapter, we have shown how prices can be adjusted to include external costs (in our case congestion costs) and to achieve welfare gains. A Pigouvian charge would, in a first-best world, be the optimal way to internalise the external costs of congestion resulting in optimal prices.

Such a first-best setting is recognised as a useful theoretical benchmark, but it is of less practical relevance. The regulator faces several barriers and constraints preventing it from setting optimal prices (e.g. technical or political in nature). Recent research has therefore focused on more realistic types of the so-called second-best pricing, in which various costs or constraints deter or prevent the setting of first-best tolls. Examples of second-best problems discussed in this chapter include the problem of networks and the interactions with other markets or sectors. Under second-best conditions, optimal prices may be said to be marginal cost based, rather than identical to marginal costs. Second-best prices are more difficult to determine and implement and may lead to considerably lower welfare gains compared with first-best prices.

ACKNOWLEDGEMENTS

This research was carried out within the NWO/Connekt VEV project on 'A Multi-disciplinary Study of Pricing Policies in Transport – An Economic Perspective'; nr. 014-34-351. Financial support is gratefully acknowledged.

REFERENCES

Arnott, R., A. De Palma, and R. Lindsey (1998). Recent Developments in the Bottleneck Model. In: *Road Pricing Traffic Congestion and the Environment: Issues of Efficiency and Social Feasibility* (K.J. Button and E.T. Verhoef, eds), Edward Elgar, Cheltenham.

Arnott, R. and A. Yan (2000). *The Two-Mode Problem: Second-Best Pricing and Capacity*, Working Paper, Boston College.

Atkinson, A.B. and J.E. Stiglitz (1980). *Lectures on Public Economics*, McGraw-Hill, London.

Baumol, W.J. (1962). On the Theory of the Expansion of the Firm, *American Economic Review*, **52**, 1078–1087.

Button, K.J. (1993). *Transport Economics*, 2nd ed., Edward Elgar, Aldershot.

Button K.J. and E.T. Verhoef (1998). *Road Pricing, Traffic Congestion and the Environment*, Edward Elgar, Cheltenham.

CE (Centrum voor Energiebesparing en Schone Technologie). (1999). *Efficiente prijzen voor het verkeer*, CE, Delft.

Coase, R.H. (1960). The Problem of Social Cost, *Journal of Law and Economics*, 3 (October), 1–44.

Dreze, J.H. and N. Stern (1987). The Theory of Cost Benefit Analysis. In: *Handbook of Public Economics II* (A.J. Auerbach and M. Feldstein, eds), North-Holland, Amsterdam, 909–989.

INFRAS/IWW (2000). *External Costs of Transport: Accident, Environmental and Congestion Costs of Transport in Western Europe*, Zurich/Karlsruhe.

Lindsey, C.R. and E.T. Verhoef (2001). Traffic Congestion and Congestion Pricing. In: *Handbook of Transport Systems and Traffic Control* (K. Button and D.A. Hensher, eds), Pergamon, Amsterdam, 77–105.

Litman, T. (1996). Using Road Pricing Revenue: Economic Efficiency and Equity Considerations, *Tranportation Research Record*, **1558**, 24–28.

Mayeres, I. and S. Proost (1997). Optimal Tax and Investment Rules for Congestion Type of Externalities, *Scandinavian Journal of Economics*, **99** (2), 261–279.

MC-ICAM (2002). *Relevant Optima and Constraints*, Deliverable 2, Project funded by the European Commission.

Oates, W.E. (1983). The Regulation of Externalities: Efficient Behaviour by Sources and Victims, *Public Finance*, **38**, 362–375.

Parry, I.W.H. and A. Bento (2001). Revenue Recycling and the Welfare Effects of Road Pricing, *Scandinavian Journal of Economics*, **103** (4), 645–671.

Parry, I.W.H. and A. Bento (2002). Estimating the Welfare Effect of Congestion Taxes: the Critical Importance of Other Distortions within the Transport System, *Journal of Urban Economics*, **51** (2), 339–365.

Pigou, A.C. (1920). *The Economics of Welfare*, MacMillan, London.

Simon, H.A. (1959). Theories of Decision-Making in Economics and Behavioural Science, *American Economic Review*, **49**, 253–283.

Stiglitz, J.E. and J. Driffill (2000). *Economics*, W.W. Norton & Company, New York.

United Nations (2001). *Sustainable Transport Pricing and Charges: Principles and Issues*, Bangkok, New Delhi.

Varian, H.R. (1999). *Intermediate Microeconomics: A Modern Approach*, W.W. Norton & Company, New York.

Verhoef, E.T. (1994). Efficiency and Equity in Externalities: A Partial Equilibrium Analysis, *Environment and Planning*, **26A**, 361–382.

Verhoef, E.T. (1996). *Economic Efficiency and Social Feasibility in the Regulation of Road Transport Externalities*, Thesis Publishers, Amsterdam.

Verhoef, E.T. (2002). *Marginal Cost Based Pricing in Transport: Key Implementation Issues from the Economic Perspective*, Unpublished paper, Amsterdam.

Verhoef, E.T., P. Nijkamp, and P. Rietveld (1996). Second-Best Congestion Pricing: The Case of an Untolled Alternative, *Journal of Urban Economics*, **40** (3), 279–302.

Verhoef, E.T. and K.A. Small (1999). *Product Differentiation on Roads: Second-Best Congestion Pricing with Heterogeneity under Public and Private Ownership*, Discussion Paper TI 99-066/3, Tinbergen Institute, Amsterdam-Rotterdam.

Vickrey, W.S. (1969). Congestion Theory and Transport Investment, *American Economic Review (Papers and Proceedings)*, **59**, 251–260.

Willig, R. (1976). Consumer's Surplus without Apology, *American Economic Review*, **66** (4), 589–597.

Threats from Car Traffic to the Quality of Urban Life
Tommy Gärling and Linda Steg (Editors)

19

BEHAVIOURAL RESPONSES TO TRANSPORT PRICING: A THEORETICAL ANALYSIS

Linda Steg, Department of Psychology, Faculty of Behavioural and Social Sciences, University of Groningen, Groningen, The Netherlands
Geertje Schuitema, Department of Psychology, Faculty of Behavioural and Social Sciences, University of Groningen, Groningen, The Netherlands

ABSTRACT

This chapter provides a psychological perspective on transport pricing. We first describe to what extent and under which conditions transport pricing may be effective in reducing car use, and consequently, in reducing the problems caused by car use. Next, we review factors influencing the acceptability of transport pricing and provide theoretical explanations for the significance of these factors. The role of policies features as well as individual characteristics will be discussed. We elaborate on the role of justice and fairness, and discuss relationships between (perceived) effectiveness and acceptability of transport pricing. Policy implications and important topics for future research will be highlighted.

INTRODUCTION

Next to technological measures, behavioural changes of individual car users are needed to reduce the problems of car use discussed in the first part of this book. Several policy strategies may be applied to manage car use (see Chapter 15 by Loukopoulos), among which are transport-pricing policies. Prices of car use may be increased (at certain times or places), while costs of the use of environmentally friendly (or: sustainable) modes of transport, such as public transport, may be reduced. Studies revealed that the latter strategy is generally not very effective in the long term. For example, lowering costs of bus use does attract new bus users, but does not result in a reduction in car use, i.e., the new passengers used to cycle or walk, or stayed at home rather than drove their car (e.g., Claassen and Kropman, 1995; Deslauriers and Everett, 1977). Furthermore, financial rewards for bus travellers appeared to be effective only in the short term; people revert to their initial travel behaviour as soon as the reward ceases (Van Knippenberg and Van Knippenberg, 1988). Increasing costs of car use is supposed to be far more effective in reducing the level of car use (e.g., Gomez-Ibanez and Small, 1994; TfL, 2004). Therefore, we focus on price increases. In the remaining of this chapter, transport pricing is defined as increasing costs of car use (and not purchase costs of cars).

Transport pricing is generally believed to be an effective and efficient way to manage problems resulting from private car use (see Chapter 18 by Ubbels and Verhoef). Especially modelling studies, but also some evaluation studies, revealed that transport-pricing policies may be highly effective in reducing car use and the problems associated with car use. However, transport pricing is not easily implemented due to the lack of public support, i.e., in general, the public thinks transport-pricing measures are not acceptable (e.g., Schlag and Teubel, 1997; Jones, 1998, 2003). This also applies to other policies: Effective policies are generally not acceptable to the public, while policies that are acceptable are generally not effective in reducing problems resulting from car use (Steg, 2003). This places policy-makers in a dilemma: Should they implement unpopular policies, or respect public opinions without solving the problems at stake? Or is it possible to design pricing schemes that are both effective and acceptable? Therefore, it is important to understand which factors affect the effectiveness and acceptability of transport policies, and more specifically, transport pricing. This facilitates the design of policies that may be acceptable and effective. Characteristics of policies (e.g., price level) as well as individual characteristics (e.g., car dependency, car attitudes) may be important in this respect.

This chapter provides a psychological perspective on transport pricing. We first discuss whether transport pricing is indeed effective in reducing car use. We will briefly review studies that have examined the effectiveness of transport pricing. Some of these studies focus on perceived effectiveness (i.e., stated preference), whereas others evaluated actual effectiveness (i.e., revealed preferences). Next, we focus on acceptability of transport-pricing policies, since acceptability is of major importance for implementing effective transport-pricing policies. We discuss factors influencing the acceptability of transport pricing and provide theoretical explanations for the significance of these factors. Furthermore, we elaborate on the relationships between acceptability and effectiveness judgements: Are effective policies indeed not acceptable to the public? Several policy recommendations are discussed, and important open questions that need to be addressed in future research are highlighted.

EFFECTIVENESS OF TRANSPORT PRICING

Various transport-pricing policies may be distinguished, such as fuel levies, road taxes, toll levies, and kilometre charges. Such user charges are generally aimed at making

Table 1: Possible Behavioural Changes Resulting from Transport Pricing

Type of Change	*Specification of Behavioural Changes*
Driving behaviour	Driving style (speed, use of brakes, changing gears)
Travel behaviour	Trip chaining Route choice Time of travel Destination choice (e.g., shopping, recreation) Number of trips (trip suppression) Mode choice (e.g., car, carpool, motorbike, public transport, bicycle, walking)
Vehicle ownership	Type of car (e.g., fuel type, size) Car ownership (number of cars, replacement of car)
Location choice	Choice of residence Choice of workplace

people drive at other times or places, or at reducing the level of car use[1]. By doing so, it is expected that problems associated with car travel discussed in the first part of this book will reduce, e.g., traffic congestion, environmental problems, and threats to urban quality of life, such as traffic safety and noise annoyance. Thus, transport-pricing policies are aimed at reducing problems resulting from car use by changing behaviour of individual car users.

Transport-pricing policies may elicit various types of behavioural changes. Table 1 gives an overview of possible changes. First, people may change their driving behaviour, i.e., adopt more energy–efficient driving styles. Second, people may consider to change their travel behaviour. They may combine trips, use different routes, change the time of travel, or visit other destinations. Furthermore, they may suppress certain trips, or travel with different modes of transport, such as public transport, cycling, walking, or carpooling. These changes are generally relatively easy to accomplish in the short term. Third, changes in vehicle ownership may occur. People may consider to buy a new car or to dispose of their car. The new car may either be smaller or more energy–efficient, but may also be larger or less economical. The latter may occur when, for example, vehicle or road taxes are abolished and incorporated in fuel prices (i.e., variabilisation of car costs), thereby reducing costs of car possession. Fourth, location choices may be affected. People may consider to move residence, or to find another job location. Location choices may have significant effects on travel behaviour because travel distances to various destinations may decrease (or increase) considerably. In general, behavioural changes are not necessarily linked to more economical (or less) car use. For example, pricing policies may well result in an increase in car use for some groups, e.g., high-income groups or individuals who receive reimbursement of their travel expenses

[1] Transport pricing may also be aimed at raising additional revenues; this will not be discussed in this chapter.

may decide to drive more often or to drive during peak hours instead of off-peak hours if congestion levels reduce. Similarly, a reduction in congestion problems may result in an increase of commuting distances for some groups, e.g., when people can afford to move residence further away from work without increasing their travel time.

Changes in car ownership and location choices (especially choice of workplace and residence) will typically occur in the long term. One may assume an ordering in these adaptation strategies (see also Loukopoulos *et al.*, 2004). People may first opt for behavioural changes with lowest behavioural costs (in terms of money, time, effort), e.g., trip chaining (e.g., combine commuting and shopping trips), choosing closer destinations (e.g., shopping close to home), or choosing different routes (e.g., travel via a route that is not tolled). When these changes are not possible or not sufficient, they may opt for changes associated with higher behavioural costs, such as using other modes of transport or suppressing trips. Finally, in the long term, they may consider buying another car or disposing of a car, or even to move home or to look for a job closer to home. Of course, individual differences may exist in behavioural costs of the adaptation strategies. For example, changing the time of travel may be possible for some workers but not for others. To get a complete and comprehensive overview of effectiveness of transport pricing, it is highly important to distinguish between these types of behavioural changes and to take into account short as well as long-term effects of transport pricing. People need time to adjust to new situations, and more significant behavioural adaptations may occur in the long term only. For example, Mogridge (1978) found that during the world oil crisis of the mid-1970s the rise in fuel prices had only a marginal effect on car use in the short term. However, it appeared that behavioural effects were apparent in the long term. People purchased smaller and more economical cars. More generally, price elasticities tend to be higher in the long term compared to short term (e.g., Priemus and Nijkamp, 1994; Wootton, 1999).

Different pricing strategies may result in different behavioural adaptations. For example, a rise in vehicle or road taxes will probably affect car ownership more strongly than route choice, while a kilometre charge may well result in changes in mode choice, but probably not in the purchase of a more energy–efficient car. Furthermore, time-dependent charges are more likely to result in changes in time of travel, while tolling main roads may typically result in changes in route choice.

The extent to which the behavioural changes discussed above will contribute to reduction in the various problems related to car use differs greatly. Changes in route or time of travel may be effective in reducing accessibility problems, but may be counterproductive in attempts to increase quality of urban life. For example, if heavy traffic volumes are spread out in time, community residents may suffer from high levels of traffic noise for a longer period of time. On the other hand, reductions in the level of car use, e.g., by shifting to less polluting modes of transport, changing destinations, residence or workplace, combining trips, or suppressing trips may improve environmental quality, urban quality of life, and destination accessibility. Purchasing new, energy–efficient or low-emission cars will especially result in reduction of environmental problems and increases in urban quality of life, but will hardly help control congestion problems. People may be even tempted to use their energy–efficient car more often because it is cheaper and more environmentally friendly. Many scholars have stressed the importance of the so-called rebound effect (e.g., Berkhout *et al.*, 2000). Therefore, policymakers should carefully take into account which policy goals should be reached and which behaviour changes are targeted when designing and implementing transport-pricing policies.

Next to changes in collective qualities, such as environmental problems, urban quality of life and accessibility, transport-pricing policies may also have consequences for individual quality of life (e.g., Poortinga *et al.*, 2004; Steg and Gifford, 2005; De Groot and Steg, 2006). Quality of life may be defined as the extent to which important values and needs of people are fulfilled (e.g., Diener, 1995; Diener *et al.*, 1999). Obviously, transport-pricing policies may affect the way and the extent to which people are able to fulfil important values and needs. In general, policymakers seem reluctant to implement policies aimed at reducing car use, such as increasing prices of car use, because it is believed that such measures will reduce individual's quality of life. However, various studies reveal that transport policies may have less negative effects for individual quality of life than expected (De Groot and Steg, 2006; see Steg and Gifford, 2005, for a review). Possible negative quality-of-life effects, such as infringements on freedom, reductions in comfort and increased costs, seem to be partly compensated for by improvements in environmental quality, nature, health, and safety, which are considered to be important for individual quality of life as well. A study by Schuitema and Steg (2005b) also revealed that people take into account effects on environmental quality and congestion when evaluating effects of transport-pricing policies for their lives. People's judgements of whether they would be better or worse off when considering all pros and cons of transport-pricing policies appeared to be most strongly (and positively) related to expected effects on congestion levels and environmental problems, while these judgements were only weakly correlated with the extent to which the pricing policy would affect one's own car use (i.e., car mileage). This implies that respondents expect their overall situation to improve if they think congestion and environmental problems will decrease as a result of transport-pricing policies (see also Schuitema and Steg, 2005a).

Psychologists have primarily studied the expected effects of transport pricing (i.e., they focus on stated preferences). Typically, respondents are asked to what extent they would reduce their car use (e.g., mileage, number of trips) when transport-pricing policies would be implemented, or to what extent these policies would be effective in reducing problems caused by car use. Various studies revealed that, in general, people think transport pricing will not be very effective in reducing car use (e.g., Steg, 1996, 2003; Jakobsson *et al.*, 2000; Loukopoulos *et al.*, 2004; Schuitema and Steg, 2005b) or the problems of car use (e.g., Schlag and Teubel, 1997; Jones, 2003; Schuitema and Steg, 2005a, 2005b), especially because they think car use reductions are hardly feasible

(Jakobsson *et al.*, 2002). People generally assume transport pricing will be somewhat more effective for others (e.g., Steg, 1996). Various group differences were found in perceived effectiveness of transport pricing. In general, low-income groups expect to reduce their car use more than high-income groups do (Jakobsson *et al.*, 2000). Furthermore, transport pricing will affect private and commuter trips somewhat more than business trips (Cavalini *et al.*, 1996; Schuitema *et al.*, 2003). Transport pricing may especially result in a reduction of shopping trips (Jakobsson *et al.*, 2002; Schuitema *et al.*, 2003). Furthermore, transport policies are more likely to result in changes in route choice, destination choice or time of travel, while the amount of kilometres driven or car ownership will hardly be affected (Cavalini *et al.*, 1996; Loukopoulos *et al.*, 2004). Especially, short trips by car will be reduced, e.g., because people cycle to their destination instead of driving (Schuitema *et al.*, 2003).

In sum, stated preference studies reveal that transport pricing will be more effective if people think they can reduce their car use (e.g., this is more likely for private trips and route choice), and if people cannot afford an increase in travel costs (e.g., low-income groups). Surprisingly, people indicate that in general, transport pricing will not be very effective, while various studies reveal that transport pricing may have substantial effects on car use. Prominent examples are the Singapore area licence scheme (a congestion charge) and the London congestion charge (see Small and Gomez-Ibañez, 1998; Santos, 2004; Santos *et al.*, 2004; Verhoef *et al.*, 2004). In Singapore, changes in route choice were most prevalent, but shifts to carpooling and bus use were observed as well, while in London switches to bus use (and taxi's) were more common than changes in route choice (see Santos, 2004; Verhoef *et al.*, 2004). Thus, revealed preferences do not seem to match stated preferences. Several explanations may be given for this finding. First, results of revealed preferences studies may not be easily generalised to other areas. Congestion pricing appeared to be effective in Singapore and London, which both were confronted with serious congestion problems. Moreover, in both cities, the quality of public transport was improved as well, and consequently, for many, feasible alternatives modes of transport were available. Experiencing serious problems and the availability of high-quality public transport may be important preconditions for transport pricing to be effective. Also, the level of price increase is important. In both Singapore and London, charges were relatively high. In contrast, low charges appeared to be less effective (e.g., toll rings in Norway; Tretvik, 2003; Ramjerdi *et al.*, 2004; see also Verhoef *et al.*, 2004). Also, the effectiveness of transport pricing may be substantially decreased when people can evade the measure, e.g., by driving at other times when time-dependent charges are introduced (such as in Trondheim, Norway, where time-dependent urban tolls were implemented; Meland, 1995; Ramjerdi *et al.*, 2004). Furthermore, stated preference methods may elicit strategic answers; people may indicate that transport pricing will not be effective, in the hope it will convince policymakers to not to implement transport-pricing policies. Moreover, people may not think through the consequences of transport-pricing policies, and may not thoroughly consider possible behaviour changes. In general, people are not very accurate in assessing future travel patterns (cf. Jakobsson, 2004).

Pricing policies may have unwanted side effects, resulting in less significant behaviour changes than expected beforehand. This will especially be the case if pricing policies reduce intrinsic motivation to help solve problems related to car use. This so-called 'crowding out' effect (e.g., Frey, 1997, 2003) may especially occur when price changes are high enough to justify behavioural changes. In that case, people may change their car use to save money only, and no longer because they feel intrinsically motivated to contribute to solutions of the problems caused by car use. This may be problematic because intrinsic motivation may more strongly affect behaviour than do monetary incentives (Thøgersen, 2003). Moreover, intrinsic motivation of those not affected by the pricing policies may diminish as well. Thus, pricing policies may be effective only to the extent that financial incentives at least compensate a possible decrease in intrinsic motivation (Frey, 1997; Thøgersen, 2003). In other cases, effects may be weakened, or may even be counter-productive. Transport pricing may give people the feeling that they have a right to drive their car because they have paid for it (Fehr and Falk, 2002). To the authors' knowledge, it has not been demonstrated yet whether the crowding out effect applies to transport pricing, but based on results of previous studies, such effects may be expected.

ACCEPTABILITY OF TRANSPORT PRICING

Lack of public support is an important barrier to the implementation of restrictive transport-pricing policies. Interestingly, policymakers and politicians seem to underestimate the support for policies aimed to reduce car use (e.g., Jones, 1995). A similar misunderstanding is prevalent among car users themselves, who, on average, indicate they find transport pricing more acceptable than does the general public (Steg, 1996). Nevertheless, many politicians are not keen on implementing policies that evoke resistance among the general population. However, some transport-pricing policies are more acceptable than others, and some may even be rather acceptable, at least for some groups. In general, the extent to which transport pricing is acceptable to the public is dependent on individual characteristics as well as on features of the policies themselves. In this section, we will discuss both types of features.

Individual Differences in Acceptability of Transport Pricing

Many (psychological) studies have tried to identify individual factors that affect public acceptability of transport pricing (e.g., Schlag and Teubel, 1997; Jakobsson *et al.*, 2000; Schade and Schlag, 2000, 2003; Schlag and Schade, 2000; Bamberg and Rölle, 2003; Schuitema, 2003; Loukopoulos *et al.*, 2005). The acceptability of transport pricing appears to be related to a wide range of factors. A study by Frey (2003) suggests that problem awareness may be an important precondition for policy acceptability. Acceptability of road pricing in Saas-Fee (Switzerland) appeared to be rather high

(57% supported this policy) because people were aware of the problems caused by car traffic and of possible solutions for these problems in their community. Similar results were found in a study conducted in the Netherlands: transport policies are more acceptable for people high in problem awareness (Steg and Vlek, 1997). Jones (2003) also stresses the importance of problem awareness for enhancing public acceptability for transport pricing (see also Small and Gomez-Ibañez, 1998). Interestingly, Schade and Schlag (2000) found that people who are especially concerned about the environmental problems of car use evaluate transport-pricing policies as more acceptable compared to those who are more strongly concerned about congestion.

Jakobsson *et al.* (2000) found that transport policies are evaluated as less acceptable when they are perceived to be unfair, and when they threaten people's freedom of choice (and thus have negative individual consequences). A study by Bamberg and Rölle (2003) revealed that, besides these factors, perceived effectiveness plays a role as well: acceptability is higher when people believe the policies will be effective in reducing car use (see also Steg, 1996; Schade and Schlag, 2003). Perceived effectiveness appeared to be related to problem awareness: the more people are aware of environmental problems caused by car traffic, the more they think transport pricing will be an effective instrument to reduce car use. Several other studies reported positive relationships between perceived effectiveness and acceptability of transport pricing as well. Jaensirisak *et al.* (2003) found that acceptability is higher if transport pricing benefits individuals as well as society as a whole by, e.g., reducing congestion levels and improving environmental quality. Schade and Schlag (2003) also report that transport pricing is more acceptable if it benefits the individual and reduces collective problems. Similar results were found by Schuitema and Steg (2005a), showing that acceptability of transport-pricing measures is higher if people think their life will not be affected too much, and if people think environmental and congestion problems will actually reduce. They found that the actual rise in travel costs for households was less strongly related to acceptability ratings. People who think transport pricing is acceptable especially stress the positive (collective) consequences of the policies, such as improved environmental quality and reduction in congestion levels, while those who think transport pricing is not acceptable typically focus on negative consequences for themselves (Loukopoulos *et al.*, 2004). From the above, we may conclude that people seem to resist policies that are not effective in solving problems caused by car use. On the other hand, Jakobsson *et al.* (2000) found that policies are not acceptable when they are very effective in changing one's behaviour, thereby seriously affecting individual freedom of choice. This implies that people seem to prefer policies that help solve collective problems resulting from car use that benefit themselves as well, while the policies should not restrict their own freedom of choice.

Various groups may actually benefit from the implementation of transport-pricing policies, for example if congestion levels decrease and accessibility improves. Furthermore, transport-pricing may benefit groups that hardly drive or do not drive, as far as the problems resulting from car use actually reduce, i.e., they will benefit from improved environmental and urban qualities. Also, people may find ways to evade the charges, and consequently, price increases may not affect them. Thus, some groups may be in favour of transport pricing. However, in general, support of these 'winners' is usually muted (Frey, 2003), and, consequently, politicians assume public support is low.

Trust in governments may also affect policy acceptability: if trust levels are low, public support for policies that reduce individual freedom to move is low as well (Ney *et al.*, 1997). This may well be related to perceived effectiveness: if people do not trust the government, they may think the policies may not be effective as well. Frey (2003) suggests that public participation in the decision-making processes may enhance public acceptability, because it increases public understanding of the need for reducing problems related to car use by introducing transport pricing. This is in line with literature on procedural justice, in which the focus is placed on the fairness of the processes by which decisions and policies are being made: decisions are perceived to be more fair, and consequently, acceptable, if those affected had a say in the decision-making process (Lind and Tyler, 1988; Clayton, 2000).

Processes of distributive justice, i.e., the extent to which various groups are affected by transport-pricing policies (see Clayton, 2000; Schroeder *et al.*, 2003), are also highly relevant for the acceptability of transport pricing. Indeed, fairness seems to be strongly related to acceptability judgements (Jakobsson *et al.*, 2000; Bamberg and Rölle, 2003; Schuitema *et al.*, 2004). For example, Bamberg and Rölle (2003) report a correlation coefficient of 0.80 between perceived fairness and acceptability. This suggests that perceived fairness is of crucial importance of policy acceptability. On the other hand, one may doubt whether respondents interpret fairness and acceptability as different constructs; they may reflect one common underlying construct. This indicates that measures of both constructs should be carefully developed, as to make sure respondents notice the conceptual difference.

Surprisingly, little is known yet about perceived fairness of different types of transport pricing. What is perceived to be fair will likely depend on the prevalent justice principle. Different justice principles may be prevalent, resulting in different fairness judgements. People may strive for equality or equity, or they may think the 'polluter should pay'. Equality refers to aiming for equal rights, privileges, quantities, qualities, etc. This implies that transport price increases should be the same for all (i.e., static pricing measures). Equity means everyone should be affected in the same degree. This suggests that different price levels for different groups may be evaluated as fair, e.g., price level should be lower for frequent travellers (e.g., business travellers) as to make total price increases equitable, or people driving large and heavy cars (more often high-income groups) should pay higher prices because they can afford it. 'The polluter pays' principle implies that price increases should be highest for those who most strongly contribute to the problems. Congestion charging is a prominent example here. Fairness judgements may also be directly related to the extent to which transport policies affect the individual relative to other people: if people think that they themselves are affected disproportionately, policies may perceived to be unfair. Another (distributive) justice principle that proved to be relevant in the environmental context is environmental justice, which respects the rights of the environment, ecosystems, other species, and future generations (Clayton, 2000). Policies may be perceived as more fair when they protect and guarantee the rights of nature, the environment, and future generations. Environmental justice may be well seen as a special type of distributive justice, by taking into account the extent to which non-human species and future generations are affected by policies. Little is known yet on which justice principle is prevalent for which groups and under which circumstances. Studies typically include quite general measures of fairness, reflecting how fair the introduction of the measure would be (often a single-item measure). More research is certainly warranted into the perceived fairness of different transport-pricing measures and the role of various justice principles in this respect.

This approach makes sense when considering acceptability judgements as attitudes towards transport pricing (e.g., Jakobsson *et al.*, 2000; Schade and Schlag, 2003), i.e., many theorists assume attitudes are based on evaluating perceived consequences of behaviour or situations (e.g., Fishbein and Ajzen, 1975; Ajzen, 1985). To put it differently, attitudes towards transport pricing are believed to be dependent on expected consequences of transport pricing. The more positive (and the less negative) consequences people expect from transport pricing, the more favourable are people's attitudes towards transport pricing, and the more positive people's acceptability ratings.

Alternatively, acceptability judgements may be defined as a specific type of (environmental) behaviour. This approach has been advanced by Stern and his colleagues (Stern *et al.*, 1999; Stern, 2000). They argue that support or acceptability of environmental policies, such as transport pricing, may be defined as non-activist behaviours in the public sphere. Non-activist behaviour affects environmental quality indirectly, by influencing public policies, which may have large effects on environmental qualities because public policies may change the behaviour of many people at once. This implies that theories aimed to explain environmental behaviour may be of particular interest to understand factors affecting public acceptability of transport pricing. Since transport pricing generally implies that people have to give up personal advantages of car use to benefit collective interests, especially theories that explain why people make short-term sacrifices in order to safeguard collective interests seem relevant in this respect, such as the norm-activation model (NAM) (Schwartz, 1977; Schwartz and Howard, 1981) and the value-belief-norm theory of environmentalism (VBN theory) (Stern, 2000).

The NAM (Schwartz, 1977; Schwartz and Howard, 1981) was originally developed to explain altruistic behaviour, but has often been applied in the environmental context (e.g., Dunlap and Van Liere, 1978; Hopper and Nielsen, 1991; Vining and Ebreo, 1992; Bamberg and Schmidt, 2003; Gärling *et al.*, 2003). According to NAM, behaviour occurs in response to personal norms that are activated when individuals are aware of adverse consequences to others or the environment (awareness of consequences or AC beliefs) and when they think they can adverse these consequences (ascription of responsibility or AR beliefs). Stern and colleagues (Stern *et al.*, 1999; Stern, 2000) proposed the VBN theory, which is in essence an extension of the NAM. Like the NAM, they propose that environmental behaviour results from personal norms, i.e., a feeling of moral obligation to act pro-environmentally. These personal norms are activated by beliefs that environmental conditions threaten things the individual values (awareness of consequences or AC beliefs) that in turn affect beliefs that the individual can act to reduce this threat (ascription of responsibility or AR beliefs). The VBN theory proposes that AC and AR beliefs are dependent on general beliefs on human–environment relations (e.g., Dunlap and Van Liere, 1978; Dunlap *et al.*, 2000) and on relatively stable value orientations. The causal chain proposed in the VBN theory moves from relatively stable and general values, to beliefs about human–environment relations, which in turn are believed to affect specific beliefs on consequences of environmental behaviour and the individual's responsibility for these problems and for taking corrective actions, which affect personal norms, and, subsequently, behaviour.

Stern (2000) explicitly states that the VBN theory can be applied to explain acceptability judgements. VBN theory has been successful in explaining various environmental behaviours, among which willingness to sacrifice and acceptability of environmental policies (Stern *et al.*, 1999; Steg *et al.*, 2005). Of special interest here is the study by Steg *et al.* (2005), which revealed that acceptability of energy policies was indeed related to personal norms, while personal norms were stronger the more people felt responsible for problems related to energy use, which was in turn related to awareness of consequences of problems related to energy use. Further, as expected, awareness of consequences appeared to be related to general beliefs about human–environment relations and to value orientations. The importance of problem awareness has also been stressed by other scholars, as discussed earlier. Other studies also suggest that policies are more acceptable when people are more aware of the (environmental) problems at stake, when they feel more responsible for these problems and when they feel a stronger moral obligation to contribute to the solution of these problems (e.g., Stern *et al.*, 1999; Schade, 2003). Further research is needed to test whether variables identified in the VBN theory are related to acceptability of transport pricing.

Policy Features That Affect the Acceptability of Transport Pricing

In this section, we discuss policy features that may affect the acceptability of transportation pricing policies, and explain the relation to individual factors influencing acceptability judgements as identified in the previous section, notably perceived effectiveness of policies, the extent to which individual freedom to move is affected, and perceived fairness of policies. Three policy features are discussed: (1) the level of price changes, (2) the extent to which price changes are differentiated, e.g., in time, place, for user groups, and (3) the way revenues are allocated.

Obviously, the level of price changes affects people's freedom to move. Marginal price increases will hardly affect people or may even not be noticed, while many people (especially low-income groups) may not be able to cope with extensive price increases. Not surprisingly, significant price increases are evaluated as less acceptable than are small increases (Schuitema *et al.*, 2003), especially if the policy is not differentiated and people cannot easily evade the measure. Moreover, significant price increases are not believed to be fair, because low-income groups will be afflicted disproportionately. On the other hand, small price increases may not be acceptable as well (Rienstra *et al.*, 1999; Schade and Schlag, 2000), because in that case travel costs will increase without problems resulting from car use being reduced. This implies that price increases are most acceptable when they are sufficiently high to reduce problems resulting from car use, without affecting one's own preferred travel behaviour.

Furthermore, acceptability may depend on the extent to which pricing policies are differentiated. First, pricing measures may be static, i.e., when price increases are similar for all drivers. Examples are flat kilometre or toll charges. Second, price increases may be variable, i.e., they may differ in time (e.g., higher tolls during peak hours) or place (e.g., higher price levels at specific routes, such as cordon charging), according to the level of the problem (e.g., congestion charging) or vehicle type (e.g., higher charges for large and heavy cars), and for different user groups (e.g., different rates for private cars and lorries). Third, dynamic price increases may be implemented, i.e., price levels may change in time or place, according to actual travel conditions (e.g., congestion levels). In case of dynamic pricing policies, at a given time actual price levels may be uncertain. As people prefer certain outcomes above uncertain ones (Kahneman and Tversky, 1984), we may hypothesise that dynamic pricing is less acceptable. In practice, however, actual price levels under dynamic pricing may be quite stable in as far as an equilibrium in travel flows is reached.

Some types of differentiations may be more fair than others, because they affect various groups differently. For example, congestion pricing will affect drivers in congested areas only, and higher tolls during peak hours will strongly affect commuters. Thus, freedom to move of some groups may be seriously affected, while other groups may not be confronted with price increases at all. In as far as people think some groups are affected disproportionately, they may think the policy is unfair.

Finally, revenue allocation may affect acceptability of pricing policies. Several studies revealed the way revenues are used may have important consequences for acceptability of transport pricing. In general, price increases appear to be more acceptable when revenues are spent in a way that benefit the individual directly, e.g., by reducing taxes for owning or using a car (e.g., Jones, 1991; Verhoef, 1996; Schade and Schlag, 2000; Harrington *et al.*, 2001; Lyons *et al.*, 2004) or when revenues are allocated within the same domain, e.g., by improving public transport (Schade and Schlag, 2000; Lyons *et al.*, 2004). In contrast, people more strongly oppose transport pricing when revenues are allocated to general public funds (Verhoef, 1996; Schade and Schlag, 2000).

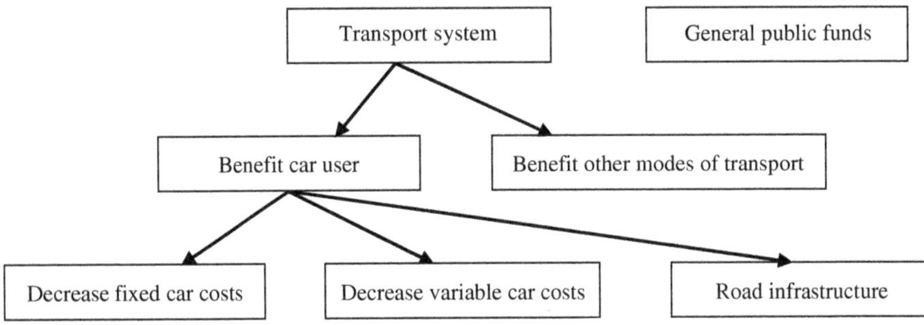

Source: Adapted from Steg and Schuitema (2003).

Figure 1: Typology of Types of Revenue Use

Figure 1 gives a typology of different types of revenue use (Steg and Schuitema, 2003). First of all, revenues may be allocated within the transport system or to general public funds. Car users evaluate the former to be more acceptable than the latter (Lyons *et al.*, 2004). Within the transport system, revenues may be allocated to benefit car users or to benefit people using other (viz., more environmentally friendly) modes of transport, such as public transport or bicycles. As may be expected, car users prefer the former above the latter (Schuitema and Steg, 2005b). Finally, car users can benefit from revenues by decreasing fixed car costs (e.g., reducing road or vehicle taxes), by reducing variable costs of car use (e.g., reducing petrol duties), or by investments in road infrastructure (e.g., build new roads). Transport pricing is more acceptable if revenues are allocated to reduce fixed and variable costs of car use rather than investing revenues in road infrastructure. In fact, investments in road infrastructure was perceived as unacceptable as allocating revenues to general public funds (Schuitema and Steg, 2005b). This seems to be in contradiction with earlier studies that revealed that investing revenues in road infrastructure was evaluated as rather acceptable. For example, a study by Verhoef (1996) revealed that investment in road infrastructure is one of the most acceptable ways of revenue use. This may be due to the method used to elicit acceptability judgements. Typically, respondents are asked to what extent they think different types of revenue use are acceptable, without explicitly linking revenue use to a specific pricing policy, and, consequently, without making any reference to the fact that respondents' themselves would be charged (e.g., Verhoef, 1996). In contrast, Schuitema and Steg (2005b) asked respondents to judge the acceptability of a specific transport-pricing policy. In addition to this, they explicated how revenues of this policy would be allocated. Thus, respondents realised they had to pay themselves for improvements in infrastructure. This implies that many people may find investments in infrastructure quite acceptable, but only if they do not realise they have to pay for it themselves. These findings highlight the importance of the study design; different designs of the research task may be interpreted differently, and may elicit different preference judgements. Another explanation for these contradictory findings is the research method followed. Earlier studies followed within subject designs in which every respondent judges all types of revenue allocation (i.e., comparisons are made between mean acceptability scores of various types of revenue use of all respondents). In contrast, Schuitema and Steg (2005b) followed a between-subject design in which each respondent evaluates one type of revenue only (i.e., comparisons are made between judgements of different respondents). In general, between-subject designs are less conspicuous then are within subject designs (see also Hendrickx and Nicolaij, 2004). This finding certainly warrants further study.

In general, it appears that transport-pricing policies are more acceptable if revenues are used to decrease fixed and variable costs of car use and/or car possession. In such cases, on an aggregate level, overall changes in costs of car use may be small or costs may not even change at all. Obviously, this may affect the overall effectiveness of the particular policy. However, costs will likely increase for some groups, and decrease for other groups, depending on the type of transport pricing and revenue allocation. For example, using revenues for decreasing fixed car costs may have relatively more negative consequences for low-income groups than for high-income groups (Verhoef *et al.*, 2004). Still, on an aggregate level, people's freedom to move may hardly be restricted. This may be one of the reasons why people generally prefer these types of revenue use. Returning revenues back to car users may also be perceived to be more fair than the other types of revenue use: the payer will receive something in return. Revenues may be used to compensate those who are affected most, but this will not be an easy task to accomplish. Some will still lose, for example those with no capacity to change, long commutes, few prospects for changing jobs, or tight budgets (Richardson and Bae, 1998).

SUMMARY AND CONCLUSION

Transport-pricing policies are aimed at changing car use, and consequently, reducing problems caused by car traffic. Transport pricing may elicit various behavioural changes that may affect collective qualities differently. Therefore, policymakers should base their selection and design of transport-pricing policies on the policy goal at stake and behavioural changes to be reached. In general, transport-pricing policies will be more effective if price increases are significant, and if feasible alternatives are available.

Policymakers seem generally reluctant to implement stringent transport-pricing policies, for it is assumed that such policies may significantly reduce individual's quality of life. However, empirical evidence for this assumption is lacking. In fact, studies have revealed that quality of life will not be strongly affected when transport pricing is implemented. One reason for this is that people do not only consider individual consequences of transport pricing, but they also take into account effects on environmental qualities and congestion. Overall, people expect to be better off when pricing policies would reduce environmental and congestion problems. This may well be related to acceptability of such measures, which is confirmed by the positive relationship between perceived effectiveness and acceptability.

Lack of public support is an important barrier for the implementation of transport-pricing policies. Acceptability is dependent on various individual perceptions and motivations. First, acceptability is dependent on the expected consequences of transport-pricing policies. Transport pricing is more acceptable when people think these policies would actually reduce environmental and congestion problems and when it benefits the individual more (which may be related to improved collective quality). This implies that acceptability may increase when the expected individual and collective benefits are clearly communicated. Process and distributive justice are also closely related to acceptability judgements; they affect to what extent policies are perceived to be fair. However, little is known yet about the role of fairness, e.g., why are policies perceived to be fair or not, and which justice principles are prevalent for different groups?

Second, acceptability may be related to moral and environmental concerns. Studies in the environmental field suggest that acceptability of policies is higher if people feel a moral obligation to act in the common good. This will be especially the case if they are aware of relevant problems and when they feel responsible for these problems, which are in turn related to general environmental beliefs and values. This implies that acceptability may increase if people are aware of the problems caused by car use, and when they feel responsible to do something about it. Further research is needed to see whether these results may be generalised to traffic and transport issues.

Various policy characteristics may be related to acceptability of transport pricing. Obviously, significant price increases will be less acceptable than minor ones. However, minor price-increases appear to be not acceptable as well because in that case problems resulting from car use will not be reduced while prices of travel increase; as indicated earlier, perceived effectiveness is an important precondition for acceptability. The extent to which transport-pricing policies are differentiated may also affect acceptability. This is closely related to processes of distributive fairness discussed earlier. Fairness of transport policies may be enhanced via revenue allocation. Revenue allocation appeared to strongly related to acceptability judgements. In general, transport-pricing policies are more acceptable if revenues benefit individual car users rather than the general public. Thus, acceptability of transport pricing may be enhanced by clarifying how revenues are used, and by compensating those who are affected most by the policies, as far as this would not decrease effectiveness of policies. Allocating revenues to road infrastructure appeared to be highly acceptable in some studies, but not acceptable at all in other studies. The design of research tasks and research method followed may be the reason for these conflicting results. Further sophisticated research is needed to clarify the role of revenue use.

REFERENCES

Ajzen, I. (1985). From Intentions to Actions: A Theory of Planned Behavior. In: *Action-Control: From Cognition to Behavior* (J. Kuhl and J. Beckman, eds), Springer, Heidelberg, 11–39.

Bamberg, S. and D. Rölle (2003). Determinants of People's Acceptability of Pricing Measures – Replication and Extension of a Causal Model. In: *Acceptability of Transport Pricing Strategies* (J. Schade and B. Schlag, eds), Elsevier Science, Oxford, 235–248.

Bamberg, S. and S. Schmidt (2003). Incentives, Morality or Habit? Predicting Students' Car Use for University Routes with the Models of Ajzen, Schwartz and Triandis, *Environment and Behavior*, **35** (2), 264–285.

Berkhout, P.H.G., J.C. Muskens, and J.W. Velthuijsen (2000). Defining the Rebound Effect, *Energy Policy*, **28** (6/7), 425–432.

Cavalini, P.M., L. Hendrickx, and A.J. Rooijers (1996). Verschillen tussen groepen automobilisten met betrekking tot hun gevoeligheid voor beleidsmaatregelen [Group differences in car users' responsiveness to policy measures], *Milieu*, **3**, 138–147 (in Dutch).

Claassen, A. and J. Kropman (1995). *Verkenningen tussen theorie en praktijk. Over de bijdrage van de sociale wetenschappen aan de beleidspraktijk op het gebied van mobiliteit* [*The Contribution of Social Sciences to Transport Policy Making*], Instituut voor Toegepaste Sociale Wetenschappen, Nijmegen (in Dutch).

Clayton, S. (2000). Models of Justice in the Environmental Debate, *Journal of Social Issues*, **56** (3), 459–474.

De Groot, J. and L. Steg (2006). Impact of Transport Pricing on Quality of Life, Acceptability, and Intentions to Reduce Car Use: An Explorative Study in Five European Countries, *Journal of Transport Geography*, **14**(6), 463–470.

Deslauriers, B.C. and P.B. Everett (1977). Effects of Intermittent and Continuous Token Reinforcement on Bus Ridership, *Journal of Applied Psychology*, **62** (4), 369–375.

Diener, E. (1995). A Value Based Index for Measuring National Quality of Life, *Social Indicators Research*, **36**, 107–127.

Diener, E., E.M. Suh, R.E. Lucas, and H.L. Smith (1999). Subjective Well-Being: Three Decades of Process, *Psychological Bulletin*, **125** (2), 276–302.

Dunlap, R.E. and K.D. Van Liere (1978). The 'New Environmental Paradigm': A Proposed Measuring Instrument and Preliminary Results, *Journal of Environmental Education*, **9**, 10–19.

Dunlap, R.E., K.D. Van Liere, A.G. Mertig, and R.E. Jones (2000). Measuring Endorsement of the New Ecological Paradigm: A Revised NEP Scale, *Journal of Social Issues*, **56** (3), 425–442.

Fehr, E. and A. Falk (2002). Psychological Foundations of Incentives, *European Economic Review*, **46** (4/5), 687–724.

Fishbein, M. and I. Ajzen (1975). *Belief, Attitude, Intention and Behavior: An Introduction to Theory and Research*, Addison-Wesley, Reading.

Frey, B.S. (1997). *Not Just for the Money. A Theory of Personal Motivation*, Edward Elgar, Cheltenham.

Frey, B.S. (2003). Why are Efficient Transport Policy Instruments so Seldom Applied? In: *Acceptability of Transport Pricing Strategies* (J. Schade and B. Schlag, eds), Elsevier Science, Oxford, 63–75.

Gärling, T., S. Fujii, A. Gärling, and C. Jakobsson (2003). Moderating Effects of Social Value Orientation on Determinants of Proevenvironmental Intention, *Journal of Environmental Psychology*, **23** (1), 1–9.

Gomez-Ibanez, J.A. and K.A. Small (1994). *NCHRP Synthesis of Highway Practice 210: Road Pricing for Congestion Management: A Survey of International Practice*, TRB, National Research Council, Washington, DC.

Harrington, W., A.J. Krupnick, and A. Alberini (2001). Overcoming Public Aversion to Congestion Pricing, *Transportation Research A*, **35**, 87–105.

Hendrickx, L. and S. Nicolaij (2004). Temporal Discounting and Environmental Risks: The Role of Ethical and Loss-Related Concerns, *Journal of Environmental Psychology*, **24**, 409–422.

Hopper, J.R. and J.Mc.C. Nielsen (1991). Recycling as Altruistic Behavior: Normative and Behavioral Strategies to Expand Participation in a Community Recycling Program, *Environment and Behavior*, **23**, 195–220.

Jaensirisak, S., A.D. May, and M. Wardman (2003). Acceptability of Road User Charging: The Influence of Selfish and Social Perspectives. In: *Acceptability of Transport Pricing Strategies* (J. Schade and B. Schlag, eds), Elsevier, Amsterdam, 203–218.

Jakobsson, C. (2004). Accuracy of Household Planning of Car Use: Comparing Prospective to Actual Car Logs, *Transportation Research F*, **7**, 31–42.

Jakobsson, C., S. Fujii, and T. Gärling (2000). Determinants of Private Car Users' Acceptance of Road Pricing, *Transport Policy*, **7**, 153–158.

Jakobsson, C., S. Fujii, and T. Gärling (2002). Effects of Economic Disincentives on Private Car Use, *Transportation*, **29**, 349–370.

Jones, P. (1998). Urban Road Pricing: Public Acceptability and Barriers to Implementation. In: *Road Pricing, Traffic Congestion and the Environment. Issues of Efficiency and Social Feasibility* (K.J. Button and E.T. Verhoef, eds), Edward Elgar, Cheltenham, UK, 263–284.

Jones, P. (2003). Acceptability of Road User Charging: Meeting the Challenge. In: *Acceptability of Transport Pricing Strategies* (J. Schade and B. Schlag, eds), Amsterdam: Elsevier, 27–62.

Jones, P.M. (1991). Gaining Public Support for Road Pricing through a Package Approach, *Traffic Engineering and Control*, **32**, 194–196.

Jones, P.M. (1995). Road Pricing: The Public Viewpoint. In: *Road Pricing: Theory, Empirical Assessment and Policy* (B. Johansson and L.G. Mattsson, eds), Kluwer, Boston, 159–180.

Kahneman, D. and A. Tversky (1984). Choices, Values and Frames, *American Psychologist*, **39**, 341–350.

Lind, E.A. and T.R. Tyler (1988). *The Social Psychology of Procedural Justice*, Plenum Press, New York, NY.

Loukopoulos, P., C. Jakobsson, T. Gärling, C.M. Schneider, and S. Fujii (2004). Car-User Responses to Travel Demand Management Measures: Goal Setting and Choice of Adaptation Alternatives, *Transportation Research D*, **9**, 263–280.

Loukopoulos, P., C. Jakobsson, T. Gärling, C.M. Schneider, and S. Fujii (2005). Public Attitudes Towards Policy Measures for Reducing Private Car Use, *Environmental Science and Policy*, **8** (1), 57–66.

Lyons, G., G. Dudley, E. Slater, and G. Parkhurst (2004). *Evidence-Base Review – Attitudes to Road Pricing*, Centre for Transport & Society, Bristol, UK.

Meland, S. (1995). Generalized and Advanced Urban Debiting Innovations. The GAUDI Project 3. The Trondheim Toll Ring, *Traffic Engineering and Control*, **36**, 457–471.

Mogridge, M.J. (1978). The Influence of the Oil Crisis on the Growth in the Ownership and Use of Cars, *Transportation*, **7**, 45–67.

Ney, J.S., Jr., P.D. Zelikow, and D.C. King (eds) (1997). *Why People Don't Trust Government*, Harvard University Press, Cambridge, MA.

Poortinga, W., L. Steg, and C. Vlek (2004). Values, Environmental Concern, and Environmental Behavior: A Study into Household Energy Use, *Environment and Behavior*, **36**, 70–93.

Priemus, H. and P. Nijkamp (1994). *Beheersing van automobiliteit: feit of fictie* [*Managing Car Use: Fact or Fiction*], Delftse Universitaire Press, Delft (in Dutch).

Ramjerdi, F., H. Minken, and K. Østmoe (2004). Norwegian Urban Tolls. In: *Road Pricing: Theory and Evidence* (G. Santos, ed.), Elsevier, Amsterdam, 237–249.

Richardson, H.W. and C.-H.C. Bae (1998). The Equity Impacts of Road Congestion Pricing. In: *Road Pricing, Traffic Congestion and the Environment. Issues of Efficiency and Social Feasibility* (K.J. Button and E.T. Verhoef, eds), Edward Elgar, Cheltenham, UK, 247–262.

Rienstra, S.A., P. Rietveld, and E.T. Verhoef (1999). The Social Support for Policy Measures in Passenger Transport. A Statistical Analysis for the Netherlands, *Transportation Research D*, **4** (3), 181–200.

Santos, G. (2004). Urban Road Pricing in the UK. In: *Road Pricing: Theory and Evidence* (G. Santos, ed.), Elsevier, Amsterdam.

Santos, G., W.W. Li, and T.H. Koh (2004). Transport Policies in Singapore. In: *Road Pricing: Theory and Evidence* (G. Santos, ed.), Elsevier, Amsterdam, 209–236.

Schade, J. (2003). European Research Results on Transport Pricing Acceptability. In: *Acceptability of Transport Pricing Strategies* (J. Schade and B. Schlag, eds), Elsevier Science, Oxford, 109–123.

Schade, J. and B. Schlag (2000). *Acceptability of Urban Transport Pricing*, VATT, Helsinki.

Schade, J. and B. Schlag (2003). Acceptability of Urban Transport Pricing Strategies, *Transportation Research F*, **6**, 45–61.

Schlag, B. and J. Schade (2000). Public Acceptability of Traffic Demand Management in Europe, *Traffic Engineering and Control*, **41** (8), 314–318.

Schlag, B. and U. Teubel (1997). Public Acceptability of Transport Pricing, *IATSS Research*, **21** (2), 134–142.

Schroeder, D.A., J.E. Steel, A.J. Woodell, and A.F. Bembenek (2003). Justice within Social Dilemmas, *Personality and Social Psychology Review*, 7 (4), 374–387.

Schuitema, G. (2003). Pricing Policies in Transport. In: *Human Decision Making and Environmental Perception. Understanding and Assisting Human Decision Making in Real-Life Settings* (L. Hendrickx, W. Jager, and L. Steg, eds), Department of Psychology, University of Groningen, Groningen, 83–102.

Schuitema, G. and L. Steg (2005a). Effects of Revenue Use and Perceived Effectiveness on Acceptability of Transport Pricing Policies, Paper presented at the 45th Congress of the European Science Association, Amsterdam, The Netherlands, August 23–27.

Schuitema, G. and L. Steg (2005b). Factors That Affect the Acceptability of Pricing Policies in Transport, Paper presented at the 7th Nordic Environmental Social Science (NESS) Research Conference, Göteborg, Sweden, June 15–17.

Schuitema, G., L. Steg, and C. Vlek (2003). Prijsbeleid: Effectief en aanvaardbaar voor veranderen van autogebruik? [Pricing Policies: Effective and Acceptable for Changing Car Use?], Paper presented at the Colloquium Vervoersplanologisch Speurwerk, Antwerpen, Belgium, November 20–21 (in Dutch).

Schuitema, G., L. Steg, C. Vlek, and J.A. Rothengatter (2004). Factors Affecting the Acceptability of Pricing Policies in Transport, Paper presented at the 3rd International Conference on Traffic and Transport Psychology (ICTTP), Nottingham, UK, September 5–9.

Schwartz, S.H. (1977). Normative Influences on Altruism. In: *Advances in Experimental Social Psychology* (L. Berkowitz, ed.), Vol. 10, Academic Press, New York, 221–279.

Schwartz, S.H. and J.A. Howard (1981). A Normative Decision-Making Model of Altruism. In: *Altruism and Helping Behaviour: Social, Personality, and Development Perspectives* (J.P. Rushton, ed.), Erlbaum, Hilldale, NJ, 189–211.

Small, K.A. and J.A. Gomez-Ibañez (1998). Road Pricing for Congestion Management: The Transition from Theory to Policy. In: *Road Pricing, Traffic Congestion and the Environment. Issues of Efficiency and Social Feasibility* (K.J. Button and E.T. Verhoef, eds), Edward Elgar, Cheltenham, UK, 213–246.

Steg, E.M. (1996). *Gedragsverandering ter vermindering van het autogebruik. Theoretische analyse en empirische studie over probleembesef, verminderingsbereidheid en beoordeling van beleidsmaatregelen* [*Behaviour Change for Reducing the Use of Motor-Cars. Theoretical Analysis and Empirical Study on Problem Awareness, Willingness-to-Change and Evaluation of Policy Measures*], Doctoral Thesis, Faculty of Psychological, Pedagogical, and Sociological Sciences, University of Groningen, The Netherlands (in Dutch, with extensive English summary).

Steg, L. (2003). Factors Influencing the Acceptability and Effectiveness of Transport Pricing. In: *Acceptability of Transport Pricing Strategies* (J. Schade and B. Schlag, eds), Elsevier, Amsterdam, 187–202.

Steg, L., L. Dreijerink, and W. Abrahamse (2005). Factors Influencing the Acceptability of Energy Policies: Testing VBN Theory, *Journal of Environmental Psychology*, 25, 415–425.

Steg, L. and R. Gifford (2005). Sustainable Transport and Quality of Life, *Journal of Transport Geography*, **13** (1), 59–69.

Steg, L. and G. Schuitema (2003). Een psychologisch perspectief op prijsbeleid in verkeer en vervoer [A Psychological Perspective on Transport Pricing], *Tijdschrift Vervoerswetenschap* [*Journal of Transport Science*], **39** (4), 23–28.

Steg, L. and C. Vlek (1997). The Role of Problem Awareness in Willingness-to-Change Car Use and in Evaluating Relevant Policy Measures. In: *Traffic and Transport Psychology. Theory and Application* (J.A. Rothengatter and E. Carbonell Vaya, eds), Pergamon, Oxford, 465–475.

Stern, P.C. (2000). Toward a Coherent Theory of *Environmentally* Significant Behavior, *Journal of Social Issues*, **56** (3), 407–424.

Stern, P.C., T. Dietz, T. Abel, G.A. Guagnano, and L. Kalof (1999). A Value-Belief-Norm Theory of Support for Social Movements: The Case of Environmentalism, *Human Ecology Review,* **6**(2), 81–97.

TfL (Transport for London). (2004). *Congestion Charge Central London: Impacts Monitoring*, Second Annual Report, Transport for London, London.

Thøgersen, J. (2003). Monetary Incentives and Recycling: Behavioural and Psychological Reactions to a Performance-Dependent Garbage Fee, *Journal of Consumer Policy*, **26**, 197–228.

Tretvik, T. (2003). Urban Road Pricing in Norway: Public Acceptability and Travel Behaviour. In: *Acceptability of Transport Pricing Strategies* (J. Schade and B. Schlag, eds), Elsevier Science, Oxford, 77–92.

Van Knippenberg, D.L. and C.W.F. Van Knippenberg (1988). *Beïnvloeding van vervoermiddelkeuze in het woon-werkverkeer door middel van een tijdelijke gedragsverandering* [*Influencing Mode Choice for Commuting via Temporarily Behaviour Changes*], Verkeerskundig Studiecentrum, Rijksuniversiteit Groningen, Haren, the Netherlands (in Dutch).

Verhoef, E., C. Koopmans, M. Bliemer, P. Bovy, L. Steg, and B. Van Wee (2004). *Vormgeving en effecten van prijsbeleid op de weg. Effectiviteit, efficiëntie en acceptatie vanuit een multidisciplinair perspectief* [*Design and Effects of Road pricing. Effectiveness, Efficiency and Acceptability from a Multidisciplinary Perspective*], Vrije Universiteit/SEO, Amsterdam; Technische Universiteit Delft, Delft; Rijksuniversiteit Groningen, Groningen (in Dutch).

Verhoef, E.T. (1996). *Economic Efficiency and Social Feasibility in the Regulation of Road Transport Externalities*, Thesis Publishers, Amsterdam.

Vining, J. and A. Ebreo (1992). Predicting Recycling Behavior from Global and Specific Environmental Attitudes and Changes in Recycling Opportunities, *Journal of Applied Social Psychology*, **22** (20), 1580–1607.

Wootton, J. (1999). Replacing the Private Car, *Transport Review*, **19**, 157–175.

Tommy Gärling and Linda Steg (Editors)

20

SOCIAL MARKETING OF ALTERNATIVE TRANSPORTATION MODES

John Thøgersen, Department of Marketing and Statistics, Aarhus School of Business, Aarhus University, Aarhus, Denmark

ABSTRACT

Researchers and practitioners working with travel demand management (TDM) seem to be increasingly interested in social marketing as a means for promoting non-car modes of transportation. However, as is true for social marketing in general, there is little clarity of the social marketing approach associated with TDM. Hence, it is not surprising that the effectiveness of this means is subject to considerable uncertainty. This chapter outlines the field and the definitions of social marketing as well as reviews the practical experience and the research on social marketing in TDM as a basis for evaluating the usefulness of social marketing in this area.

INTRODUCTION

Owing to the growing societal harms produced by car traffic, as described in the first part of this book, countries, regions, and municipalities all over the world are searching for and experimenting with ways to reduce car traffic. Some of the "tougher" measures to this end, such as restrictions on driving and economic penalties, are met with widespread and persistent political and public resistance (see, e.g., Gärling and Loukopoulos, Chapter 17 in this volume; Steg and Schuitema, Chapter 19 in this volume). This is an important reason why marketing techniques for inducing voluntary behavior change are becoming increasingly popular in travel demand management (TDM) (e.g., Brög *et al.*, 2003; Victoria Transport Policy Institute, 2004). Another reason is documented effects, some of which have been reviewed in this chapter.

Owing to current attempts to reduce car driving being motivated by social goals, such as reducing congestion and the emission of greenhouse gasses, marketing campaigns in this area are increasingly social marketing campaigns. Below, I specify what I mean by social marketing and summarize how social marketing techniques have been used in TDM. I also describe three specific applications of social marketing in this area. Finally, I briefly outline important limitations of this approach.

A CONCEPTUAL ANALYSIS OF SOCIAL MARKETING

Marketing techniques have been used to promote public transit services for a long time (Deka, 1996). Experiences from these commercial applications can be useful also in a TDM context, but here the perspective extends what is commercially profitable. When marketing is used to promote social objectives, rather than for private ends, it is called social marketing (e.g., Kotler and Zaltman, 1971; Andreasen, 1994; Kotler and Roberto, 1989).

Kotler and Zaltman introduced the social marketing concept in 1971. At the beginning, social marketing was mainly applied in family planning, but it quickly spread to the area of public health and later to other areas. Currently, the main application areas for social marketing are health, safety, environment, and community involvement (Kotler *et al.*, 2002). Social marketing programs have also been applied to travel mode choice (which I will return to) and a variety of other transport-related behaviors (e.g., preventing drunken driving and speeding and promoting the use of seat belts, and even, in Sweden in 1967, promoting the change of the traffic rules from driving on the left to driving on the right side, see Andreasen, 1991; Kotler *et al.*, 2002). Social marketing belongs to the so-called "soft" measures of intervention, together with, for example, communication and education. However, social marketing differs from other "soft" measures as the ultimate goal is always to influence behavior, and not just audiences' knowledge or attitudes (Andreasen, 1991, 1994).

Social marketing principles and techniques have been tested in a large number of empirical studies (cf., Fox and Kotler, 1980; Malafarina and Loken, 1993; and the Social Marketing Quarterly). In spite of this, and in spite of the publication of widely cited conceptual pieces (e.g., Kotler and Zaltman, 1971; Andreasen, 1993, 1994) and book-length treatments of social marketing (Fine, 1981; Kotler and Roberto, 1989; Andreasen, 1995; Kotler *et al.*, 2002), what constitutes a social marketing approach to social change is, however, still in dispute (Lefebvre, 2001).

Here, I adopt one of the most widely cited definitions of social marketing, proposed by Andreasen (1994):

> "Social marketing is the adaptation of commercial marketing technologies to programs designed to influence the voluntary behavior of target audiences to improve their welfare and that of the society of which they are part."

Like other influential definitions of social marketing (e.g., the one proposed by Kotler and Zaltman, 1971), this one emphasizes that social marketing is more than advertising or communication. Communication is needed in social marketing campaigns to create or increase awareness about a (new) social idea and perhaps even to persuade audiences to accept it, but the bottom line of marketing programs is to change behavior, and awareness or attitudes do not always transform into behavior. The idea that people can be persuaded by communication alone to accept a behavior that conflicts with their own felt needs and wants, is rejected as futile by social marketers. Instead, social marketers are concerned with designing solutions to the targeted social problem (i.e., "products") that are perceived as desirable and gratifying by their potential adopters. They are also concerned with providing convenient distribution channels and offering solutions that are generally characterized by a favorable balance between perceived costs and benefits. Market research is used to ascertain the needs, the wants, and the perceived barriers of those targeted by the program and to monitor and evaluate their satisfaction and the needs for adjustments after implementing the program.

How a social marketing program should be designed in order to maximize the chances of success depends on the targeted behavior. Andreasen (1991) suggests that three characteristics or dimensions are particularly important when assessing behavior in this context:

1. The involvement of the actor (low–high);
2. Whether it is a one-off or a continuing behavior; and
3. Whether it is performed by individuals or groups.

For instance, a social marketing campaign with the purpose of reducing harms produced by car traffic might focus on persuading people to buy a more energy-efficient car (e.g., Kurani and Turrentine, 2002). In this case, it would target a high-involvement decision made infrequently by individuals (or individual families). Although, most people would undoubtedly like their car to be energy efficient, buying one can be impeded by lack of knowledge and understanding of what makes a car energy-efficient, which cars are energy efficient, and/or the advantages of having an energy-efficient car. Actually buying one may also be impeded by a higher price, lower capacity for carrying passengers and luggage, less comfort and safety, and a less-powerful engine (e.g., A. Gärling and Thøgersen, 2001; Kurani and Turrentine, 2002). To change behavior successfully, the social marketing campaign needs to target important impediments, be they misconceptions or real disadvantages of the promoted behavior.

Another social marketing campaign might aim to persuade drivers to increase their use of alternative travel modes for commuting, such as walking, cycling or public transportation. This campaign would target a continuing behavior performed by large groups of people every day, but usually on an individual basis, and normally with little involvement in the decision on travel mode (unless something interferes with their chosen mode).

Changing continuing behavior involves breaking old habits, learning new ones, and freezing the new pattern of behavior, which is often more difficult than changing a one-shot action (e.g., Verplanken *et al.*, 1994; Verplanken *et al.*, 1997; Møller and Thøgersen, in press; but see Bamberg *et al.*, 2003). Low involvement means that people are usually not interested in information about the issue, although it could help them make better choices (Aarts *et al.*, 1997; Verplanken *et al.*, 1997). Especially, the disinterest in information reduces the effectiveness of mass communication. Hence, a special effort is needed to create sufficient situational involvement to make target commuters receptive to information about alternative travel modes and willing to try alternatives. Further, mechanisms that enable and make it convenient for people to translate their motivation into action are particularly important for changing this type of behavior (e.g., Andreasen, 1991; Thøgersen, 1997).

As already indicated, social marketing campaigns purporting to reduce the harms from car traffic can target different behaviors. One suggested approach is de-marketing of the car, that is, to attempt to reduce its attractiveness (Wright and Egan, 2000). Another approach would be to promote alternatives to traveling, for example, stay home and watch TV instead of going to the theater or teleworking instead of commuting (e.g., Jensen *et al.*, 2003; T. Gärling, 2005; see also http://www.secure-teletrips.com). However, the most common approach – and probably also the most feasible in most cases – is to promote alternative modes of transportation. Below, I focus especially on the latter approach.

REVIEW OF SOCIAL MARKETING APPLICATIONS

The last couple of decades have witnessed an increasing number and a wide variety of applications of social marketing principles in TDM. In some cases, a social marketing program has been designed to promote a specific alternative to single occupancy cardriving, such as ridesharing or carpooling (e.g., McClintock, 2000; Chun, 1993; Kearney and De Young, 1995; Smith and Beroldo, 2002; Glazer *et al.*, 1986), cycling (e.g., Cleary and Ryley, 2001; Gaffron, 2003; Rose *et al.*, 2003) or public transit (e.g., Bachman and Katzev, 1982; see also Deka, 1996; Enoch and Potter, 2002; Stradling, 2002). In other more ambitious cases, the goal has been to match available alternatives (individual modes or a combination) to the individual driver's needs (e.g., Lohmann-Hansen *et al.*, 2001; Brög *et al.*, 2003)[1]. Often, special events are organized – or the campaign is linked to one – in order to create public interest, PR, and situational involvement (e.g., Thayer, 1992; Rose *et al.*, 2003). Events may be planned (e.g., a bike-to-work week) or unplanned, but foreseen (e.g., smog alert days), and they may be single (e.g., marking the end of a period with sub-standard service, for instance, due to the renovation of a railway line) or recurrent (e.g., a yearly car-free day).

[1] See also the Boulder, Colorado, case at http://www.toolsofchange.com.

To create public interest, event-based social marketing may rely mostly on a single marketing parameter, usually a special offer (e.g., Tetraplan, 2004) or the campaigners may create a package of propositions for broader appeal, often including a strong social element (Rose *et al.*, 2003).

Below, I describe three cases in some detail. The cases are chosen to illustrate the breadth of social marketing campaigns in the TDM area. The effects of the campaign are reasonably well documented and publicly available. It has been less important whether or not campaigners have conceptualized their campaign in social marketing terms.

Case 1: Individualized Marketing

The "individualized marketing" approach was developed by the German consultant company Socialdata GmbH (Brög *et al.*, 2003). After preliminary tests in the early 1990s in Kassel and Nuremberg, Germany, followed by an international demonstration project consisting of 45 local projects in 13 European countries, the first large-scale application of this approach was carried out in South Perth, Australia, in 1997 (the "TravelSmart" campaign).

The "individualized marketing" approach is marketing oriented, as it is based on a thorough customer survey aiming at uncovering the reasons behind an individual's mode choice for every trip, awareness of alternatives, and perceptions about barriers that prevent the individual from choosing alternative modes of transportation. Another marketing oriented characteristic is that "the customer base" is segmented, and different segments are targeted differently. Further, the "individualized marketing" approach utilizes a number of techniques from the marketing toolbox, notably personal communication and sales promotion (in the form of a free "sample"), in addition to information material. An important limitation of the "individualized marketing" approach is, however, that it focuses on promoting existing solutions only. It contains no element of product development where the marketer is open to the possible need for new solutions that are adapted to customer needs and wants. In this respect, it is sales oriented, rather than being marketing oriented (cf., Andreasen, 1991; Kotler, 1991).

It is a unique feature of the "individualized marketing" approach that it includes an attempt to reach *all* households in the targeted area with direct and personal communication. Contacted households, who are willing to participate, are segmented based on behavior (use of various travel modes), interest (in the promoted modes), and knowledge. No further resources are wasted on those not interested or those already using alternative modes to the car and having sufficient knowledge about such alternatives (except for a small gift to the participants). Those already using alternative modes, but expressing a need for further information, receive the required information. Those *not* using alternative modes, but being interested, in doing so receive requested information and are offered a home visit by a specialized advisor regarding the use of public transportation or bicycling and/or walking. If relevant, they also receive a free "test ticket" for a month so that they can familiarize themselves with public transit in the area.

In the South Perth case, 40% of contacted households fell in the "not using, but interested" category[2]. About 10% of these (i.e., 4% of contacted households) received a home visit by Perth bus staff offering advise about public transport as well as free tickets, and 42% (i.e., 17% of contacted households) received home visits by an advisor regarding cycling and/or walking. Eight percent of contacted households received personalized timetables to public transit and about 40% received packages of more general information about alternative travel modes. Based on all residents in the area (not just the participants), the program achieved a 14% reduction of car-as-driver trips, with the total number of trips remaining constant. Car-as-passenger trips (i.e., carsharing) increased by 9%, public transportation by 17%, cycling by 61%, and walking by 35%. Moreover, these changes were still sustained more than two years after the program ended.

The "individualized marketing" approach is obviously not cheap. However, based on the South Perth pilot study, it was calculated that the individualized-marketing approach had a cost–benefit ratio of 1:13 (James, 2002). Hence, there is reason to believe that it may be well worth the costs, also in other contexts.

After South Perth, more than 100 large-scale applications of the individualized marketing approach have been conducted in Europe and Australia, in addition to over 60 pilot projects, including one in Portland, Oregon (Brög *et al.*, 2003). The short-term results of these additional applications are similar to those reported from South Perth. However, most of them are too new for available information about long-term effects.

Case 2: The Nottingham Cycle-Friendly Employers' Project[3]

The Nottingham Cycle-Friendly Employers' (NCFE) project was one of the major projects implemented in the late 1990s under the UK Department of Transport's Cycle Challenge project. Although comparatively ambitious, the project shares elements and experiences with a large number of other projects aimed at promoting cycling, especially for commuting to work. Although it was not conceptualized as such, the NCFE project contains many of the defining characteristics of a social marketing campaign.

---

[2] The presentation of the South Perth case is based on Brög et al. (2003), James (2002), and the municipality's web site http://www.dpi.wa.gov.au/travelsmart.

[3] This summary of the Nottingham Cycle-Friendly Employers' (NCFE) project is based on Cleary and McClintock (2000) and Department for Transport Local Government and the Regions (2001).

The NCFE project commenced in 1996 and was scheduled to run for two years (but both the implementation and the evaluation took longer than envisaged). Its objective was to increase cycling for commuting journeys and official work trips. Eight large employers in the Greater Nottingham area were involved as project partners, four of which were institutions of higher education. They together employ over 32,000 people, and if students are included the numbers using the sites increase to around 77,000. The project was partly financed by a grant from the UK Department of Transport (GBP 225,000).

Indeed, this project was essentially a (social) marketing campaign, as important "offers" provided as part of the project were based on consultations with potential "customers" – that is, employees of the project partners – about their needs, wants, and barriers, using a combination of travel surveys, e-mail, and discussion groups. However, not all project partners planned such consultations from the outset, which was pointed out as a major source of delays in the campaign evaluation (Cleary and McClintock, 2000). The campaign combined a number of means, including substantive changes in facilities for cyclists, publicity and information material, promotional events, and social interventions.

According to interviewed cyclists, the most important substantive changes were improved workplace showering and changing facilities for cyclists, and secure cycle parking. The publicity and information material endorsed environmental and health benefits of cycling as well as the usefulness and appropriateness of this means of transport for local journeys. Among the promotional events were bike-to-work days and bikers' breakfasts. The promotional events functioned both as means for catching attention and as social events. Most of the participating employers also organized bicycle user groups, i.e., groups of enthusiastic cycling employees, which were viewed both as a means of social influence and as a source of advice and feedback about the program.

As concluded by Cleary and McClintock (2000, p. 122), it is "not altogether clear cut ... (which) ... precise number of new cycle commuters was generated directly as a result of the project". Cleary Hughes Associates was contracted to evaluate the project in 1999 by means of a questionnaire survey, which was sent to samples of cyclist and noncyclist commuters at each of the partner employers. Below, I summarize the most important results from the survey (Department for Transport Local Government and the Regions, 2001), knowing that behavioral self-reports, and especially retrospective ones, are not very reliable. Self-reports about awareness and opinions are much less error-prone.

- Forty-two percent of cyclists reported that they cycled to work more frequently after the NCFE project than before. Forty-nine percent cycled the same amount and 9% cycled less. Amongst those who said they cycled to work more frequently, 30% said this was as a result of the new facilities, 30% as a result of moving house or job, and 30% for health reasons.

- Sixty-seven percent of cyclists were aware of the improved facilities that resulted from the NCFE project.
- Newsletters and promotional events were by far the most successful ways of spreading information about the improvements.
- Sixteen percent of cycle owners said that they used their bikes for journeys at work (short official journeys), with 7% doing so on a weekly basis.
- For cyclists, provision of showers and/or lockers was seen as the most important worksite improvement to encourage cycling by 46% of respondents, followed by cycle parking (36%), and financial incentives (10%). For non-cyclists, cycle parking is most important for 36% of respondents, followed by showers and/or lockers (29%), and financial incentives (25%).
- Thirty-two percent of noncyclists would consider cycle commuting in future. Of these, 38% said that they did not cycle to work because of a lack of facilities.

Case 3: Zero Fare on the Svendborg Line

Occasionally, users of public transit experience periods of substandard service due to renovation and/or maintenance work. Delays, cancellations, and substitution of one's regular travel mode with an inferior one (e.g., bus instead of train), inevitably lead to customer dissatisfaction, and some dissatisfied customers may change to private car driving. This was the situation, for example, facing Danish railways company DSB's Svendborg-Odense line (Svendborg line for short) in late 2003 (Høberg, 2003). For months, a major modernization project had led to irregularities in the service and to the use of buses instead of trains on the line. Hence, DSB made an extra effort to promote the new and modernized Svendborg line when the modernization project was complete. They decided to offer their customers a free gift, which was expected to make an impression: A zero fare in the whole month of January 2004.

DSB management hoped that the free month would make customers feel compensated for the substandard service they had had to put up with the year before, and that lost customers would be won back this way (Ellesøe and Flensburg, 2004). They also hoped that the initiative would attract a lot of public attention, create interest and curiosity, and convince new customers to try the Svendborg line. Furthermore, they hoped that some of these new customers would continue using the line after the promotion period (Ritzau, 2004). All these expectations were borne out, according to a thorough evaluation of the campaign issued by the Danish Ministry of Transport and Energy (Tetraplan, 2004).

In addition to DSB's regular passenger counts, a questionnaire survey was carried out, covering a random sample of about 2,500 passengers on the Svendborg line in the second half of January 2005, and those who gave their permission and their e-mail addresses were contacted again two months later for a reinterview. According to the passenger counts, the number of passengers on the Svendborg line in January 2004 (the free month) more than doubled compared with the month of January the previous two years. It was estimated that the number of new (i.e., "trial") users amounted to 8% of the population in the region where the line is situated. In February 2004, where the fare was back to normal, the number of passengers was 25% higher than in February the year before and 15% higher than two years before. It was estimated that about a third of the increase could be attributed to the free month, while two-thirds were due to improved service on the new and modernized Svendborg line. Eighteen percent of the passengers in the free month said they would have taken the car had the train not been free. Consistent with this, the Danish Road Directorate registered a (small) decrease in the traffic on the Odense-Svendborg highway from Fridays to Sundays in the free month. Eighty percent of those passengers trying the line only in the free month reported that they traveled to destinations serviced by the Svendborg line less than once a week. Hence, they used the line for an occasional journey or just out of curiosity. Still, offering these passengers a free trial is not necessarily a waste. Owing to the experience, they may be more inclined to use the line regularly, should their circumstances change in the future. Generally, trial users were more satisfied with the service on the modernized Svendborg line than the regular users, indicating a positive trial experience.

DSB's communication to their customers about the issue was multiplied by good press coverage, in local as well as national media, and this is part of the reason for the high interest and trial created by the zero-fare month. Owing to the novelty and conspicuousness of the approach, the free month was seemingly perceived as newsworthy by the news media, as expected by DSB management.

The most important results from the reinterviews, two months after the campaign, in March, are the following.

The passengers in January, who did not travel with the Svendborg line in March, were mainly people who rarely traveled to the destinations served by the line. They traveled in January either because of the free trip or because of rare circumstances. Generally, the 10% of the reinterviewed who were new customers, using the line in March as well as in January, on average used the line less frequently than the regular users. The reason was partly that they used the line for less regular travel purposes, partly that they used the line for commuting to work less often than the regular customers. Hence, the new customers tended to use the train as a supplementary travel option, or they were still in a trial phase two months after starting to use the line. However, it is a promising observation that, except with regard to the number of departures, new customers in general expressed greater satisfaction with the service than the regular users.

Because of the success of the zero-fare campaign, DSB decided to repeat it on the modernized Grenaa-Aarhus line in May 2005, with the same short-term effect as on the Svendborg line in terms of the number of passengers in the free month (Plougsgaard, 2005).

Discussion on the Findings

As illustrated by the three cases, social marketing can be an effective approach to social change, also in the area of travel mode choice. The core distinguishing characteristic of this approach is its foundation in the conviction that fulfilling the target "customer's" needs and wants is a prerequisite of voluntary behavior change. Hence, the collection of information about actual and potential customers' needs and wants is crucial for designing a successful marketing campaign. Usually, information is collected by means of market research, but as illustrated by the third case feedback from customers to front personnel and/or the customer services department is an additional, important source of information.

Social marketing differs from commercial marketing in many important ways. For instance, there are often fewer opportunities to modify offerings (Andreasen, 1991). This means that the possibilities of designing offerings that are desirable to target adopters can be severely restricted. For example, it has been suggested that an important reason why fewer and fewer people use public transit is its low status and that the solution, therefore, is to develop transit offerings that appeal to status conscious commuters (e.g., Everett and Watson, 1987). Although there is substantial evidence backing the first proposition (e.g., Steg *et al.*, 2001), there are obvious limits to how far one can go to remedy this without compromising (some of) the social goals that motivate the promotion of alternatives to car driving (e.g., the need to reduce the emission of greenhouse gasses).

This is not to say that it is impossible to make alternative travel modes more appealing within the given constraints. As demonstrated by campaigns, such as the NCFE project, it is often possible to improve the balance between perceived costs and barriers for alternative modes. Sometimes the most important improvements may concern peripherals of the travel (e.g., lockers and shower facilities and safe cycle parking in the NCFE case) rather than the travel mode itself.

Still, considering the restrictions that do exist when it comes to developing more appealing, but still sustainable travel options, it is a valuable lesson from the "individualized marketing" campaigns that it is at least sometimes possible to obtain substantial changes in the modal split away from the car, although the campaign is restricted to promoting already existing alternatives only.

Another important lesson from the "individualized marketing" campaigns is that impressive results in terms of travel mode changes can be produced by combining a number of specific measures, none of which have proven particularly successful when applied individually. For instance, the general result from studies which have systematically evaluated the use of sales promotion in the form of a free month test ticket for public transit (e.g., Fujii and Kitamura, 2003; Thøgersen and Møller, 2004) is that the use of public transit increases when the free ticket is in effect, but that it returns to baseline when drivers have to pay the normal fare again. Another element included in the individualized marketing approach, providing personalized timetables, has proven to be ineffective even in the short run when given alone to randomly assigned drivers (Østergaard and Schougaard, 1997) and so have general appeals (Hutton and Ahtola, 1991) and information about the societal benefits from using alternative options (Staats *et al.*, 1996).

Actually, two things distinguish campaigns based on the individualized marketing approach from these latter campaigns and tests.

Firstly, the individualized marketing approach combines various means of influencing behavior rather than relying on a single one. It is well documented that the combined effect of influence means, such as information and an economic incentive, can be much stronger than the additive effect of the individual means (e.g., Stern, 1999).

Secondly, the individualized marketing approach tailors the offering to the situation and wants of individual "customers". A major advantage of tailoring the offering is that it is much more cost effective to allocate (costly) incentives and customized information to households that actually need and want them. In addition, by tailoring the offering one reduces the risk of drowning audiences in superfluous information.

As mentioned in the introduction, producing situational involvement – and the use of attention-catching techniques for this purpose – is a necessary component of a social marketing campaign targeting a low involvement-continuing behavior, such as everyday travel mode choice. It is an important common trait of the three cases described here that they all involved the application of special (but different) techniques for catching the audiences' attention. The individualized marketing approach uses direct communication to catch attention (see Seethaler, 2004). DSB's zero-fare campaign relied on a conspicuous special offer, and the NCFE campaign organized special events, such as bike-to-work weeks and bicyclers' breakfasts, among other things. The fact that everyday travel mode choice is group behavior means that there are often opportunities for organizing social events, which may be instrumental in catching attention, and which may, in addition, initiate and build on social dynamics to spread the word and to support and reinforce individuals' decision to use alternative travel modes (Rose *et al.*, 2003).

DISCUSSION AND CONCLUSIONS

There is plenty of evidence documenting the effectiveness of social marketing techniques in TDM. Social marketing campaigns in this area vary a lot in scope and in level of ambition. Successful social marketing is based on a thorough understanding of "customers' " needs, wants, and perceived barriers; it uses a combination of means to create an attractive offering tailored to the needs, wants, and perceived barriers of individual segments of consumers; and it applies proven techniques for catching attention to the offerings.

However, there are limits to what social marketing can accomplish. There may be individuals or communities for whom private car driving has a strong symbolic value and therefore the norms dictate that a private car is "the only proper travel mode for a decent person". Social marketing is not particularly well suited to change values and norms (Andreasen, 1991). Here, a propaganda aimed at demarketing the car may be required (perhaps for a long time) before it is meaningful to implement a campaign promoting alternatives. Also, previous experiences indicate that social marketing campaigns alone can reduce car driving by only 10%–15%. Larger reductions require more fundamental structural changes, either in the form of improvements for alternative travel modes or restrictions on or penalties for car driving (e.g., Frederick and Kenyon, 1991).

REFERENCES

Aarts, H., B. Verplanken, and A. Van Knippenberg, (1997). Habit and Information Use in Travel Mode Choices, *Acta Psychologica*, **96**, 1–14.

Andreasen, A.R. (1991). Social Marketing. In: *Strategic Marketing for Nonprofit Organizations* (P. Kotler and A.R. Andreasen, eds), 4th ed., Prentice-Hall, Englewood Cliffs, 402–429.

Andreasen, A.R. (1993). Presidential Address. A Social Marketing Research Agenda for Consumer Behavior Researchers. In: *Advances in Consumer Research* (M. Rothschild and L. McAlister, eds), Vol. 20, Association for Consumer Research, Provo, 397–404.

Andreasen, A.R. (1994). Social Marketing: Its Definition and Domain, *Journal of Public Policy & Marketing*, **13**, 108–114.

Andreasen, A.R. (1995). *Marketing Social Change* Jossey-Bass, San Francisco.

Bachman, W. and R. Katzev (1982). The Effects of Non-Contingent Free Bus Tickets and Personal Commitment on Urban Bus Ridership, *Transportation Research A*, **16**, 103–108.

Bamberg, S., D. Rölle, and C. Weber (2003). Does Habitual Car Use Not Lead to More Resistance to Change of Travel Mode? *Transportation*, **30**, 97–108.

Brög, W., E. Erl, and N. Mense (2003). Individualised Marketing Changing Travel Behaviour for a Better Environment, Paper presented at the TRIP Research Conference: The Economic and Environmental Consequences of Regulating Traffic, Hillerød, 2–3 February.

Chun, D. (1993). Ridesharing and the Consumer: A Tale of two marketing strategies, *Transportation Research Record*, **1390**, 60–65.

Cleary, J. and H. McClintock (2000). Evaluation of the Cycle Challenge Project: A Case Study of the Nottingham Cycle-Friendly Employers' Project, *Transport Policy*, 7, 117–125.

Deka, D. (1996). Public Transit in the Social Marketing Framework, *Journal of Public Transportation*, **1**, 65–86.

Department for Transport Local Government and the Regions. (2001). *The Nottingham Cycle Friendly Employers Project*, Traffic Advisory Leaflet No. 9/01, Department for Transport Local Government and the Regions, Traffic Advisory Unit, London.

Ellesøe, M. and T. Flensburg (2004). *Trafikøkonom: Gør busserne gratis [Traffic Economist: Make the Busses Free.]*. Politiken, 19.01.04.

Enoch, M. and S. Potter (2002). Marketing and the British Bus Industry, *Municipal Engineer*, **151**, 49–56.

Everett, P.B. and B.G. Watson (1987). Psychological Contributions to Transportation. In: *Handbook of Environmental Psychology* (D. Stokols and I. Altman, eds), Vol. 2, Wiley, New York, 987–1008.

Fine, S.H. (1981). *The Marketing of Ideas and Social Issues,* Praeger, New York.

Fox, K.F.A. and P. Kotler (1980). The Marketing of Social Causes: The First 10 Years, *Journal of Marketing*, **44** (Fall), 24–33.

Frederick, S.J. and K.L. Kenyon (1991). Difficulties with the Easy Ride Project: Obstacles to Voluntary Ridesharing in the Suburbs, *Transportation Research Record*, **1321**, 94–101.

Fujii, S. and R. Kitamura (2003). What Does a One-Month Free Bus Ticket Do to Habitual Drivers? An Experimental Analysis of Habit and Attitude Change, *Transportation*, **30**, 81–95.

Gaffron, P. (2003). The Implementation of Walking and Cycling Policies in British Local Authorities, *Transport Policy*, **10**, 235–244.

Gärling, A. and J. Thøgersen (2001). Marketing of Electric Vehicles, *Business Strategy and the Environment*, **10**, 53–65.

Gärling, T. (2005). Changes of Private Car Use in Response to Travel Demand Management. In: *Traffic and Transport Psychology: Theory and Application* (G. Underwood, ed.), Elsevier, Amsterdam, 551–572

Glazer, J., A. Koval, and C. Gerard (1986). Part-Time Carpooling: A New Marketing Concept for Ridesharing, *Transportation Research Record*, **1082**, 6–15.

Høberg, J. (2003). Dsb Er i Krise På Fyn [*DSB Face a Crisis on Funen.*], *Jyllands-Posten,* 31 October.

Hutton, R.B. and O.T. Ahtola (1991). Consumer Response to a Five-Year Campaign to Combat Air Pollution, *Journal of Public Policy & Marketing*, **10**, 242–256.

James, B. (2002). Travelsmart—Large-Scale Cost-Effective Mobility Management. Experiences from Perth, Western Australia, *Municipal Engineer*, **151**, 39–48.

Jensen, L.M., C. Jensen-Butler, B. Madsen, J. Millard, and L. Schmidt (2003). A Web-Based Study of the Propensity to Telework Based on Socio-Economic, Work Organisation and Spatial Factors, Paper presented at the TRIP Conference: The Economic and Environmental Consequences of Regulating Traffic, Hillerød, February.

Kearney, A.R. and R. De Young (1995). A Knowledge-Based Intervention for Promoting Carpooling, *Environment and Behavior*, 27, 650–678.

Kotler, P. (1991). *Marketing Management. Analysis, Planning, Implementation, and Control*, 7th ed., Prentice-Hall, Englewood Cliffs, NJ.

Kotler, P. and E.L. Roberto (1989). *Social Marketing: Strategies for Changing Public Behavior*, The Free Press, New York.

Kotler, P., N. Roberto, and N. Lee (2002). *Social Marketing: Improving Quality of Life*, 2nd ed., Sage, Thousand Oaks.

Kotler, P. and G. Zaltman (1971). Social Marketing: An Approach to Planned Social Change, *Journal of Marketing*, 35 (July), 3–12.

Kurani, K.S. and T.S. Turrentine (2002). *Marketing Clean and Efficient Vehicles: A Review of Social Marketing and Social Science Approaches*, Institute of Transportation Studies, University of California, Davis, CA.

Lefebvre, R.C. (2001). Theories and Models in Social Marketing. In: *Handbook of Marketing and Society* (P.N. Bloom and G.T. Gundlach, eds), Thousand Oaks, Sage, 506–518.

Lohmann-Hansen, A., H. Lahrmann, and J.C.O. Madsen (2001). *Cykelbus'ter Projektet i Aarhus: Fra Bil Til Cykel Eller Bus Med Positive Virkemidler – Projektevaluering* [*The Aarhus Cykelbus'ter Project: From Car to Bicycle or Bus by Positive Means–Project Evaluation.*], Transportrådets notatserie No. 01-01, Transportrådet, København.

Malafarina, K. and B. Loken (1993). Progress and Limitations of Social Marketing: A Review of Empirical Literature on the Consumption of Social Ideas. In: *Advances in Consumer Research*, (M. Rothschild and L. McAlister, eds), Vol. 20, Association for Consumer Research, Provo, 397–404.

Møller, B.T. and J. Thøgersen (in press). Car-Use Habits: An Obstacle to the Use of Public Transportation? In: *Road Pricing, the Economy, and the Environment* (C. Jensen-Butler, B. Madsen, O.A. Nielsen, and B. Sloth, eds), Elsevier, Amsterdam.

Østergaard, S. and J. Schougaard (1997). Direkte Markedsføring i HT [*Direct Marketing by HT.*]. In: *Trafikdage på Aalborg Universitet, 1997* (H. Lahrmann and L.H. Pedersen, eds), Transportrådet og Trafikforskningsgruppen, Aalborg Universitet, Aalborg, 567–572.

Plougsgaard, H. (2005). Gratis Nærbane er Populær [*Free Local Rail is Popular.*], *Jyllands-Posten,* 2 May 2005.

Ritzau. (2004). Gratis Tog Succes for DSB [*Free Train Success for DSB.*], *Politiken*, 10 February.

Rose, G., H. Marfurt, and P. Harbutt (2003). Using a Ride to Work Day Event to Promote Travel Behavior Change, Paper presented at the 26th Australasian Transport Research Forum, Wellington, New Zealand, October.

Ryley, T. (2001). Translating Cycling Policy into Cycling Practice, *World Transport Policy and Practice*, 7 (2), 38–43.

Seethaler, R.K. (2004). Using the Six Principles of Persuasion to Promote Travel Behaviour Change – Preliminary Findings of a Travelsmart Pilot Test, Paper presented at the 26th Conference of Australian Institutes of Transport Research (CAITR), Melbourne.

Smith, V. and S. Beroldo (2002). Tracking the Duration of New Commute Modes Following Service from a Ridesharing Agency: Longitudinal Study, *Transportation Research Record*, **1781**, 26–31.

Staats, H.J., A.P. Wit, and C.Y.H. Midden (1996). Communicating the Greenhouse Effect to the Public: Evaluation of a Mass Media Campaign from a Social Dilemma Perspective, *Journal of Environmental Management*, **45**, 189–203.

Steg, L., C. Vlek, and G. Slotegraaf (2001). Instrumental-Reasoned and Symbolic-Affective Motives for Using a Motor Car, *Transportation Research F*, **4**, 151–169.

Stern, P.C. (1999). Information, Incentives, and Proenvironmental Consumer Behavior, *Journal of Consumer Policy*, **22**, 461–478.

Stradling, S. (2002). Transport User Needs and Marketing Public Transport, *Municipal Engineer*, **151**, 23–28.

Tetraplan. (2004). *0-takst På Svendborgbanen. Effekten af at Gøre Brugen Gratis i Januar 2004 [0 Charge on the Svendborg Line. The Effects of Making the use Free in January 2004.]*, Trafikministeriet, København.

Thayer, M. (1992). Effectiveness of a Statewide Ridesharing Promotion: California Rideshare Week, *Transportation Research Record*, **1338**, 94–101.

Thøgersen, J. (1997). Facilitating Recycling. Reverse-Distribution Channel Design for Participation and Support, *Social Marketing Quarterly*, **4** (1), 42–55.

Thøgersen, J. and B. Møller (2004). *Breaking Car-Use Habits: The Effectiveness of Economic Incentives*, Paper presented at the 3rd International Conference on Traffic and Transport Psychology, ICTTP 2004, Nottingham, September.

Verplanken, B., H. Aarts, and A. Van Knippenberg, (1997). Habit, Information Acquisition, and the Process of Making Travel Mode Choices, *European Journal of Social Psychology*, **27**, 539–560.

Verplanken, B., H. Aarts, A. Van Knippenberg, & Van C. Knippenberg, (1994). Attitude Versus General Habit: Antecedents of Travel Mode Choice, *Journal of Applied Social Psychology*, **24**, 285–300.

Victoria Transport Policy Insitute. (2004). TDM Encyclopedia: TDM Marketing, Information and Encouragement Programs, Retrieved 15 January, 2004, from http://www.vtpi.org/tdm/tdm23.htm, 4 June.

Wright, C. and J. Egan (2000). De-Marketing the Car, *Transport Policy*, 7, 287–294.

Tommy Gärling and Linda Steg (Editors)

21

INTELLIGENT TRANSPORT SYSTEMS FOR VEHICLE DRIVERS

Karel Brookhuis, Department of Psychology, Faculty of Behavioural and Social Sciences, University of Groningen, Groningen, The Netherlands; Delft University of Technology, Faculty of Technology, Policy and Management
Dick de Waard, Department of Psychology, Faculty of Behavioural and Social Sciences, University of Groningen, Groningen, The Netherlands; Delft University of Technology, Faculty of Technology, Policy and Management

ABSTRACT

The primary objective of Intelligent Transport Systems (ITS) is to enhance efficiency and safety in traffic and transport, for a large part by eliminating (the effects of) traffic accidents as far as possible. The benefits of ITS implementations are potentially considerable, not in the least if a significant decrease in accidents can be established, decreasing human suffering, economical cost and pollution. However, there are also potential problems to be expected, since the task of driving a motor vehicle under ITS is changing from slight adaptations to fundamental transformations with considerable effects. Many of these effects pertain to changes in driver behaviour.

INTRODUCTION

In the Netherlands (preferably unlimited) mobility is considered to be of major importance for keeping up the quality of life in Dutch Society. Mobility, however, cannot be unrestrained, it should expand in a sustainable manner, i.e. not counteracting societal or environmental conditions. In 2004, specific laws have been passed accordingly, concerning what is called Sustainable Mobility, in fact adopted to be the "road to the future". In the governmental memorandum "Sustainable Mobility" a role of some importance is given to Sustainable Safety, a programme that was already launched in 1992 (Koornstra *et al.*, 1992). With the introduction of the memorandum Sutainable Mobility actually the second phase of Sustainable Safety is marked, which is potentially an important step forward in the field of traffic safety and efficiency, amongst others since Intelligent Transport Systems (ITS) as a new set of measures is explicitly introduced in the Sustainable Safety programme as well (Wegman and Aarts, 2005). The expectations of ITS' effects in traffic and transport are highly positive. Transport efficiency, throughput, will be enhanced, accidents will be avoided, driver errors caught and attenuated, "it makes us better drivers"; in short, the whole transport system will be greatly improved, ... is the word.

However, problems are to be expected too, of different kind and severity. For instance, a seemingly uncontrolled proliferation of systems is noticeable, mature and immature systems try to conquer the market indiscriminably, consequently, selection is utterly difficult. Besides, it is not clear who should do the selection, and on the basis of what, to begin with. Other problems concern ITS applications that can have all sorts of (unintended or even unwanted) side-effects, for instance adverse effects, such as driver attention distraction, that are difficult to predict, determine and avoid. Nevertheless, such systems, often technology driven, will increasingly claim a position in traffic and transport.

BACKGROUND

The costs of road traffic accidents for society are enormous both in terms of human suffering and economic loss. In Europe alone around 50,000 people are killed in traffic accidents each year, while more than 1,500,000 are injured. Traffic congestion, i.e. the regular ones and those following traffic accidents, is a daily nuisance and economic concern, predominantly present in the economically most sensitive places. Directly attributable to accidents are about 12.5% (one in eight) of all traffic jams in The Netherlands (AVV, 2001). In Europe, at least € 70 billion are spent each year on medical treatment of injured people, the cost of congestion is many times that amount, while many thousands of person-years of work are lost. Safety is a major factor in economy.

But, safety is primarily a "human factors" case. Driver impairment is the first cause of accidents on (European) motorways. Based on a literature survey, Smiley and Brookhuis (1987) stated that at least 90% of all traffic accidents are to be attributed to human failure, for instance, through fatigue, inattention or drowsiness at the wheel. According to Vallet (1991), it is generally a loss of alertness, which is the principal cause of fatal accidents (34%). Alertness is usually considered a "multi-factorial collection" of causes whereas single-factor causation is generally difficult to extract. Nevertheless, some suggest that alcohol is up to at least 20% of all accidents the "prime causative factor" during the weekend nights, while fatigue as *contributing* "single factor" is estimated to be responsible for up to around 25% of all accidents (Brookhuis *et al.*, 2003b).

What is now generally known as ITS is in fact to be considered as the collection of systems and subsystems to solve the currently increasing problem that we face in traffic, i.e the traffic density, the accidents and concomitant increase in traffic congestion with all its consequences. ITS appears in many forms and shapes, from big central systems for managing traffic flows in built-up areas to intelligent individual driver-support systems. It is the latter, mostly described as Advanced Driver Assistance Systems (ADAS), that we are mainly concerned with in this chapter. The purpose of ADAS may be directed so that primary driving performance is supported or even that negative effects of driver's decreasing alertness are detected, i.e. the consequences of driver impairment by alcohol or fatigue are detected and attenuated. The benefits of ADAS implementations are potentially great in that case, because with increased automation driver error will be reduced or even eliminated, variability is decreased and efficiency is enhanced, since (cf., Congress, 1994):

- Many more vehicles can be accommodated, on regular highways but especially on dedicated lanes;
- High-performance driving can be conducted with less regard to vision, weather and environmental conditions;
- Drivers using ADAS applications can be safe and efficient drivers (e.g. elderly or inexperienced drivers).

INTELLIGENT TRANSPORT SYSTEMS

Mobility Management Systems

An integral part of Sustainable Mobility should be Mobility Management, intending to influence choices preceding and during individual transport, such as transport needs, modal split and timing, all with the aid of ITS applications. In this way, Mobility Management may change exposition and with that traffic safety at the same time. Firstly, Information and Communication Technology (ICT)-facilities, such as facilities for teleworking, which is at the heart of the need for moving, may well decrease transportation needs and can bring down the number of trips. Secondly, once a decision for transport is made, ITS can help to make a modal shift from car to public transport, and vice versa of course. The so-called chain mobility on the basis of – or following – reliable (!) traffic and travel information can support travellers to optimise their travel schemes on the one hand, and help traffic and travel facilities' managers to distribute travellers and traffic participants across the available "room" in a broad sense, on the other hand (Bovy and Van der Zijpp, 1999, Dicke *et al.*, 2004). In a far-reaching version of Mobility Management, electronic, on-line paid use of traffic and travel facilities may motivate distribution among all three choices (transport needs, modal split and timing, see Wegman and Aarts, 2005). Instant electronic paying for all forms of travel might shift or even diminish travel needs for a start (see Chapter 18, Ubbels and Verhoef, this volume). Steering and control between modes of travel, i.e. by on-line payment dependent on time and location, could greatly influence modal shift from car to other modes of transport, for instance, by making car driving highly expensive at times and/or at locations. The technique for that is available or about to be, but implementation and introduction waits political will (c.f. Marchau *et al.*, 2005).

The road systems' manager would like to be able to organise the distribution of traffic flow in a dynamic way, to anticipate varying traffic densities on different parts of the road network. Safety is not central in the matrix of factors in this problem field but works as a sort of precondition at best, not completely in line with the principles of Sustainable Safety. One way of attaining the necessary behavioural change is through providing electronic travel and traveller information that facilitates distribution of traffic and traffic flows (Dicke *et al.*, 2004). This type of information may be provided both pre-trip or on-trip, in various forms and shapes. Pre-trip information utilisation is increasing at immense speed, mostly through Internet, but also services via mobile phone in that sense are starting to find their way into the society. Well-known on-trip examples are dynamic parking indications, intelligent traffic lights, for instance, mutually communicating to realise "green waves", dynamic route information panels (DRIP), and it's modern version, the graphical or full-colour information panel.

A different type of on-trip information provision, in the vehicle itself, is by navigation systems. Based on digital maps and global positioning satelite (GPS), drivers are guided through traffic quickly and safely since they can devote their attention to the surrounding traffic and traffic environment, the system reducing their uncertainty. In the next generations, navigation systems will undoubtedly have much more information available, such as local speed limits, or even dynamic speed advice based on dynamic information from mutual communication or the road side about relevant matters such as cars (congestion) ahead, fog or ice on the road, etc. One of the major advantages will be a reduction in speed differences, a decrease in variance. Enhancement of traffic homogeneity will improve traffic flow, efficiency, safety and environment.

Advanced Driver Assistance Systems

ADAS as a subset of ITS covers a continuum, from information systems keeping the driver in full control over the vehicle, receiving supporting information only, to a fully automated transport system (even if only partly realised, see ADASE project deliverables, 1998). Only when on a dedicated traffic lane, the vehicle can be operated under fully automated control, which is very similar to the automatic pilot in aeroplanes (Congress, 1994), banning the human factor. ADAS concepts on this continuum include among others (advanced) route guidance systems, radio data systems–traffic message channel (RDS–TMC), blind spot detectors, Lane Deviation Warning Assistance, brake assistance, Intelligent Speed Adaptation, Adaptive Cruise Control, Autonomous Intelligent Cruise Control, platoon driving, etc. Some of the technology is available in the market, or ready to be marketed, some is being developed but as a prototype still under test. Even a prototype fully automated traffic lane exists in the USA near

San Diego, including a limited number of vehicles running on the lane, for testing and demonstration purposes.

History

ITS has a considerable history. In Europe, several car manufacturers and research institutes started the Prometheus initiative, around 1986. A series of projects were carried out under this umbrella, most of them aiming at practical solutions to urban traffic problems. The European Union initiated the Dedicated Road Infrastructure for Vehicle safety in Europe (DRIVE) programme shortly thereafter, in which a considerable number of projects tackled practical problems as well as fundamental issues. An example of the latter is the Generic Intelligent Driver Support (GIDS) project, the largest project in DRIVE-1 (Michon, 1993), ahead of its time and still relevant. The overall objective of this ambitious project was "to determine the requirements and design standards for a class of intelligent driver-support systems which will conform with the information requirements and performance capabilities of the individual drivers". On the one hand, this class of systems will aid the driver's detection and assessment of road and traffic hazards, on the other, they will provide guidance on the driver's ability to deal with specific hazards.

SAFETY ASPECTS

Since traffic safety is the main factor in transport efficiency, within Sustainable Mobility the Sustainable Safety programme should serve as a natural characteristic, as a precondition of Mobility Management. Distribution of traffic among routes, modal shift and even time of travelling should be based on safety principles. Driving at night, for instance, may be highly efficient because of low-traffic density and distributed road use, but the odds against daytime driving are high, 5.4 times the likelihood of getting involved in an accident (Connor *et al.*, 2002, Van Winsum, 1997).

In Figure 1, the relative "unsafety" of driving at any moment of the 24-hour period is depicted, as a function of the odds against the safest daytime period, i.e. between 9:00 a.m. and 12:00 p.m. set at 1 (Ouwerkerk, 1986).

As said, generally, (traffic) safety is an important factor in transport efficiency as accidents are a major factor in traffic jams (AVV, 2001). The distribution of contributory factors to accidents can be hierarchically categorised as representing driver (error and impairment), environment and vehicle factors contributing to accidents (Shinar, 1998). The incidence of these factors is revealing. The incidence of alcohol in accidents was established at an average of 3.8% in total, which is very close to that for all accidents in England where a driver is known to be over the drink-drive limit (4.2%). For the

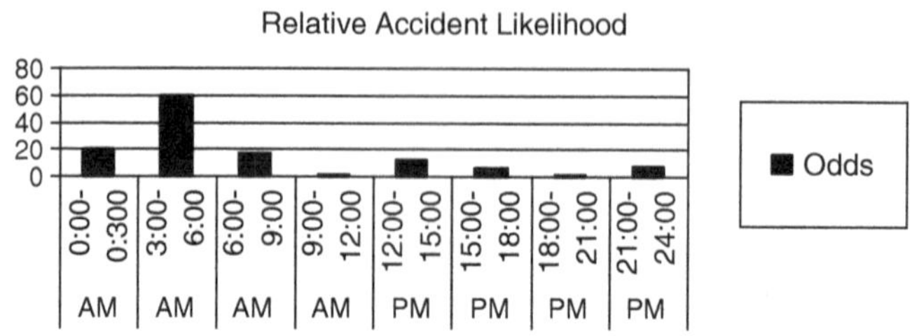

Source: Adapted from lorry driver survey data (Ouwerkerk, 1986).

Figure 1: The Odds, i.e. Likelihood of Involvement in Accidents by Lorry Drivers, Relative to the Safest Period of the Day, Just Before Noon (after Ouwerkerk, 1986)

other impairment-related factors the situation is not so straightforward. Impairment due to fatigue was recorded as a factor in only 0.8% of accidents by Shinar (1998), whereas in-depth studies and a large volume of anecdotal evidence shows that the contribution of this factor is more like up to at least 25%. This under-representation of fatigue-related accidents is now well recognised and results largely from the absence of direct evidence of sleepiness or tiredness being a factor (Brookhuis *et al.*, 2003b). There is no quantitative measure of this effect on drivers. If drivers survive an accident caused by sleepiness, they are unlikely to admit it; if they do not survive there is often very little direct physical evidence, except for in some cases perhaps the absence of skid marks on the road suface. Other factors, e.g. vehicle defects, which are sometimes erroneously cited as causes, usually need to be eliminated before fatigue becomes apparent.

ITS DEVELOPMENT, INTRODUCTION AND IMPLEMENTATION

The development and introduction of ITS in road traffic and transport has been pushed by efficiency and comfort mainly to date, and not by safety, which is a reason for some concern (Wegman, 2004). For systems that do have safety as primary objective, directly, high expectations with respect to savings in casualties are fostered by many (cf. Carsten and Tate, 2005). For the OECD countries, a reduction in casualties of some 40% is expected by safety-related ITS alone (OECD, 2003). Still, this would mean only a beginning of a match with rail and air traffic where safety has been central and incorporated much earlier. Nevertheless, implementation of most of the known safety systems is waiting for different reasons. Some of them are not yet fully developed, i.e. tested to functionality, safety, reliability and what have you; others are not yet introduced because of liability, effect- or cost-effectiveness uncertainty, lack of political courage and potential unwanted behavioural effects.

Driver comfort appears to be a strong impetus for the development of electronic driving aids in the first place, at least from a marketing point of view. Car manufacturers are keen on driver comfort and invest considerable effort in the development and improvement of comfort-enhancing electronic aids. Another and maybe the most important reason why car manufacturers present a host of electronic driving aids as comfort systems is because of impending liability issues with safety systems. If a system that claims to enhance safety fails for one reason or the other, manufacturers are likely to be held responsible, or at least liable (Van Wees, 2004).

Well-known examples of this type of applications are the earlier mentioned navigation or route guidance systems, adaptive cruise control (ACC) systems and the intelligent speed adapter (ISA). Originally developed for safety enhancement, many of these types of systems are now found as comfort systems in the brochures of car manufacturers.

Navigation Systems

Among the first systems on the market were the navigation systems; prototypes passed a number of tests (and improvements, see Michon, 1993) and were successfully placed on the consumer market, notwithstanding the fact that they were quite expensive in the beginning. These systems are at least very comfortable in case of new destinations, but at the same time prevent horribly hazardous behaviour like reading maps while driving.

Adaptive Cruise Control

ACC is an extension of the widely-existing cruise control systems. They provide assistance to drivers by automating parts of the longitudinal driving task in terms of operational control of headway and speed. Using frontal radar, ACC automatically increases the following distance to a vehicle ahead by reducing fuel flow and/or actively braking the vehicle if set to a cruise speed faster than the speed of a vehicle ahead. A meta-analysis (Dragutinovic *et al.*, 2005) suggests that different operational settings of ACC that have consequences for the allocation of different tasks to drivers as well as for the level of support provided by the system, are significant for the effects of ACC on driving behaviour, i.e. at least for ACC effects on mean driving speed and mean time headway. According to Van Arem (2003), ACC could decrease specific types of accidents by 10%.

Intelligent Speed Adapter

ISA is an example of in-vehicle ITS system that is designed and developed to reduce speeding. It can be introduced as a policy instrument to improve speed-limit compliance and with that traffic safety. Joksch (1993) found that the probability that a driver is killed increases with increased speed fits regression models with exponents as high as four, meaning that each km/h faster exponentially increases the fatality risk. An average reduction of as little as 2–5 km/h could lead to a reduction of 10–30% in injury accidents (see Brookhuis and De Waard, 1999). Field research has demonstrated that ISA has the intended effect with respect to speeding and environment (e.g. Várhelyi and Mäkinen, 2001), specifically in densely populated areas. However, previous research also indicated that the acceptance of this instrument is rather limited. Adopting a policy that may increase driver acceptance before the introduction of the instrument is therefore recommended. Molin and Brookhuis (2007) suggest that explaining clearly how ISA can contribute to attaining various personal and societal goals may be a viable policy to increase ISA acceptance.

Before the actual marketing of any system, user needs research (or marketing research) is indispensable and also studies on acceptance, but certainly on safety effects are still necessary after implementation since all kinds of adverse effects are imminent. Consumer acceptance is dependent upon such requirements as system safety, but also validity (does the system function as it should) and benefit (is there a positive cost-benefit balance). Finally, environmental issues are not decisive in this area yet, but will gain weight consistently.

ADVERSE EFFECTS

Since the primary functionality of ADAS is to facilitate the task performance of drivers by providing real-time driving support, this type of system is also described by the term "co-driver systems" or "driver-support systems". The different operation modes, i.e. advisory, taking over part(s) of the driving task, semi-automatic to fully automatic may have different consequences for the driving task, traffic safety and the environment. Although the purpose of a driver-support system is to have a positive effect on traffic, that is on safety, efficiency and the environment, adverse effects have been demonstrated as well. Firstly, the provision of information potentially leads to a situation where the driver's attention may be diverted from traffic (see Carsten and Brookhuis, 2005). Both visual distraction and cognitive (secondary) task load induce impairment on driving performance. Secondly, taking over (part of) the driving task by a co-driver system may well produce behavioural adaptation (Dragutinovic *et al.*, 2004). A meta-analysis of the effect of ACC systems showed that with compulsive systems average speed was increased. Thirdly, because of automation, drivers may be less alert or attentive and might gradually lose skills (Bainbridge, 1983, De Waard *et al.*, 2004). Automation may cause complacency, i.e. unjustified over-reliance on ADAS, and increased reaction time. As a result, either the driver might not (or too late) be aware of a sudden hazard, or is not ready for an adequate reaction. A few examples of disadvantages of ADAS will be worked out in the next paragraphs.

Distraction

Although adequate and timely traffic and travel information helps to decrease travel time and cost, whilst the drivers do not have to direct their attention to other sources of travel information, potential negative side-effects by certain aspects of specifically new types of traffic information provision systems are imminent. New information sources add to the information load of drivers, potentially counteracting the potential benefits of decreased distraction and workload. For instance, navigating in an unfamiliar traffic network is a demanding driving task that may well be supported by a route guidance system. However, if the operation of the system entails studying small screens down at the dashboard, the positive effects are annihilated by the negative.

Behavioural Adaptation

Automation in any degree and form necessarily produces behavioural adaptation, i.e. changes in behaviour that can be intended and wanted but also not intended and unwanted (Dragutinovic *et al.*, 2005b). It may increase reaction time, for instance. In case of monitoring of a system function, reaction time to events in a driving task can be restricted to less than one second, while if more than one functions have to be monitored and other tasks are attended to, awareness of the situation has to be refreshed with increased frequency. At the same time, possible malfunction of an application and its origin have to be determined which might take many seconds. In this way, an attempt to reduce workload is actually likely to lead to an increased workload (Hancock and Parasuraman, 1992). There are also several studies that have shown that monitoring of systems for malfunctions during prolonged periods of time induce high levels of workload, despite the fact that information processing requirements for these tasks are low in itself. Humans are poor "process monitors" (e.g. Molloy and Parasuraman, 1996) and enforced vigilance in the operational environment is very stressful (Hancock and Parasuraman, 1992). An adaptable interface between human and machine would be better, an interface that dynamically allocates tasks to these two and maintains an appropriate and tolerable load on the operator. Research on adaptive task allocation shows that temporarily returning control to the human operator at the right moment, i.e. when the driver is ready for it, has favourable effects on detection of automation failures during subsequent automation control (Parasuraman *et al.*, 1996). For adaptive automation to be successful, a high level of user awareness has to be maintained to allow judgement of the appropriate behaviour of the automation (Parasuraman *et al.*, 2002). First applications of adaptive automation were found in the military aviation domain. The general idea is to present the right information in the right format at the right time. This is at the same time one of the risks of adaptive automation: a changing interface that can even be perceived as inconsistent (see also Hoogeboom and Mulder, 2004). Many of these aspects are subject to research in the early new millennium.

Human Supervision

The classic goal of automation is to replace human manual control, planning and problem solving by automatic devices. However, these systems still need human beings for supervision and adjustment. It has been suggested that the more advanced a control system, the more crucial is the contribution of the human operator (Bainbridge, 1983).

The point made by Bainbridge (1983) is as follows: normal operation is performed automatically, abnormal conditions are to be dealt with manually. Unfortunately, as a result of automation, experience is limited, while in case of abnormal conditions (i.e. something is wrong with the process) unusual actions will be required. Also, human problem solving is not optimal under time-pressure.

Supervising of (present) automatic processes is based on skills that formerly manual operators have, and that future generations of operators (/drivers) cannot be expected to have (Bainbridge, 1983). Pilots also indicated that although automation reduced workload, it also had a negative effect on flying skills. They considered manually flying of a part of every trip important to maintain these skills (McClumpha *et al.*, 1991).

Complacency

When a system fails to work or is in a state that failure is possible, feedback should be provided in order to let the driver know that s/he cannot rely on the system. The main reason for this is that automated systems can and will lead to what has been called complacency (Wiener and Curry, 1980). Complacency is an attitude of (over)reliance on an automated system. A number of studies found signs of complacency (Desmond *et al.*, 1998, De Waard *et al.*, 1999, De Waard *et al.*, 2004, Brookhuis and De Waard, 2005), others reported deterioration in driving performance such as increased weaving. In a test of reaction time to a system failure cue, Knapp and Vardaman (1991) found support for complacency, i.e. the reaction time to this cue increased compared to normal task performance. Ward *et al.* (1995) also found evidence for complacency, and also poor lane position control and failure to yield to other traffic was more frequently observed in drivers driving a car with Autonomous Intelligent Cruise Control (AICC) compared to drivers driving a normal car. However, Moray and Inagaki (2000) have criticised the existence of complacency. Their main point is that the operator (here: the driver) should not be blamed for not detecting signals when the system is badly designed and optimal driver behaviour is not defined. It is true that in many fully automated systems, the operator/driver is thought of as safety guard and is supposed to be continuously alert for system failure. This role is not surprising since in almost all cases the driver is legally liable. Automation in general will be considered to support the driver, and not to take-over tasks and responsibilities. In this context, it is very important that if an automated system fails, it should fail saliently (e.g. De Waard *et al.*, 1999). Also, what the driver can expect from the system and the division of responsibilities between human and machine should be clear in advance. Key questions with respect to automation include user acceptance, whether drivers will trust automated vehicles, whether they actually will reclaim control if required, what driving mode they prefer and whether they finally accept supervising an automated vehicle instead of driving.

STAKEHOLDER NEEDS AND ACCEPTANCE

Needs and acceptance in road transport in a general sense stem from unsatisfactory conditions. The reason could be in (physical) working conditions, causing fatigue and health problems, or traffic safety and/or inefficiency. In the past, many studies have been done in this area into the driving conditions of professional drivers. Ouwerkerk (1986), for instance, in a survey among studies as far as known at that time reported that more than 50% of the professional drivers admitted to have been falling asleep at least once and/or having had near-accidents. These numbers were found for both lorry- and bus-drivers. In the present time of electronic driving aids, solving the need for support to avoid these problems, and accidents, is thought to be realised, however, the increased complexity of the "cockpit" increases the likelihood of failure by the driver too. In case of electronic system, malfunction of at least one of the system's components, either by "spontaneous" failure or by design errors (Janssen *et al.*, 1992), severe problems may arise. The installation of support systems that take over parts of the driving task thus requires additional alertness of the driver, while at the same time driver alertness in general and attention for the driving task per se is decreasing in case of automation of the driving task (Brookhuis and De Waard, 2005).

Acceptance of ITS is a basic matter in ITS implementation, dependent upon the mode of operation. For instance, takeover of control in case of short headway to a lead car was less appreciated than warnings or suggestions of the appropriate action in a test of different types of Collision Avoidance Systems (Nilsson and Alm, 1991). Although drivers expect a positive safety effect by this type of anti-collision systems and other forms of ITS, they have at the same time reservations against it. Handing over control to a device and the automated braking function are evaluated as negative aspects of ITS systems (Hoedemaeker, 1996; Hoedemaeker and Brookhuis, 1999). Complete take-over in specific circumstances has been tested as well. The SAVE project (EU DG XIII TR1047) aimed to develop a system that takes over vehicle control in case of real emergency, such as sudden illness. An international questionnaire survey carried out in the SAVE project indicated that the driver population is reluctant to release vehicle control, but is willing to accept it in emergency situations (Bekiaris *et al.*, 1997). De Vos and Hoekstra (1997) focused on behavioural aspects of leaving the fully automated highway traffic lane. Apart from studying the exit procedure, they also found that a short headway was considered less comfortable and less accepted than a larger headway to vehicles-in-front. In follow-up projects funded by the EU, many of these aspects are tackled with the aim to help introducing ITS (e.g. Wiethoff, 2001).

As stated before, electronic driving aids operate in three different modes, from non-committal information supply via active support at driver's will to leaving control to technology completely. Not surprisingly, also the surveys demonstrated that acceptance varies with driver control (Bekiaris *et al.*, 1997); taking the driver out of the loop is considered a problem by many (potential) end-users. The views with respect to the different modes may well be different for different stakeholders such as the authorities, the general public and the end-users including professionals.

One of the most important aims in the development of electronic driving aids is to integrate the information to drivers. Not only through combining different types of messages located in the road environment, but also by their integration with information inside the vehicle, through different types of devices. It is clear that a poorly designed system inside vehicles can adversely affect the different actors, or more generally, the social benefits associated with the system (Bekiaris *et al.*, 1997). In-vehicle ITS devices that are limited to supplying information are most likely to meet a priori acceptance. Examples are route guidance systems, traffic jam warning systems and advanced driver monitoring systems.

Several reservations that have consequences for acceptance by the relevant stakeholders, hold for any of the three modes. A few studies found support for complacency, i.e. excessive reliance on automated (ITS) systems, others reported deterioration in driving performance. This and other forms of behavioural adaptation, or compensation as it is called in a wider field, are factors that should be taken into account when investigating the conditions for introduction of ITS (Verwey *et al.*, 1996).

CONCLUSION

A specific concern with the development of driver-support systems that are intended to reduce accidents is that it is very difficult to forecast the reduction of accidents, savings of death and disability that might result from the introduction of newly developed systems. Although there is an urgent need to know what the effects are of introducing a specific system before it enters the market, for new systems no data exist on which estimates of the benefits and the risks of system-specific traffic behaviour can be based. That necessarily takes time. The only type of effects that can be studied directly are effects on behavioural aspects of the driving task per se, in an experimental setting. These aspects should be selected on the basis of known adverse effects on traffic safety, such as insufficient safety margins in lateral and longitudinal positioning (cf., Rumar, 1988). Hence, each individual ITS application should be subjected to a test on behavioural effects before marketing, in order to pinpoint both beneficial and unwanted side-effects at the behavioural level. For this, however, exact criteria are preliminarily assessed but still have to be further developed (Brookhuis *et al.*, 2003a). Eventually, the first known traffic-safety effects are confined to an extrapolation from these test results (see Carsten and Brookhuis, 2005).

The potential of in-vehicle systems is great, provided ITS will be widely accepted and introduced in the (near) future. For this, (sub)systems will all have to be made as fail-safe as possible. Whenever the system fails, safety is to be determined by the provisions taken to avoid serious accidents and in case of an accident the measures to minimise the consequences for the passengers. Acceptability of ITS is highly dependent upon solid demonstration of these features. Acceptability is also found to be dependent of the form in which ITS applications are implemented. For the end-user the benefits should be clear and preferably directly noticeable. For this reason, comfort-enhancing features stand a better chance than safety-enhancement properties. Most drivers consider themselves at least better drivers with respect to safe behaviour than average. Strict requirements for ITS applications by all stakeholders are safe (and valid) operation and reliability, false alarms are not acceptable for end-users particularly.

Introduction of a fully automated highway system is technically possible now, but public introduction on a large-scale is at least waiting for safety provisions and public acceptance, and also on proper legislation that clearly establishes responsibility and liability. ITS America initially planned USA full deployment for 2005, comparable initiatives in Europe and Japan were announced to certainly close in as soon possible. None of these plans will be realised within short notice, however. In the meantime, many (current, at the moment of writing) projects such as the EU-projects AWAKE and HASTE (both running until 2005) try to pave the way for introduction of ITS and above all acceptance of ITS by stakeholders and community.

ADAS and Mobility Management systems as well, should be developed in such a way that they reduce the negative aspects of road transport, instead of producing them. The direct negative consequences for driving performance, resulting in accidents, congestion and pollution, should be attenuated by providing information to and guidance for safer, more efficient and environmental friendlier use of the road transport infrastructure in general. In principle, such applications should reduce accident incidence, allow higher traffic intensity and decrease average travel times. Environmental impacts, such as emissions and fuel consumption are subsequently reduced by avoiding unnecessary manoeuvering or minimising superfluous driving. The provision of adequate, accurate and timely information will significantly improve traffic safety and with that traffic efficiency, as they assist drivers to avoid hazardous situations as well.

REFERENCES

ADASE (1998). *Advanced Driver Assistance Systems in Europe*, Project Reports D3.1, D3.2, D3.3 and D4.1 to the European Commission.

AVV (2001). *Verkeersgegevens. Jaarrapport 2000*, Ministerie Verkeer en Waterstaat, Verkeer and Vervoer, Adviesdienst.

Bainbridge, L. (1983). Ironies of Automation, *Automatica*, **19**, 775–779.

Bekiaris, E., S. Petica, and K.A. Brookhuis (1997). Driver Needs and Public Acceptance Regarding Telematic In-Vehicle Emergency Control Aids. In: *Proceedings of the 4th World Congress on Intelligent Transport Systems*, Ertico (on CD), Brussel.

Bovy, P.H.L. and N.J. Van der Zijpp (1999). The Close Connection between Dynamic Traffic Management and Driver Information Systems. In: *Behavioural and Network Impacts of Driver Information Systems* (R. Emmerink and P. Nijkamp, eds), Ashgate Publishing Company, Aldershot, 355–370.

Brookhuis, K.A. and D. De Waard (1999). Limiting Speed, Towards an Intelligent Speed Adaptor (ISA), *Transportation Research, Part F, Psychology and Behaviour*, **2**, 81–90.

Brookhuis, K.A. and D. De Waard (2005). ADAS' Acceptance and Effects on Behaviour: the Consequences of Automation? In: *Traffic and Transport Psychology. Theory and Application* (G. Underwood, ed.), Elsevier, Amsterdam, 273–278.

Brookhuis, K.A., D. De Waard, and S.H. Fairclough (2003a). Criteria for Driver Impairment, *Ergonomics*, **46**, 433–445.

Brookhuis, K.A., D. De Waard, J.H. Kraaij, and E. Bekiaris (2003b). How Important is Driver Fatigue and What Can We Do about It? In: *Human Factors in the Age of Virtual Reality* (D. de Waard, K.A. Brookhuis, S.M. Sommer, and W.B. Verwey, eds), Shaker Publishing, Maastricht, 191–207.

Carsten, O.M.J. and K.A. Brookhuis (eds) (2005). The Relationship Between Distraction and Driving Performance: Towards a Test Regime for In-Vehicle Information Systems, *Transport Research Part F: Psychology and Behaviour* (Special Issue), **8**, 75–196.

Carsten O.M.J., and F.N. Tate (2005). Intelligent Speed Adaptation: Accident Savings and Cost-Benefit Analysis, *Accident Prevention and Prediction*, **37**, 407–416.

Congress, N. (1994). The Automated Highway System: An Idea whose Time Has Come, *Public Roads On-Line*, (Summer 1994, A Journal of Highway Research), **58**, 1–7.

Connor, J., R. Norton, S. Ameratunga, E. Robinson, I. Civil, R. Dunn, R. Bailey, and R. Jackson (2002). Driver Sleepiness and Risk of Serious Injury to Car Occupants: Population Based Case Control Study, *British Medical Journal*, **324**, 11–25.

Desmond, P.A., P.A. Hancock, and J.L. Monette (1998). Fatigue and Automation-Induced Impairments in Simulated Driving Performance, *Transportation Research Records*, **1628**, 8–14.

De Vos, A.P. and W. Hoekstra (1997). *Behavioural Aspects of Automatic Vehicle Guidance (AVG); Leaving the Automated Lane*, Report TM-97-C010, TNO Human Factors Research Institute, Soesterberg, The Netherlands.

De Waard, D., K.A. Brookhuis, E. Fabriek, and P.C. Van Wolffelaar (2004). Driving the Phileas, a New Automated Public Transport Vehicle. In: *Proceedings ICCTP2004*, online: http://www.psychology.nottingham.ac.uk/IAAPdiv13/ICTTP2004papers2/ITS/deWaard.pdf

De Waard, D., M. Van der Hulst, M. Hoedemaeker, and K.A. Brookhuis (1999). Driver Behavior in an Emergency Situation in the Automated Highway System, *Transportation Human Factors*, **1**, 67–82.

Dicke, M., F.L. Bouwers, E. Koenders, R. Kok, R. Van der Veer, and K.A. Brookhuis (2004). Human Factors in the Design of Travel and Traffic Information. In: *Human Factors in Design* (D. De Waard, K.A. Brookhuis, and C.M. Weikert, eds), Shaker Publishing, Maastricht, 95–111.

Dragutinovic, N., K.A. Brookhuis, M.P. Hagenzieker, and V.A.W.J. Marchau (2005a). Behavioural Effects of Advanced Cruise Control Use – A Meta-Analytic Approach, *European Journal of Transport and Infrastructure Research*, **5**, 267–280.

Dragutinovic, N., K.A. Brookhuis, M.P. Hagenzieker, and V.A.W.J. Marchau (2005b). ACC Effects on Driving Speed – A Second Look. In: *Human Factors in Design, Safety, and Management* (D. De Waard, K.A. Brookhuis, R. Van Egmond, and Th. Boersema, eds), Shaker Publishing, Maastricht, the Netherlands, 381–387.

Hancock, P.A. and R. Parasuraman (1992). Human Factors and Safety in the Design of Intelligent Vehicle-Highway Systems (IVHS), *Journal of Safety Research*, **23**, 181–198.

Hoedemaeker, M. (1996). *Behoeften, rijstijlen en meningen ten aanzien van Automatische Voertuig Besturing [Needs, Driving Styles and Opinions Towards Automated Driving]*, Report, Technical University Delft, Delft, The Netherlands.

Hoedemaeker, M. and K.A. Brookhuis (1999). Driving with an Adaptive Cruise Control (ACC), *Transportation Research, Part F*, **3**, 95–106.

Hoogeboom, P.J. and L.J.M. Mulder (2004). Physiological Indices for the Estimation of Momentary Changes in Cognitive Workload and Mental State. In: *Human Factors in Design* (D. De Waard, K.A. Brookhuis, and C.M. Weikert, eds), Shaker Publishing, Maastricht, the Netherlands, 147–160.

Janssen, W.H., M. Wierda, and A.R.A. Van der Horst (1992). *Automatisering van de rijtaak en gebruikersgedrag: een inventarisatie en research agenda [Automation of Driving and User Behaviour]*, Report TNO-IZF 1992 C-16, TNO-TM, Soesterberg, The Netherlands.

Joksch, H.C. (1993). Velocity Change and Fatality Risk in a Crash – A Rule of Thumb, *Accident Analysis and Prevention*, **25**, 103–104.

Knapp, R.K. and J.J. Vardaman (1991). Response to an Automated Function Failure Cue: An Operational Measure of Complacency, *Proceedings of the Human Factors Society 35th Annual Meeting*, HFES, Santa Monica, CA, USA, 112–115.

Koornstra, M.J., M.P.M. Mathijssen, J.A.G. Mulder, R. Roszbach, and F.C.M. Wegman (eds) (1992). *Naar een duurzaam veilig wegverkeer*, Stichting Wetenschappelijk Onderzoek Verkeersveiligheid SWOV, Leidschendam.

Marchau, V.A.W.J., R.E.C.M. Van der Heijden, and E.J.E. Molin (2005). Desirability of Advanced Driver Assistance from Road Safety Perspective: The Case of ISA, *Safety Science*, **43**, 11–27.

McClumpha, A.J., M. James, R.G. Green, and A.J. Belyavin (1991). Pilots' Attitudes to Cockpit Automation, *Proceedings of the Human Factors Society 35th annual meeting*, HFES, Santa Monica, CA, USA, 107–111.

Michon, J.A. (ed.) (1993). *Generic Intelligent Driver Support*, Taylor and Francis, London.

Molin, E.J.E. and K.A. Brookhuis (2007). Modelling Acceptability of the Intelligent Speed Adapter, *Transportation Research Part F*, **10**, 99–108.

Molloy, R. and R. Parasuraman (1996). Monitoring an Automated System for Single Failure: Vigilance and Task Complexity Effects, *Human Factors*, **38**, 311–322.

Moray, N. and T. Inagaki (2000). Attention and Complacency, *Theoretical Issues in Ergonomics Science*, **1**, 354–365.

Nilsson, L. and H. Alm (1991). *Collision Avoidance Systems – Effects of Different Levels of Task Allocation on Driver Behaviour*, Report DRIVE V1041/GIDS/MAN3, University of Groningen, Traffic Research Centre, Haren, The Netherlands.

OECD (2003). Report *'Road Safety – Impact of New Technologies'*.

Ouwerkerk, F. (1986). *Relationships between Road Transport Working Conditions, Fatigue, Health and Traffic Safety*, Report VK-86-01, University of Groningen, Traffic Research Centre, Haren.

Parasuraman, R., M. Mouloua, and R. Molloy (1996). Effects of Adaptive Task Allocation on Monitoring of Automated Systems, *Human Factors*, **38**, 665–679.

Parasuraman R., T.B. Sheridan, and C.D. Wickens (2002). A Model for Types and Levels of Human Interaction with Automation, *IEEE Transactions on Systems, Man, and Cybernetics – Part A: Systems and Humans*, **30**, 286–297.

Rumar, K. (1988). Collective Risk, But Individual Safety, *Ergonomics*, **31**, 507–518.

Shinar, D. (1998). *Psychology on the Road: The Human Factor in Traffic Safety*, Wiley, New York.

Smiley, A. and K.A. Brookhuis (1987). Alcohol, Drugs and Traffic Safety. In: *Road Users and Traffic Safety* (J.A. Rothengatter and R.A. de Bruin, eds), Van Gorcum, Assen, 83–105.

Vallet, M. (1991). Les dispositifs de maintien de la vigilance des conducteurs de voiture. In: *Le maintien de la vigilance dans les transports.* (M. Vallet, ed.), Paradigm, Caen, 1–5.

Van Arem, B. (2003). *Kan de auto beter zelf rijden?* PhD Thesis, AIDA Knowledge Centre, Utwenthe.

Van Wees, K.A.P.C. (2004). *Intelligente Voertuigen, veiligheidsregulering en aansprakelijkheid – Een onderzoek naar juridische aspecten van, Advanced Driver Assistance Systems*, PhD Thesis, TRAIL Thesis T2004/10, TRAIL Research School, TU Delft.

Van Winsum, W. (1997). *A Validation Study of a PC-Based Test of Safety Aspects of In-Vehicle Information Systems: A Test of a Map Display Version of a RDS-TMC Task*, Research TM-97-C057, TNO Human Factors Research Institute, Soesterberg.

Várhelyi, A. and T. Mäkinen (2001). The Effects of In-Car Speed Limiters – Field Studies, *Transportation Research, Part C*, **9**, 191–211.

Verwey, W.B., K.A. Brookhuis, and W.H. Janssen (1996). *Safety Effects of In-Vehicle Information Systems*, Report TM-96-C002, TNO Human Factors Research Institute, Soesterberg, the Netherlands.

Ward, N.J., S.H. Fairclough, and M. Humphreys (1995). The Effect of Task Automation in the Automotive Context: A Field Study of an Autonomous Intelligent Cruise Control System, *Unpublished Paper Presented at the International Conference on Experimental Analysis and Measurement of Situation Awareness*, Daytona Beach, Florida, USA, November 1–3.

Wegman, F.C.M. (2004). Naar een tweede generatie duurzaam veilige maatregelen: aanzet tot een discussie over de toekomst van Duurzaam Veilig, Presented at the Dutch National Transport Safety Congress, Leidschendam.

Wegman, F.C.M. and L. Aarts (eds) (2005). *Doorgaan met Duurzaam veilig*, Stichting Wetenschappelijk Onderzoek Verkeersveiligheid/Institute for Road Safety Research SWOV, Leidschendam.

Wiener, E.L. and R.E. Curry (1980). Flight Deck Automation: Promises and Problems, *Ergonomics*, **23**, 995–1011.

Wiethoff, M. (2001). How Could Intelligent Safety Transport Systems Enhance Safety. In: *Proceedings of the Critical Infrastructure Conference* (J. Althoff, ed.), Delft University Press, Delft.

Threats from Car Traffic to the Quality of Urban Life
Tommy Gärling and Linda Steg (Editors)

22

EFFECTIVENESS OF TRANSPORT POLICIES IN REDUCING CAR TRAVEL

Phil Goodwin, Centre for Transport and Society, University of the West of England, Bristol, UK

ABSTRACT

Policies intended to solve problems caused by excessive traffic usually seek to change travel behaviour. These changes are assessed with extensive empirical evidence, including inference from the extent of behavioural change that occurs for other reasons such as age and income, case studies of increasing or reducing road capacity, econometric analysis of the effects of fuel price and public transport fares, and observation of impacts of soft measures such as travel plans and improved information. The results show a dynamic process of adjustment over periods of some years, with impacts that are larger, but more complex, than often assumed.

INTRODUCTION

This chapter addresses the empirical issue of what actually happens in practice when policies aimed at reducing car use are implemented. Is it, in fact, possible to change travel behaviour? The underlying proposition in the chapter is that by understanding how travel behaviour *does* change, we are better equipped to understand how it *can be* changed.

TRAVEL BEHAVIOUR CHANGES

As Figure 1 shows, since 1950 the number of cars in Britain increased nearly 1,200%. Today, 75% of households have access to a car, compared to only 14% in 1950. The increase in car ownership has been the main impetus to increasing traffic, over the same period traffic increased by over 800%.

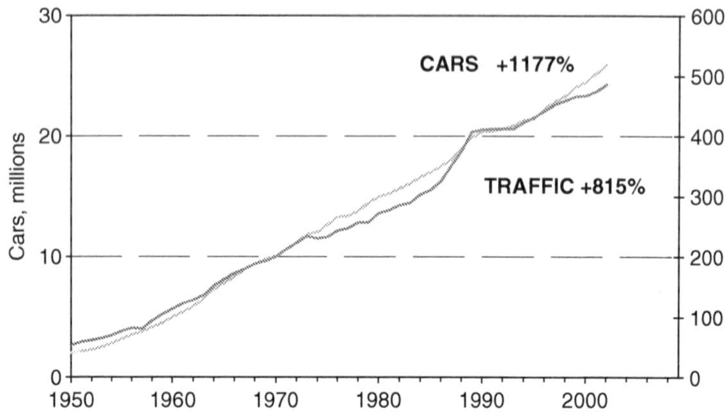

Figure 1: Cars and Traffic in Great Britain

In Figure 2, we see the well-established pattern that car ownership has allowed people to travel further distances. In the UK, 50 years ago the average person travelled 4,000 km per year. Today it is three times as far, over 12,000 km, even without including international air travel. During this period, total passenger kilometres travelled by car increased by over 900%, while rail increased by 26%. All other modes have declined: bus travel by 50%, cycle by 83%, and walking (on which statistics have only recently been recorded at the same level of detail) probably by more. However, these trends are aggregate results of many different changes in travel behaviour by individuals.

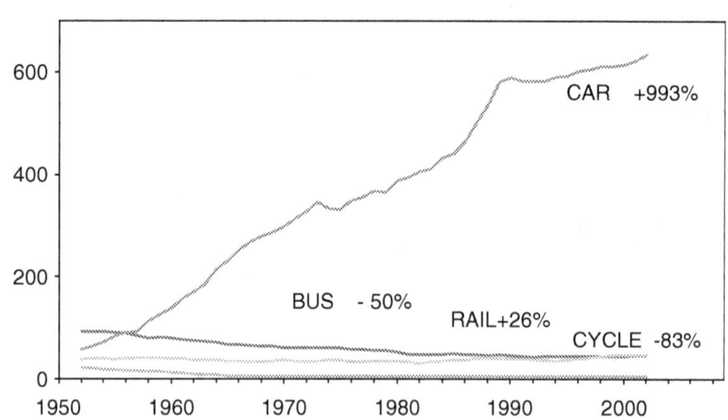

Figure 2: Passenger Travel in Great Britain (Billion per Kilometre)

The trends appear to show change only in one direction, but the individual changes do not go all in the same direction, which is of importance when considering policies that seek to alter the direction of trends. This is usually ignored because empirical work has largely been based on aggregate time-series data, which can only detect net changes or disaggregate cross-section data (which contain no information about changes at all). What we need is panel data, in which the behaviour of individuals, or in some cases groups of individuals, is actually tracked over time.

The first source of panel data for such analysis was for South Yorkshire, starting with a study by Goodwin and colleagues (Goodwin, 1983) and followed by a series of reports by Stokes and others the last of which was by Stokes (1995, summarised in Goodwin, 1997). The key conclusion was that apparently relatively slow and steady increases in car ownership across the whole population were actually made up of many households losing cars and many (usually more) gaining them, with the largest group showing no change (see Figure 3). This phenomenon has been described as "churn", a term used in political science to describe voting behaviour, whereas a small net shift from one party to another (the "swing") may be composed of very large numbers of people changing in opposite directions.

In five separate panel surveys between 1981 and 1991, a similar pattern was observed. Car ownership would rise by about 3% or 4% over a two- or three-year period, but this would be made up by about 13% or 14% of people in households increasing the number of cars in their household, and 8–10% reducing the number of cars in the household. Figure 3 shows the figures from 1988 to 1991, and is fairly typical of the other surveys. The individual changes are much larger than can be seen in snapshots of aggregate data. Figure 4 shows that a similar result is found for the number of trips

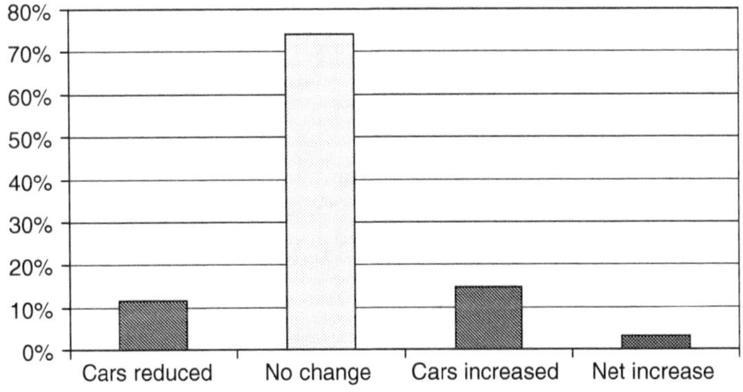

Figure 3: Proportion of People in Households Whose Car Ownership Increased or Reduced

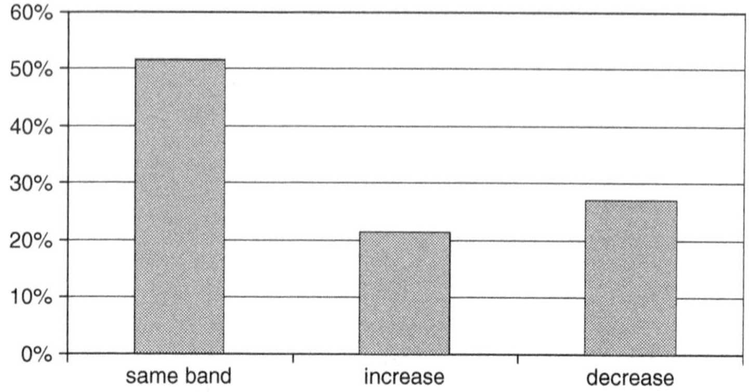

Figure 4: Percentage of People in Groups with Similar Numbers of Trips per Person

people made. Re-analysis of data from London panel surveys in 1976 showed that over a six-month period there were some major changes. For over 5,000 people surveyed, there was a very small reduction in the average number of journeys by all modes made during the diary week. But even just over this six-month period, 52% stayed in the same category of number of trips made, 21% increased by at least one class, and 27% decreased by at least one class. This survey was done during a period of rather volatile public transport fares, where policies to reduce fares were overturned by legal intervention.

Substantial churn was also found in the total time spent travelling. This is of interest because in aggregate terms time spent travelling is remarkably constant over time (60–65 minutes for an average person), and from time to time it has been suggested that the travel time budget is a sort of universal constant. But that only applies at the aggregate level – for individuals, there is no sign of stability or constancy at all: It changes quite radically with major life transitions, and also, apparently, randomly.

The results from South Yorkshire suffered from the typical feature of transport panels, namely that they usually cover a small geographic area and a short-time period. Longer-time periods of analysis were possible using the British Household Panel Survey (BHPS) that contains information on car ownership and commuting mode and time, along with a large number of socio-economic and demographic variables important in determining travel behaviour. Details are given in Dargay (2003) and Dargay and Hanly (2004). The BHPS data showed an increase in the number of cars per household, over the past decade, at an annual average rate of 0.2%. This small net-increase conceals the fact that a relatively large number of households (15.8%) change the number of cars they own between any two years. Slightly less than half of these (7.6%) reduce the number of cars

Table 1: Main Commuting Mode – Percentage of Commuters

	Rail	*Tube*	*Bus*	*Motorcycle*	*Car Driver*	*Car Passenger*	*Cycle*	*Walk*
Average year	3.3	1.4	6.2	1.5	66.9	7.4	3.3	9.7
At least *n* years								
1	7.8	3.9	16.5	4.7	83.1	25.0	9.4	22.1
2	5.6	2.6	11.5	3.1	79.3	13.9	6.2	16.4
3	4.9	1.9	8.4	2.2	76.3	9.4	3.9	12.9
4	3.4	1.6	6.7	1.5	73.4	7.2	3.4	10.6
5	2.9	1.2	5.4	0.9	70.8	5.3	2.8	8.5
6	2.5	0.9	4.4	0.7	67.8	4.5	2.3	7.4
7	2.0	0.6	3.6	0.7	63.9	3.5	1.7	6.5
8	1.6	0.4	2.7	0.5	58.7	2.6	1.3	5.6
9	1.2	0.4	1.8	0.4	53.4	1.9	1.2	4.3
10	1.1	0.2	1.4	0.2	42.4	0.9	0.9	2.4

they own, including 1.9% who give up car ownership totally. A similar volatility is noted for commuting mode, often thought of as relatively habitual and resistant to change. Nearly 18% of commuters changed main travel mode between any two years, as shown in Table 1.

We track main modes of transport for the journey to work for a sample of people for whom data are available over a 10-year period. For example, for the average year, about 7.4% of journeys to work were made by car passengers. Of these, less than 1% were car passengers in every year out of the ten. But almost three times as many (25%) commuted mainly as car passengers for at least one year out of the decade. It is a similar picture for the other modes – for rail, for example, nearly three times as many people are dependent on rail for a period of their life, than appears in any one year.

What this implies is that although only a small proportion of the population may use a particular mode in any given year, they are not the same individuals each year, so that over a period of years the number of individuals using the mode in one year or another is much greater. Of course, if we include trips for purposes other than commuting, we would find the proportion of individuals using any given mode over a longer-time period is even greater.

In summary, at the aggregate level, average travel behaviour has changed enormously during the period of one lifetime – mostly, apparently, in the same direction. But underlying that, individual behaviour changes more than the average, and includes substantial proportion of people changing in the opposite direction to the average. For whatever reason, behaviour is evidently very volatile. Virtually all established appraisal, even at disaggregate level, ignores churn and volatility in assessing the scale of change in behaviour.

EXOGENOUS DRIVERS OF CHANGE: DEMOGRAPHIC FACTORS AND INCOME

Simply to establish that change happens is important in a policy debate where one argument is that nothing moves. But that is not useful information until we know something more about what drives those changes. We start with the factors that are normally not considered to be transport policy instruments at all, demographic trends, and movements in income.

This analysis used the technique of "pseudo-panels" which had not previously been used in the transport field (Dargay, 2001, 2002). Pseudo-panels use cross-section data collected at different points in time to construct something resembling a panel – not of individuals, but of individuals sharing the same characteristics. Such groups of individuals, generally called cohorts, are then followed in each of the annual datasets, and the average values for car ownership, income, etc. are treated as observations in a panel.

Extending work reported in the first term, we constructed a pseudo-panel dataset able to show some changes in life cycle and income effects, from the annual UK Family Expenditure Surveys over a period of 25 years. We define the cohorts by the year of birth of the head of household in five-year bands. These cohorts are then followed over time using the annual datasets.

Figure 5 shows car ownership over time for cohorts defined by the age of the household head (this providing a relatively unambiguous identifier, albeit one of less social relevance than used to be the case). This age is given on the horizontal axis and car ownership on the vertical. The lines represent the different cohorts, with the birth-year bands shown adjacent. The initial data point for each cohort is obtained from the first survey in which an observation for the cohort is available, generally 1970, while the final data point is obtained from the last survey containing a comparable observation, generally 1995. For example, take the cohort whose head was born between 1931 and 1935. His or her mean age was 37 in the 1970 survey and 62 in the 1995 survey. The average household in this cohort owned about 0.75 cars when the head was 37 years of age. Household car ownership increased until the head approached the age of 50, reaching a maximum of 1.2 cars, thereafter declining to 1 car by the age of 62.

Two effects are discernable in the plot: a *life-cycle effect* – car ownership increases until the head is in his or her early 50s, and then declines; and a *generation effect* – at every synthetic age car ownership is higher for more recent cohorts than for earlier ones. A similar pattern of life cycle and generation effects can also be seen for car use. This pattern can partially be explained by differences in household income over the life cycle and differences in real income between generations. Household income increases up until the head reaches his or her late 40s and declines thereafter. Similarly, at each age,

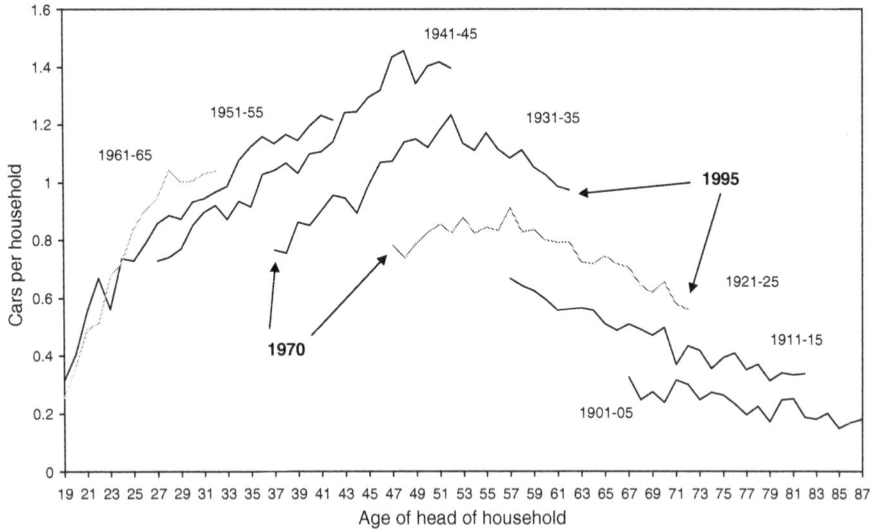

Figure 5: Household Car Ownership by Cohort

real incomes are higher for more recent generations due to general real income increases over time.

The change in number of adults in the household over the life cycle is clearly an important determinant of household income, car ownership, and car travel. As young adults form households, income increases, and first, then perhaps second, cars are purchased and car travel increases. This is compounded as the children grow up and learn to drive – often contributing to the household income and obtaining cars of their own. Later, both car ownership and use decline, as adult children leave home taking their car with them or through the disposal of second cars, and finally predominantly through the death of a spouse.

In contrast to the aggregate trends shown earlier, this approach does provide for individual household level car ownership and car travel to go down as well as up over time. Although this is largely a systematic and sensible relationship with changes in income and household composition over the life cycle, there is evidence that the relationship between car ownership and car travel, and income may not be symmetric (see Figure 6).

The vertical axis shows car ownership per household and the horizontal axis is total weekly expenditures. The earliest cohort shown, at the lower left of the figure, is representative of pensioner households. The head ages from 59 to 80 over the observed time period and both car ownership and income are declining. The most recent cohort shown,

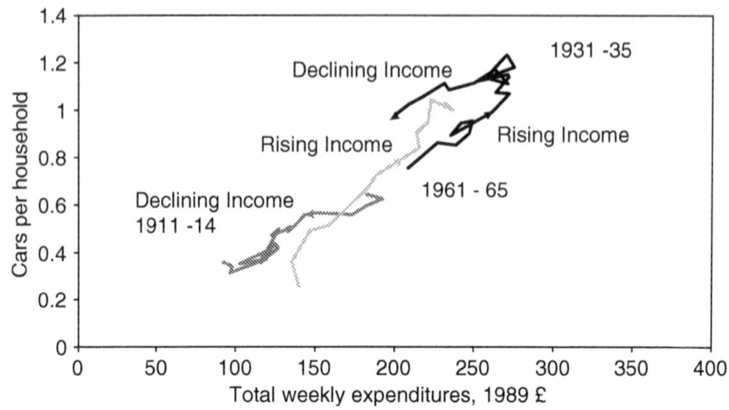

Figure 6: Relationship between Car Ownership and Income

in the middle, is an example of a relatively young household, with the head ageing from 20 to 30. Both income and car ownership are increasing rapidly. Comparing these two cohorts, it is apparent that the slope of the line indicating the car ownership–income relationship is greater for the increasing income case than it is for the decreasing income case. Rising income when young leads to increased car ownership, but falling income when old does not lead to an equally great reduction in car ownership. Both effects are shown in the case of a single, middle-aged, cohort at the upper right of the figure. Here, we follow the cohort as the head ages from 35 to 60. Between the ages of 35 and 50, household income and car ownership are increasing; while after the age of 50 or so, both income and car ownership begin to decline. But the same path is not followed. As income declines, car ownership declines, but to a lesser degree than it rose as income increased. For each income level, we have two rather than one level of car ownership depending on whether income is increasing or decreasing. Thus, there is no unique car ownership use–income relationship, but a hysteresis[1] loop. One explanation for this would be that households have become accustomed to the advantages of car travel. Such car dependency is not easily reversed, so there is a tendency to maintain car ownership in spite of falling income. Such resistance to change is unlikely, however, to be permanent, so that the timing of eventually reducing car ownership may be determined by the car finally ending its useful life or requiring disproportionate expense to keep on the road.

To summarise, we identify the powerful effects of demographic influences and income. This in itself is not new – everybody accepts they are powerful. What is new is the

[1] A term from the study of magnetism, relating to the energy of a system where particles are subjected to an electrical field, and gain, then lose, magnetic properties (see Goodwin, 1976).

nature of that influence, and in particular that the relationships take place over very long periods of time – decades. The feature of asymmetry, or hysteresis, is important. It means that even though change is universal, it is more difficult to reverse a long-established trend than to accelerate it. In our time this has taken the form of it being easier to increase the number of cars than to reduce the number, but there is a logical possibility that if the dominant trend was to reduce car ownership (for whatever reason), then hysteresis would work in the opposite direction.

THE IMPACT OF POLICY INITIATIVES

Policy initiatives, whether affecting the provision of infrastructure or the conditions under which it is used, normally are intended either to provide capacity for demand growth or to reduce or reverse that growth. If hysteresis applies, the two would not be opposite and equal, and the appraisal of moving from "A" to "B" is not equivalent to moving in the reverse direction. We demonstrated that existing travel patterns cannot be overwhelmingly dominated by unchangeable habits, but it is also suggested that such resistance to change still has a degree of power.

Here, we must confront an analytical problem in the tradition of transport studies. The longest established methods of analysing the effects of a new policy or facility (whether for forecasts or for retrospective interpretation) is to use a model which is itself based on observing differences in choices made by different people at a point in time, not on changes made by the same person over time. The presumption is that relationships based on *differences* can be used to determine *changes*. That is not always true and it can be very misleading. If hysteresis is a possibility, observing that people with cars make fewer public transport trips than people without cars cannot tell us what happens if a non-car owner buys a car, and if a car owner stops being so, since these may be different from each other. So what we have done is to seek contexts where such a change has happened, and observe or measure the change in travel behaviour that follows. At best, this is done in real time with continuing observation, but retrospective before-and-after enquiry can also provide useful insights, albeit with some care when retrospective memory-based reports are elicited.

Park-and-Ride

One unexpected example of complex behavioural responses to a new system emerged from investigations of the effects of bus park-and-ride schemes (see Parkhurst, 1995, 2000, for more detail). The work started with two schemes in operation, in Oxford and York, and was eventually extended to ten other towns and cities. The policy intention behind park-and-ride scheme was reasonably straightforward. By placing car parks at the edge of the urban area and linking them to the town centre using superior bus services, targeted specifically at motorists, it would be possible to capture cars before they entered the congested urban area.

Analyses were based on a combination of retrospective and prospective inquiries supported by counts and market information over time. The overall impact needs to be separated into at least four interacting processes of behavioural response. First, around half of park-and-ride users were people who would have driven all the way to the town centre but were instead intercepted at the car park and travelled the last stage of their journeys on the park-and-ride bus. Thus, second, an increase in bus traffic occurs, in providing the additional park-and-ride bus services. But overall some traffic in the urban area is usually avoided. Third, some users indicated that their trips to the particular town were in some way extra compared to the period before park-and-ride had been made available. This was sometimes the case because a new job, or other major life event, had led them to alter travel patterns, or alternatively as a result of realising that park-and-ride turned out to be a cheaper way to travel than using either car only, or bus only, to access the city centre, for example if the park-and-ride scheme is subsidised. And fourth, an often sizeable minority of users had changed mode of travel from public transport to park-and-ride, rather than from car to park-and-ride. As the park-and-ride sites were located at the edge of the city, some of the trips now made by car as far as the park-and-ride site would be relatively long ones, of perhaps 15–20 km. This means that only relatively few public transport users with a car available choosing to transfer to park-and-ride would be necessary for the reduction in traffic in the town centre to be more than compensated for by the growth in traffic outside the urban area. Thus, in total there was a net increase in the total amount of vehicle kilometre travelled (Figure 7). Reductions in inner urban traffic were offset by larger increases in outer and non-urban traffic, the net increase being mostly in the range 5–20 km per intercepted car.

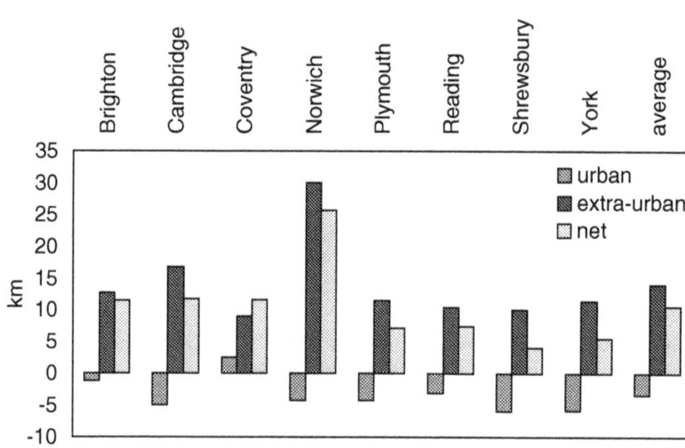

Figure 7: Overall Traffic Effects per Intercepted Car

The result that investing in park-and-ride facilities for motorists reduced the use of public transport was not the intention of either the local authorities or the bus companies. Since the first intimations of this effect were published at the beginning of the research programme (and then progressively asserted with more confidence as the data base increased), it was possible to observe some effects of the research itself on policy development. After an initial resistance, awareness that there are disadvantages to park-and-ride as well as advantages has grown. Whilst the problems have not disappeared, park-and-ride schemes are now generally being better integrated with policies to discourage car use and in some cases innovative approaches to park-and-ride provision are being pursued, such as, on occasion, provision of the car parks on existing bus routes, and locating the car parks further from the towns they serve, so more of the car trip can be intercepted.

Road Traffic Increases Due to Road Building

The next example is that of induced traffic due to road building, which is summarised briefly because it has been widely discussed, notably in the work reported in SACTRA (1994), and subsequently in a number of articles, local studies, and international comparisons, adding up to some hundreds of observations of cases. This growing literature is summarised in Goodwin and Noland (2003). An overview of the evidence is shown in Table 2.

The empirical case is that induced traffic exists. Its size, of course, varies according to circumstances. On average, road improvements had induced about 10% of the base traffic in the short run, and about 20% in the long run, and there were some schemes with induced traffic at double this level. The biggest levels of induced traffic were on the alternative routes that the schemes were intended to relieve, for example, bypasses aimed at removing traffic from roads through a town (but unaccompanied by restrictions) saw an increase of traffic over the predictions not only on the bypass itself, but on the town roads also. If there is little congestion to start with, or expected, then the induced traffic from its relief will naturally be small. But in the average conditions where road capacity increases are considered in the UK, existing or expected congestion is usually rather high. As a rule of thumb, we can say that when extra road capacity initially reduces congestion, which saves some time for drivers, something between half of the time saved, and all of the time saved, will be ploughed back into more, or longer distance travel, which erodes the benefits and creeps back – sometimes rushes back – towards the conditions of congestion observed before. The remaining question to be solved is the extent to which the extra traffic is composed of the same people travelling more, or new people, but this distinction itself becomes unclear when dealing with forecasts of years or decades.

When considering the effect of this research on established methods of forecasting and appraisal, we can say that in one sense research has solved a dispute. The formal

Table 2: Empirical Evidence on Whether New Roads Generate Extra Traffic

Definition	"Traffic" is produced by the combined effects of choices to take more frequent trips, take longer routes, shifts in method of transport used, new choices of where to work or shop (leading to longer average journey distances), and indirect changes due to making undeveloped land more accessible, changes in employment, and the consequential land-use changes resulting from all these choices
Mechanism	Extra capacity could lead to induced traffic not due to the capacity itself, but to the reduction in the time and/or money cost of travel it produces. Therefore, the relevant context is primarily that where extra capacity reduces congestion or increases accessibility to undeveloped land, hence speeding up journey time, reducing fuel costs, and possibly increasing the returns on development. Therefore, the context where road expenditures might influence the volume of traffic is when the expenditure is substantially on road capacity expansions, *and* that capacity expansion leads to reductions in time and cost or spurs new development because of better accessibility
Differential growth rates	Since the 1950s, that traffic growth rates have generally been slowest where congestion is worst, and fastest where existing capacity is still spare, or new capacity is provided. From 1980 to 1990, measured traffic on major roads in UK built-up areas grew by 20%, on major roads in non-built-up areas by 58%, and on motorways by 73%. The M25 Motorway round London experienced particularly high growth rates in traffic – much higher than the forecasts made for it – almost as soon as it was opened
Econometric estimation of elasticities of demand	Some 500 citations in the economics and engineering literature produce strong evidence on the effect of transport price changes, which can be converted to equivalent time changes by the use of the concept of the "value of time". These suggest that an average speed change, saving 10% of journey time, would cause around a 5% increase in traffic volume in the short term, and perhaps twice this in the long term
	Direct statistical measurement of the relationship between additions to actual road capacity, expressed as lane miles of road, and the impact on total vehicle travel give an average result of 0.3% to 0.5% increase in traffic for each 1% increases in lane miles, and perhaps somewhat higher in the long run
Comparisons of predicted and out-turn traffic levels	Traffic counts generally taken about one year after opening for 151 schemes, with data on the alternative routes, sometimes called "relieved routes" for 85 of these, showed a wide variation, on average the observed traffic on the improved roads being about 10% higher than had been forecast assuming no induced traffic. The discrepancy for the relieved roads was greater, on the order of 16%
Before and after traffic counts on road improvement schemes	More detailed information for 11 before-and-after studies, over a longer period, showed growth rates in the corridors observed in every case substantially greater than the roads used as controls, and greater than background growth rates, with an unweighted average of "unexplained" growth of 25% from 20 counts cited, with a range from 7% to 66%. This discrepancy in traffic growth between expanded and other roads systematically increased over time. The unweighted average of the growth rates over less than a year was 9.5%, increasing to 33% after five years

recommended procedures are that induced traffic should be estimated and included in evaluation before proceeding with a scheme. But even now some highways schemes are still being assessed on the basis of an argument that induced traffic can be ignored. There has been growing emphasis since 2003 by the UK Department for Transport on the concept of "locking-in" – referring to the importance of having complementary policies, including pricing, traffic control, or other measure aimed at reducing induced traffic, to make sure that improvements in traffic conditions are not eroded by the extra traffic that the improvements themselves attract. This is a useful and original policy response to induced traffic. The significance of the induced traffic discussion is that it provides evidence of changes that are big enough to be seen, and to be important, at the aggregate level. It was not possible to argue that detailed changes by individuals would get lost in the great volumes of traffic; rather the aggregate effects of these disaggregate changes are big enough to radically affect total volumes of traffic, and therefore influence appraisal of policies aimed at providing for it. It is simply the existence of induced traffic that undermines the legitimacy of the simplest version of "predict-and-provide" rule – for logical reasons, not political ones. Since the provision influences the prediction, one cannot proceed in a single line from prediction to provision, but must change the prediction as well, in a recursive process. This is now accepted in principle, though not always in practice.

Road Capacity Reductions

Following on from the SACTRA report, which showed that increasing road capacity could induce traffic, there was strong interest in the opposite question. What happens if road capacity is reduced, by taking space away from cars? As demonstrated above, not all relationships are precisely reversible. But even if they are only approximately reversible, you would expect reductions in capacity to have some effect on travel choices.

A review of the evidence on this question is reported in Cairns *et al.* (1998), and updated in Cairns *et al.* (2002). The evidence base was formed from examples where road capacity already had been reduced. Many changes to town centres, for instance, large-pedestrianisation schemes affecting the whole centre, or small schemes affecting only a couple of streets, all involve taking some road capacity away from traffic. Road space for cars can also be reduced by bus priority schemes, street-running light rail systems, cycle lanes, and pavement widening. There are a large number of other reductions in capacity due to earthquakes, bridge maintenance, road works, and other unintended events that, though not being policy initiatives, still provide evidence on behavioural responses, especially in the short term. Altogether detailed information was available for over 70 locations in 11 different countries together with partial information from over a hundred other studies.

When it is planned to take road space away from cars, there are often dire predictions of "traffic chaos". However, in retrospect these predictions rarely, if ever, prove accurate. Where congestion is already bad, it does often stay bad, sometimes with short-term disruption, and problems on specific local streets. However, wide-scale, long-term major inability to cope is simply not reported.

This phenomenon of a repeated difference between expectations and outcome represents a problem for appraisal of such policies, especially where various simplified methods are used of estimating traffic effects, such as fixed trip matrix traffic assignment models, not allowing for full adaptation. This is often the case because the small scale of the scheme does not seem to justify the use of expensive appraisal methods.

With some caution (the caveats being fully spelled out in the published reports), it is possible to collate a simplified presentation of the empirical evidence, shown in Figure 8a (the original data) and 8b (where the updated date allowed a distinction to be made between two major types of capacity reallocation). Each bar corresponds to an individual case study of road capacity reduction, and shows the overall change in traffic after road capacity was reduced.

In most cases, overall traffic levels went down on the streets where capacity was reduced, and in the order of half of this, traffic reappeared on neighbouring streets. Across all case studies, the average traffic reduction overall was 22%, with a median of 11%. Thus, in half the cases, more than 11% of the traffic previously using the affected street or area could not be found on the local network afterwards. The effects of changes in town centre shopping streets were, on average, substantially greater than the changes observed for bus lanes.

The data we had were not sufficient to test directly for hysteresis, that is to determine whether the decrease in traffic caused by a reduction in capacity was equal and opposite to the increase in traffic caused by an increase in capacity. However, the orders of magnitude seem similar with observations of 20% and more being not uncommon in both cases. This suggests that hysteresis may not, here, be substantial. As a minimum, we can assert that appraisal of capacity reductions must allow for "disappearing" traffic for reasons that are as compelling as those which require that capacity increases must allow for induced traffic.

Price

In recent years, the most influential result on discussions of the effect of price on car use has been the experience of congestion charging, with the London scheme, though not the first, attracting most attention. The initial results are shown in Figure 9, taken from the Transport for London website (http://www.tfl.gov.uk/tfl/cclondon/cc_publications-library.shtml), where there are updated results from time to time. There was a swift

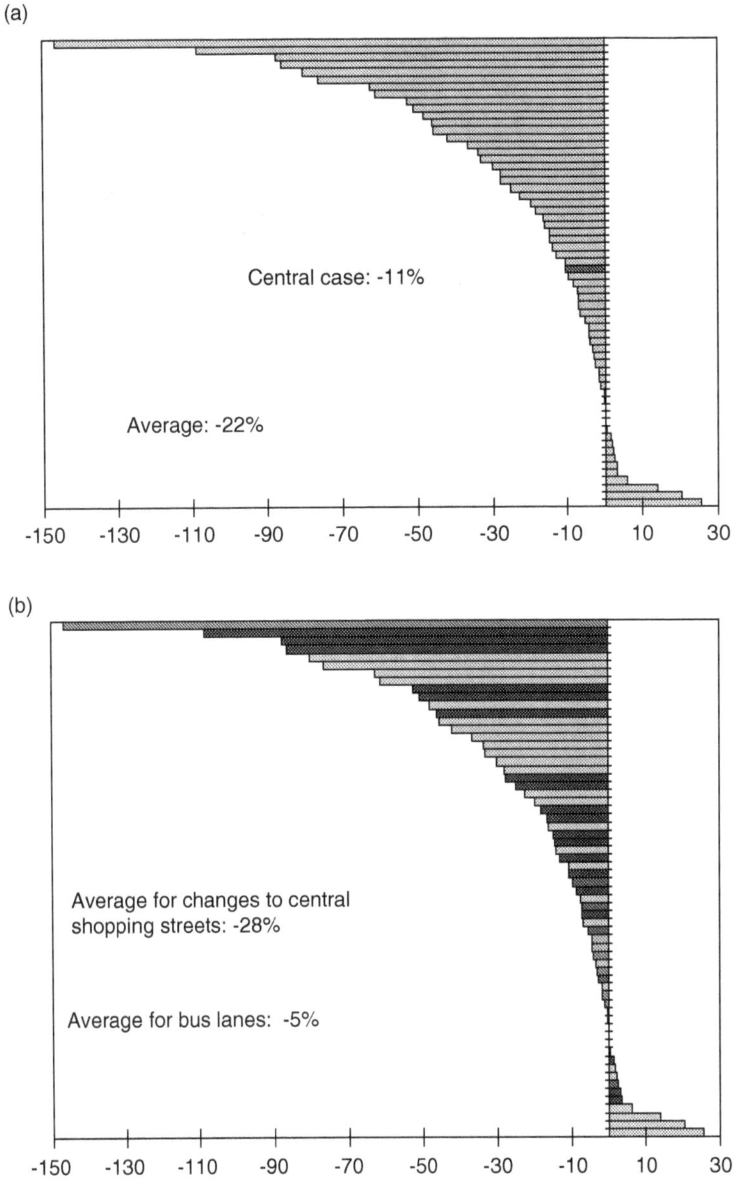

Figure 8: Changes in Traffic

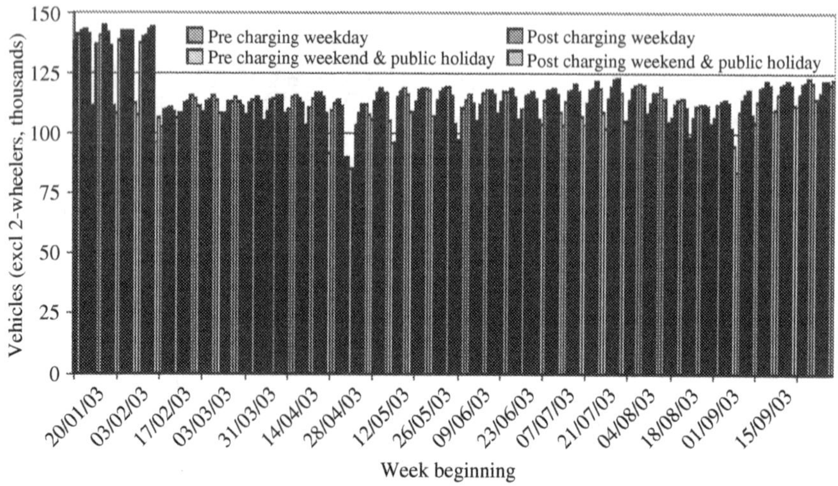

Figure 9: Traffic Entering the Charging Zone During Charging Hours (and on Weekends 7.00 am to 6.30 pm) on a Representative Selection of Major Entry Points

reduction in traffic, within which the reduction of car traffic from those without discounts or exemptions was about 30%, significantly higher than had been expected. This result needs to be considered in the context of the circumstances of commuting to central London, where a large majority already use public transport.

The effectiveness of pricing strategies depends on a wide range of conditions and details, but the common thread for this chapter relies on assessment of value of price elasticity of demand – that is, is demand highly sensitive to price, or only slightly, or not at all. Elasticity is a dimensionless constant, very easy to interpret. An elasticity of –0.5, for example, means that if price goes up to 10%, demand will go down by 5%. It is always driven by individual disaggregate choices by specific people for specific journeys, but usually expressed for the whole market.

For decades, it was thought that these measures were reasonably well established, and rather small. The bus-fare elasticity, for example, was –0.3. If you put bus fares up by 10%, one loses 3% of the market, but still make extra revenue. And the fuel price elasticity was smaller. The Department for Transport and its predecessors assumed, for a while, a figure of about –0.1 to –0.15, that is an increase in fuel price of 10% would reduce traffic volume a little, 1% or so, but not enough to make a difference to anything in practical terms. But the elasticities typically used for policy analysis had been generally estimated using the assumption that at an observed time, demand had already fully adjusted to recent changes in price, which we have called an equilibrium approach.

For technical reasons not discussed here, this has a tendency to produce biased results in contexts where long-run responses are different from short-run responses, and for this reason equilibrium modelling was rejected in favour of a dynamic approach. To observe how travel behaviour changes over time, it is necessary to use longitudinal data, for example, aggregate time-series data for the country or a region; a combination of cross-section time-series data including pseudo-panels, or true panel data, and model specifications that allow for the build-up of effects over time (thus incorporating lags, habit or other sources of asymmetric response). We report here two specific sets of results, first for bus patronage, and then for car ownership.

Bus fares. The study of bus-fare elasticities is reported by Dargay and Hanly (2002). The main objectives were to estimate elasticities which could be used in policy calculations to project the change in bus patronage nationally as a result of a given average fare change. We used national, regional and county data on bus patronage, fares, service and income over time, and a dynamic model. The results are shown in Figure 10.

The estimated fare elasticity for Great Britain as a whole was about –0.4 in the short run (close to the received wisdom), but –0.9 in the long run (very far from received wisdom). This means that if the average fare of all operators in a local market increases by 10%, total patronage will decline by 4% within one year. The complete response takes around seven years, by which time patronage will have declined due to the fares change itself (quite apart from changes due to other important factors such as income, service level, etc. which are included in a properly specified model, but not discussed here), by a further 5%, giving 9% in total, not taking account of changes due to other factors such as

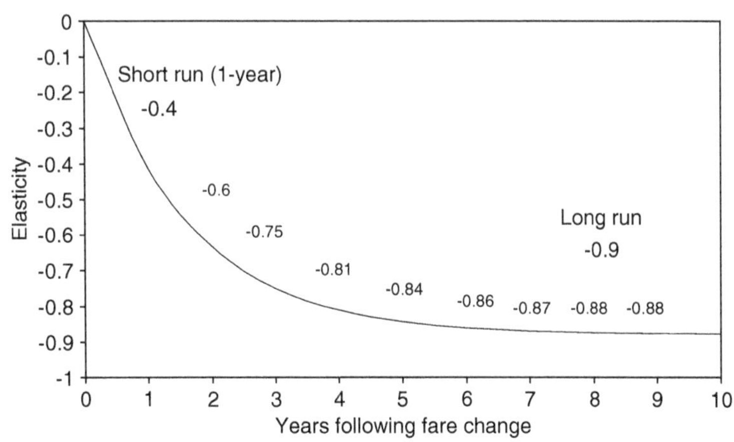

Figure 10: Dynamic Bus Fare Elasticity

Table 3: Average Elasticities

	Car Ownership		*Car Travel*	
	Short Run	*Long Run*	*Short run*	*Long run*
Car purchase costs	−0.1	−0.2	−0.3	−0.4
Fuel price	0	0	−0.1	−0.2
Income	+0.3	+0.7	+0.7	+1.0

income, car ownership, or inflation. The dynamic elasticity is illustrated in the figure. The fare elasticity increases over time, but at a declining rate, finally to reach its long-run value. Clearly, the estimated long-run elasticity of –0.9 is much higher than the elasticity of –0.3 previously taken as given. Other results of this study were that the fare elasticity increases at higher fare levels, and is greater for non-urban than urban areas. Service – in terms of vehicle kilometres – is also found to have a substantial impact on patronage.

Car ownership and car traffic. Estimates of elasticities for car ownership and car travel have been obtained by applying suitable dynamic models to the pseudo-panel data shown earlier. A number of different model specifications were estimated and the results shown in Table 3 are average values. Again we find a significant difference between short- and long-run elasticities. Car travel responds more strongly and more quickly to price and income changes than car ownership. Both car ownership and car travel are more sensitive to car prices than to fuel prices, but income is the most important determinant of both.

The modelling work also provides evidence of differences in elasticities between individuals. For example, households in rural areas are less price-sensitive than urban households regarding car ownership and use. In addition, high-income households are less-price sensitive than low-income households. Both of these findings have implications for the distributional effects of price-related policies.

We also find statistical evidence of saturation – the income elasticity declines at higher levels of car ownership and use – and of asymmetry with respect to income – the effects of income increases are not the same as of income decreases, depending on the direction of trend as discussed above.

These results were supported by a new review of the literature (Goodwin *et al.*, 2004b), which led to the conclusion that the elasticity of traffic with respect to the fuel price is of the order of –0.1 in the short run and –0.3 in the long run (see Figure 11). This is about half the elasticity of fuel consumption with respect to the fuel price, implying that a significant part of the response to fuel price increases is through the use of more fuel-efficient vehicles.

	short term	long term
fuel consumption	-0.25	-0.60
traffic volume	-0.10	-0.30

Figure 11: Elasticities with respect to Fuel Price per Litre Literature Review

Taken together, the results for cars are similar to the public transport results, namely that the elasticities were higher than had been assumed, and showed a distinct build-up of effect over time scales in the order of year years or so.

Soft Factors

There has recently been increasing interest in policies aimed at changing behaviour by changing the quality of transport alternatives, and the information and perceptions that people have about those alternatives. These measures have been called "soft factors", though the label is contested and the Department for Transport in the UK decided to use the label "smart" instead.

In a study (Anable *et al.*, 2004; Cairns *et al.*, 2004), the present and potential effects of the following soft policy measures were assessed:

- Workplace and school travel plans;
- Personalised travel planning;
- Public transport information and marketing;
- Travel awareness campaigns;
- Car clubs and car sharing schemes;
- Telework and teleconferencing;
- Home shopping schemes.

Most of these policies are relatively new – dating back 10 years or less – although the amount of local authority experience of implementing them is growing rapidly, largely because there are some very positive reports of their effects, particularly in terms of their effectiveness at cutting traffic levels.

The work involved a major literature review and critique, collation of evidence and visits to 24 of the leading initiatives we were able to identify, and interviews with those responsible for organising them. In summary, the conclusions were that these measures, pursued vigorously for a period of about 10 years, could lead to

- A 21% cut in urban peak-hour traffic;
- A 14% cut in non-urban peak-hour traffic;
- An 11% reduction in national traffic, overall.

These figures are significantly higher than the Department for Transport had previously assumed in guidance given for the multi-modal studies, where a cautious figure of 5% traffic reduction was suggested. Attention was drawn to a strong caveat we had made, derived from the evidence on induced traffic: if car use amongst one group of people is reduced, freer road conditions may attract others onto the roads, unless the benefits of the soft factor policies are carefully "locked-in" using other traffic restraint measures.

Components and Triggers of Behavioural Change

The work described above establishes that travel behaviour can change, substantially, and that this can happen both as spontaneous processes and as responses to policy initiatives building up over time. Deeper understanding depends on being able to detect the relative importance of different types of change, and the influences bearing on them. This section is more tentative, but some important general insights have emerged. In the work on reducing road capacity, the evidence suggested a wide range of responses, which can be broadly divided into three groups. In the first instance, traffic does not reduce, it intensifies, as people change their driving styles – driving closer together, getting through traffic lights quicker – in order to maximise the road space. Second, traffic spreads out over time and space, with people swapping to alternative routes, or changing their journey times by leaving earlier or later. Third, as it becomes more difficult to make such adjustments, a whole variety of changes get mentioned in surveys, which would explain why traffic can disappear from a network overall. These include changes in

- How to travel;
- Where to go;
- How often trips are made;
- Car sharing;
- Doing more than one thing on the same journey;
- Who does what within the household;
- Entirely new trips;
- Where new developments are built;
- Job location;
- Home location.

The evidence suggests that in general people do not make such changes only because of a change in road capacity. But precisely because of the natural variability in travel

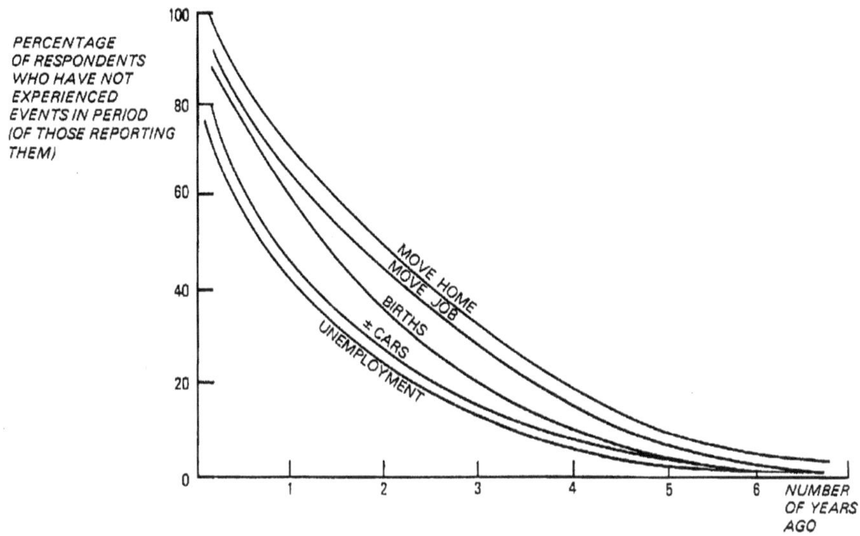

Figure 12: The Early Research: Frequency of Life Events Enabling Choice

choices for other reasons, a change in road conditions may help to tip the balance in which change is chosen. Habits and constraints, based on existing circumstances, may be so strong as to prevent changing behaviour in response to some policy initiative, because the constraints of life simply do not allow it. Even so, after a while, there will certainly be a "life-shock" of some sort, and then some change in behaviour is unavoidable. This can be done in a way that enables a response to the new circumstances, higher costs, lower costs, more reliable buses, or less road space.

The interesting thing is that it has been observed, as Figure 12 shows, that year by year fewer and fewer people will be left who have not had some such upset in their routine. The time scale – one to five years, for the majority of the population – is very close indeed to the sort of lags that the econometric results found for elasticities. Thus, the people who had the biggest changes in their lives were also the people who showed the largest response to policy effects such as, in this case, bus fares.

POLICY IMPLICATIONS

The policy monitoring work in conjunction with the econometric analyses suggest that long-term effects are bigger than short-term ones, precisely because people learn, adapt, and are able to include in their behaviour not only the change of route, but also a different pattern of life. The existence and importance of such dynamic features is one of the key findings. Ignoring the inter-temporal nature of the response to policy will lead to an underestimate of the scale of behavioural response, and hence a biased assessment of the potential for policies to have the intended impact. At the same time, the dynamic processes may also lead to other, unintended impacts, which may offset or reverse the desired effects unless such problems are recognised and remedial action taken.

It is in the nature of policy-related research that there is always room for disagreement on interpretation of evidence and in inferences about what follows. But some general conclusions are suggested.

Public transport fare elasticities. Our results imply that attempts to raise revenue by fare increases will work in the short run, but so erode the market that in the long run the extra net revenue is very small indeed, and negative in terms of congestion and environmental impacts. What can we do about the dynamics of revenue, in a world where the accountants do not really want to look beyond next year? And where, then, would the funding come from for the continued environmental improvement of public transport systems?

The effects of motoring costs. The results imply that car traffic levels do respond to these – up and down – which strengthens the transport effectiveness of road pricing, but raises less revenue than expected. This arises directly if elasticities are higher than expected, since a price-increase then produces a greater shift in behaviour and this in turn produces less revenue.

Soft measures. This work presupposes a very big increase in the priority of these measures at local level, and at the moment some local authorities do not even count the staff employed in this area as being on the proper establishment. They are the poor relations, the staff on bursaries or temporary contracts – unlike the engineers in the drawing room, whose career track is simpler, and a lot more favourable. And then, suppose that a local authority does decide to treat this area as a proper job, and does put priority on it – but does *not* manage to secure support for the complementary traffic policies to prevent induced traffic. This is a live policy problem in many areas.

Road building. "Locking-in" is the answer to one question: if we have the right road scheme, how to protect its benefits. But a programme of locking in, and demand management, and congestion charging, and soft measures, and public transport improvements: taken together, these mean that appraisal of any proposed future road schemes must take into account a different picture of travel demand (and hence location of traffic in time and space) than has been the case in the past.

Dynamic analysis. On the analytical side, at root, the concepts of dynamics are simple – simpler, in fact, than the standard transport models based on a utility maximising achieved equilibrium, whose precepts about behavioural change are, at heart, self-contradictory and elusive. But there is a problem of familiarity. Where are the consultants who will offer a dynamic forecast with as much confidence as they now offer an equilibrium forecast? Understanding of change cannot be derived from observations of states at a point in time but only from consideration of pathways, history, and the process of adaptation. That requires observations of data over time, and models using dynamic methods, which need to become more part of the standard toolbox than the off-the-shelf surveys and models now used.

CONCLUSION

In summary, we have looked altogether at about 20 different influences on travel behaviour that transport policy uses as instruments. We find that travel behaviour is very much more volatile and changeable than is usually assumed, and in particular that car ownership and use do go down as well as up, though the strength of the effects is different.

Revealed effects are larger than has generally been assumed: changes of 10–30%, at the aggregate level, in periods of small numbers of years, are quite common even in conditions that are within the bounds of "normal life", as most people perceive it, not those of extreme emergency or revolutionary change. The problem is that these changes are complex, and are easily offset by other unintended changes especially where the policy instruments are not all pulling in the same direction, which is common.

Overall, the conclusion is that policymakers and transport industries have more scope to influence travel behaviour than they think, but only if transport interventions are consistent with each other, maintained over a lengthy period, and supported by analytical methods and appraisal frameworks that are not yet commonplace.

ACKNOWLEDGMENTS

This is an edited version of Changing Travel Behaviour, being the final report of a research programme funded by the UK Economic and Science Research Council at the Transport Studies Unit (initially Oxford University, subsequently University College London), 1994–2004, coauthored with Sally Cairns, Transport Research Laboratory; Joyce Dargay, Institute for Transport Studies, University of Leeds; Mark Hanly; Graham Parkhurst, Centre for Transport and Society, University of the West of England, Bristol; Gordon Stokes, UK Countryside Agency; Petros Vythoulkas, National Technical University of Athens, Greece. The research was carried out while each author was a member of the research team of the ESRC Transport Studies Unit initially at the University of Oxford, subsequently at the University College London.

REFERENCES

Anable, J., A. Kirkbride, L. Sloman, C. Newson, S. Cairns, and P. Goodwin (2004). *Smarter Choices. 2: Case Study Reports*, Department for Transport, London.

Cairns, S., S. Atkins, and P. Goodwin (2002). Disappearing Traffic: The Story So Far, *Proceedings of the Institution of Civil Engineers, Municipal Engineer*, **151**, 13–22.

Cairns, S., C. Hass-Klau, and P. Goodwin (1998). *Impacts of Highway Capacity Reduction: Assessment of the Eidence*, Landor, London.

Cairns, S., L.C. Sloman, J. Anable, A. Kirkbride, and P. Goodwin (2004). *Smarter Choices. 1: Changing the Way We Travel*, Department for Transport, London.

Dargay, J. (2001). The Effect of Income on Car Ownership: Evidence of Asymmetry, *Transportation Research A*, **35**, 807–821.

Dargay, J. (2002). *Pseudo-Panel Analysis of Household Car Travel*, Report No 2002/04, ESRC Transport Studies Unit, University College London, London.

Dargay, J. (2003). Travel to Work: An Investigation Based on the British Household Panel Survey, Paper presented at NECTAR Conference No. 7, Umeå University, Sweden, June.

Dargay, J. and M. Hanly (2002). The Demand for Local Bus Services in England, *Journal of Transport Economics and Policy*, **36**, 73–92.

Dargay, J. and M. Hanly (2004). Volatility of Car Ownership, Commuting Mode and Time in the UK, Paper presented at World Conference on Transport Research, Istanbul, Turkey, August.

Goodwin, P. (ed.) (1983). *Subsidised Public Transport and the Demand for Travel: The South Yorkshire Example*, Gower, Aldershot, UK.

Goodwin, P. (1997). Have Panel Surveys Told Us Anything New? In: *Panels for Transportation Planning: Methods and Applications* (T. Golob, R. Kitamura, and L. Long, eds), Kluwer, Amsterdam, 79–96.

Goodwin, P., S. Cairns, J. Dargay, M. Hanly, G. Parkhurst, G. Stokes, and P. Vythoulkas (2004a). *Changing Travel Behaviour*, Final Report to ESRC, Centre for Transport Studies, University College London, London.

Goodwin, P., M. Hanly, and J. Dargay (2004b). Elasticities of Road Traffic and Fuel Consumption with Respect to Price and Income: A Review, *Transport Reviews*, **24**, 275–292.

Goodwin, P. and R. Noland (2003). Building New Roads Really Does Create Extra Traffic: A Response to Prakash et al., *Applied Economics*, **35**, 1451–1457.

Parkhurst, G. (1995). Park and Ride: Could It Lead to an Increase in Car Traffic? *Transport Policy*, **2**, 15–23.

Parkhurst, G. (2000). Influence of Bus-Based Park-and-Ride Facilities on Users' Car Traffic, *Transport Policy*, **7**, 159–172.

SACTRA (1994). *Trunk Roads and the Generation of Traffic*, Report of the Standing.

Stokes, G. (1995). *Travel Behaviour in South Yorkshire in 1993*, Transport Studies Unit, University of Oxford, Oxford.

Threats from Car Traffic to the Quality of Urban Life
Tommy Gärling and Linda Steg (Editors)

23

Societal Management of Sustainable Transportation: International Policy Review, Commons Dilemmas and Solution Strategies

Charles Vlek, Department of Psychology, Faculty of Behavioural and Social Sciences, University of Groningen, Groningen, The Netherlands

ABSTRACT

The environmental burden from motorised transport may be effectively reduced in light of the commons dilemma model. This captures the tension between individual and collective interests, and it implies that collective transport problems can be resolved through changes in individual transport behaviours. To bring model thinking and policy substance together, recent problem analyses and policy intentions by the OECD, EU and WBCSD are summarised and critically evaluated from a behavioural-science perspective. Commons-dilemma thinking is explicated, including various strategies for behaviour change. Sustainable transport development is described following a three-dimensional framework of guidelines and indicators, five policy principles, and a broad set of substantive policies. One conclusion is that the supply side of transport needs to be addressed at least as much as the demand side.

INTRODUCTION

Transport motorisation is an overwhelming development of the past 100 years, slow at first and rapidly accelerated since World War II. Transport accounts for one-quarter of the world's energy use and about half of the world's oil production (UNEP, 1999). Following their "Mobility 2030" project, the WBCSD (2004) expects that worldwide the number of light-duty vehicles will grow from the present 750 million to 1 billion around 2020 and up to 2 billion around 2050. Rees (2003) notes that cities are accommodating more and more of the world's population and that, through their massive "ecological footprint" – significantly due to motorised transport, they are turning into centres of unsustainable development. The physical infrastructure needed for modern mobility may be seen as part of a global move towards a "frictionless society", a utopian vision of humankind in which historical impediments of space and time are ultimately overcome (Flyvbjerg *et al.*, 2003).

The development of motorised transport – by road mostly, but also by air – is increasingly being perceived as unsustainable in the long run. This was already acknowledged in the early 1990s (ECMT, 1990, 1993; ISIRT, 1992; Banister and Button, 1993). More recently, among the UNEP (1999), OECD (1997, 2002) and EU (2001b, 2003; see also EEA, 2006) there is growing concern and policy exploration in view of unsustainable transport developments. This has contributed to a wealth of international activities. In November 2005, an Internet search with the entry "sustainable transport" yielded about 20 million hits. The broader entry "sustainable development" gave 108 million hits.

Massive motorised transport constitutes an impressive, multifaceted example of a common-resource or commons dilemma. This is a social situation of conflict between a set of aggregate collective interests such as public health or living environments, and numerous individual interests such as getting fast and comfortably from A to B. In pursuing their own personal interests, many individuals freely shift off limited negative effects on their common physical and social environment. By accumulation of the many small impacts the overall effect may be rather burdening, so that collective environmental qualities significantly deteriorate. Examples are urban air pollution, space occupation (yielding congestion) and noise, as well as disturbance of natural areas by road infrastructure (see Button and Rothengatter, 1993; Hensher and Button, 2003; see also Chapter 2 by Van Wee, this volume). Such social processes of environmental decline often proceed slowly and gradually, which stands in stark contrast to the immediacy and frequency of the benefits obtained by individual resource users.

Resolving a commons dilemma means finding a sustainable balance between individual and collective interests. For motorised transport, negative environmental impacts may be reduced at several system levels. Making motor vehicles lighter, cleaner and quieter is one strategy. Using them more efficiently (e.g., by combining several travel motives per trip) is another. Changing travel mode is a third strategy, which presupposes the availability of feasible alternatives. One could also try to reduce travel demand, e.g., by bringing frequent travel origins and destinations closer together. Finally one might restrict motorised transport altogether, e.g., by limiting car ownership or through a tradable-quota system of transport kilometres.

In the remainder of this chapter first a review is given of problem statements and policy intentions as formulated by several international organisations. Then a summary description is provided of the commons dilemma model, including various strategies for actors' (transport) behaviour change. The second half of the chapter is devoted to sustainable transport development, in terms of a definition, a set of policy principles and a spectrum of substantive policy strategies.

POLICY CONCERNS AND INTENTIONS ABOUT TRANSPORT DEVELOPMENT

Effective therapies usually begin with an insightful diagnosis of the problem needing to be resolved. Let us therefore, by way of problem analysis, first consider the concerns and propositions of international policy-making bodies.

OECD Actions in the 1990s

At the end of the international OECD conference *Towards sustainable transportation* in Vancouver, 1996, it was concluded (OECD, 1997, p. 56) that:

> "Sustainable transportation is achieved when needs for access to people, services, and goods are met without producing permanent harm to the global environment, damage to local environments, and social inequity. (. . .) Systems of transportation used in OECD and some other countries are unsustainable".

This statement, preceded by similar ones published by ISIRT (1992) and the ECMT (1993), clearly implies that the current transport system (much more of which is to be expected; see WBCSD, 2004) is actually producing permanent global harm, local damages and social inequities.

In its second *Environmental Assessment Report*, the European Environment Agency (EEA, 1999, pp. 30, 62, 416) elucidates some of the factors whereby motorised transport is consistently expanding:

> "Transport and mobility is jeopardising the EU's ability to achieve many of its environmental policy targets. (. . .) improved eco-efficiency is not a sufficient condition for sustainable development (. . .) In the past, economic growth and lowering transport prices have raised demand for transport. Where congestion occurred, new roads, airports and other infrastructures were constructed. (. . .) This closes the vicious circle of ever expanding transport volumes. (. . .) While many instruments are being applied to reduce transport damage, these are being overwhelmed by the rapid rise in demand for transport".

This message reflects two problem characteristics. First, the expansion is ascribed to "market forces", particularly the growing demand by transport consumers, which is continually stimulated by the producers of transport vehicles. Secondly, governments have traditionally followed a strategy of supplying physical infrastructure (roads, airports) and system regulations (taxes, safety regimes) without realising much that their supply strategy worked to confirm and stimulate ongoing transport developments – which they tended to call autonomous.

Meanwhile, the OECD (2002) has strongly invested in its international project on "Environmentally Sustainable Transport (EST)", in which the 1996 conclusion was elaborated that: "Achievement of sustainable transportation will likely involve improvements in vehicles, fuels, and infrastructure (. . .), and reductions in personal mobility and in the movement of goods, (. . .)" (OECD, 1997, p. 56).

Thus, next to a Business as Usual (i.e., current policies') scenario, a high-technology and a behavioural-change scenario were composed. In a national study tuned to the densely motorised Netherlands, Geurs and Van Wee (2000, 2003) concluded that a combination-scenario involving both far-reaching technical innovations and significant behavioural changes, would be most realistic for achieving sustainability goals.

EU Common Transport Policy since 2001

In 2001, the European Commission issued a White Paper entitled "European Transport Policy for 2010: time to decide" (EU, 2001b). With reference to the Commission's 1992 White Paper aimed at "opening up the transport market", in 2001 the Commission points at successes of market development, "except in the rail sector". It emphasises the great economic importance of the transport sector: total expenditure about 1000 billion euros, that is, more than 10% of gross domestic product, and employing 10 million people, and it reconfirms the need for further developing the Trans-European Networks (TENs) as incorporated in the Maastricht Treaty of 1992. The Commission pronounces four essential freedoms of mobility, that is, of persons, goods, services and capital.

At the same time, however, the European Commission expresses concern about the unequal growth in different modes of transport (by road, air, rail and water), the increasing congestion on the main road and rail routes and around airports, and the harmful effects of transport on the environment and on public health. Noting that the concept of sustainable development offers a new lever for adapting the EU's common transport policy, the Commission aspires to "break the link between economic growth and transport growth and make for a modal shift" (EU, 2001a, pp. 13–14). Thus it pronounces, among other things, that EU co-financing should be re-directed to give priority to rail, short-sea and inland waterway transport. The Commission considers regulating a reduction in the mobility of persons and goods and imposing a redistribution between transport modes "a simplistic solution". Instead the Commission lists three options: (a) pricing policies alone, (b) pricing and efficiency improvements of other (than road) transport modes, and (c) an integrated approach favoured in the White Paper (EU, 2001a) and comprising pricing, revitalizing alternatives, and targeted investments in the Trans-European Networks, ". . . without there being any need to restrict the mobility of people and goods" (EU, 2001a, p. 15). The Commission believes that rail transport is the key sector for achieving a shift of balance, particularly for goods transportation. The principal measures proposed in the White Paper are summarised in Table 1.

Noteworthy in the "Policy Guidelines of the White Paper" (EU, 2001a) is one sentence devoted to the use of bicycles (p. 18): "Guaranteeing road safety in towns is a precondition for, for example, developing cycling as a means of transport". Meanwhile, the European Conference of Ministers of Transport have issued a "policy note and declaration" about the promotion of cycling in urban areas (ECMT, 2004). Ministers emphasise the environmental friendliness of cycling, its cost-effectiveness and its opportunities for health and physical fitness. They recommend that countries recognise and promote cycling as an integral part of urban transport. To this end, improvements are necessary in policy coordination, in cycling infrastructure and users' safety, and in general attitudes towards cycling as a sustainable way of transport, not just something for leisure, sports or children's activities. In a special issue of *IATSS Research* on non-motorised transport Tiwari (1999, p. 70) writes about Indian cities: ". . . pedestrians, bicyclists and non-motorised rickshaws are the most critical elements in mixed traffic . . . (effective) measures may imply restricting mobility of car users to ensure mobility of bicycle users". Similarly, Pucher *et al.* (2005) and Zacharias (2003) describe the critical situation of non-motorised transport in Indian and Chinese cities, respectively.

Table 1: Principal Measures from the EU's Common Transport Policy for 2010 (EU, 2001b)

- Revitalising the railways, especially for goods transport
- Improving the quality of road transport, especially in regard to security, safety and social conditions
- Promoting transport by sea and inland waterways
- Striking a balance between growth in air transport and the environment, especially noise pollution
- Turning inter-modality of (esp. container) transport into a reality
- Improving road safety: halving the number of victims by 2010
- Effectively charging users for transport (including infrastructure)
- Recognising the rights and obligations of transport users (esp. air passengers)
- Developing high-quality urban public transport (thereby enabling a more "rational" use of the car)
- Putting research and technology at the service of clean, safe and efficient transport
- Managing the effects of globalisation (e.g., through a coherent sustainable-development policy for international air and sea transport)
- Developing medium and long-term environmental objectives for a sustainable transport system

Finally, the European Commission admits that it is still wrestling with the basic concept underlying its common transport policy: "This sustainable transport system needs to be defined in operational terms in order to give the policy-makers useful information to go on" (EU, 2001b, p. 20). A mid-term review of the White Paper is undertaken in 2006. A possibly revised version is due to appear later that year.

In addition, the EEA, already cited above, has reported about "ten key transport and environment issues for policy-makers" (EEA, 2004). Six of these are worrying ("frownies"), three are good news ("smilies"), and one gets a question mark. The ten EEA issues are summarised in Table 2. Two years later the EEA (2006) reiterates that environmental problems are increasing, specifically through the growth in passenger as well as freight transport (mostly by road), in air transport, in CO_2 emissions and in air pollution.

The WBCSD Project "Mobility 2030"

The EU's policy development about sustainable transport is reasonably well in tune with a multi-annual project conducted by the World Business Council for Sustainable Development in Geneva, Switzerland. In this Sustainable Mobility Project (SMP), over 200 experts from a broad set of twelve motor vehicle and energy industries produced a collective report entitled "Mobility 2030: meeting the challenges to sustainability" (WBCSD, 2004). Acknowledging that "promoting mobility is a key part of our companies' business", the authors declare that, in their collective view, increasing mobility and lessening transport's impact may both well be achievable.

According to the WBCSD report, a worldwide average growth in personal transport may be expected of 1.7% per year up to 2050, with especially high growth rates for Latin America (2.9%), India (2.3%), China (3.0%) and the countries of the former Soviet Union (2.0%). The worldwide total of light-duty vehicles is expected to grow from the current 750 million to 1 billion around 2020, 1.5 billion in 2040, to 2 billion in 2050. For road and rail freight transport, the average annual growth up to 2050 is

Table 2: Ten Key Transport and Environment Issues (EEA, 2004)

1. Growing transport volumes are challenging decoupling policy: ☹
2. Emissions of air pollutants from road transport are falling, despite a growth in traffic: ☺
3. Greenhouse gas (CO_2) emissions from road and air transport are increasing: ☹
4. Alternative fuel policy is starting to take effect with bio-fuels: ☺
5. Market shares of road and air travel are continuing to grow: ☹
6. Access to many basic services is dependent on car use: ☹
7. Present price structures are favouring individual transport: ☹
8. There are signs of promising developments for transport pricing: ☺
9. Infrastructure investment needs to balance economic and environmental needs: ?
10. Transport infrastructure is fragmenting natural habitats: ☹

Table 3: Proposed Societal Goals for Sustainable Transport, from "Mobility 2030" (WBCSD, 2004)

1. Reduce transport-related conventional emissions (CO, NO_x, SO_2, VOCs, PM and lead)
2. Limit transport-related greenhouse gas emissions (CO_2, NH_4) to sustainable levels
3. Significantly reduce the worldwide number of traffic deaths and serious injuries
4. Reduce transport-related noise
5. Mitigate transport-related congestion
6. Narrow mobility "divides" between rich and poor, able and disabled citizens
7. Preserve and enhance mobility opportunities for the general population

projected at 2.3% (2.5% up to 2030), with especially high growth rates for Africa (3.1%), India (3.8%), China (3.3%) and "other Asia" (3.7%). Most of this growth is considered to be related to expected increases in per capita real income (and in population size, one may presume).

The corresponding picture of environmental impacts is somewhat depressing. Transport-related conventional emissions of CO, NO_x, SO_2, VOCs and PM[1] will decline sharply in "developed"[2] countries, while they will first rise and then decline in urbanised areas of developing countries. In contrast, greenhouse gas emissions will grow significantly, especially in developing countries. Transport-related deaths and serious injuries will decline in OECD countries and some (richer) developing countries; but they would rise in (poorer) developing countries. Traffic congestion in major urbanised areas will increase everywhere in the world, while traffic noise will probably not decrease. The "ecological footprint" of transport, in terms of materials use, land use and energy use will further increase (see also Rees, 2003). Finally, the authors of "Mobility 2030" express concerns about transport security and about the equality of transport opportunities for the poor, the disabled and handicapped, and the elderly. "Factoring in all these findings, it appears to the SMP that the present system of mobility is not sustainable, nor is it likely to become so if present trends continue" (WBCSD, 2004, p. 10).

In view of these projected developments and expected increases in environmental impacts, in "Mobility 2030" seven goals are formulated for society to strive for in the near future. These are summarised in Table 3.

The means considered for reaching these goals are ranging from improvements in engine fuel efficiency to using hydrogen, bio-fuels and electricity (also from nuclear power), and from enhanced road safety programmes to noise-reducing measures applied to road surface, engine characteristics and driving style. With respect to goal 5, it is said that

[1] CO, carbon monoxide; NO_x, nitrous oxides; SO_2, sulfur dioxide; VOCs, volatile organic compounds; PM, particulate matter; CO_2, carbon dioxide (strictly not a pollutant); NH_4, methane.

[2] By "developed" is usually meant that a country is industrially advanced and has a relatively high level of infrastructure, material possessions and consumption per capita.

"transport congestion cannot be eliminated completely without destroying transport's vital role in enabling economic growth. But its effects can be mitigated substantially" (WBCSD, 2004, p. 23). Equal opportunity of mobility (the average west-European now travelling about ten times as much as the average African) would have to be realised through cost-lowering of rural transport, the development of inexpensive motor vehicles for harsh rural conditions, and ensuring mobility opportunities enabling the economic development of poor people. Maintenance and/or revitalisation of public transport systems is considered to be a key strategy in preserving general mobility opportunities (goal 7).

CRITICAL COMMENTS ON OFFICIAL TRANSPORT-POLICY VIEWS

Surveying the concerns of, among others, the OECD (1997, 2002), EU (2001b, 2003) and WBCSD (2004), as preceded by ISIRT (1992) and ECMT (1990, 1993), we may note that these organisations seem to agree that our current, "western-industrialised" transport system is unsustainable. Their policy intentions and suggestions for the future, however, invite critical comments on several major points, as follows.

First of all, there is the widespread belief that mobility and transport are essential for economic growth, and that growth is needed for improving human quality of life. Two doubts may be raised here. One is about the relationship between economic growth and transport growth, which the EU (2001a) now intends to break. The other doubt concerns the link between economic growth and human quality of life. Could enhanced and more evenly distributed quality of life be achieved with less environment-burdening economic growth? Adding the two links up: could society effectively strive for enhanced quality of life going along with less, not more motorised transport? (see also Chapter 3 by Gifford and Steg, this volume). If there is sufficient political will and society is creative enough, the answer may well be positive.

Second, the expected growth of motorised transport necessitates the construction of additional infrastructure. The OECD does not say much about this, the WBCSD is silent about it (cf. Table 3), while the EU is promoting its Trans-European Networks. New infrastructure goes at the expense of open landscapes, wild flora and fauna, and neighbouring human populations. It strengthens and confirms the road transport system and it shapes and fixates people's needs and opportunities for road transport for a very long time. The money, materials and land going into new infrastructure are no longer available for the development and promotion of other, less environment-burdening transport modes (see also Chapter 16 by Newman and Kenworthy, this volume). Being reluctant in deciding about new infrastructure for environment-burdening transport, therefore, seems to be a wise strategy.

Third, it is regularly said (e.g., WBCSD, 2004) that transport growth may well go along with a decrease in absolute levels of conventional pollution (excluding CO_2) of air, soil and water. On the one hand, this reflects a limited notion of pollution that seems to exclude environmental resource use, noise, accidents and space occupation. On the other hand, with cleaner, quieter and more energy–efficient motor vehicles, the total environmental burden from transport may not decrease at all, due to the rebound effect (see Berkhout *et al.*, 2000; Alcott, 2005), whereby such motor vehicles are actually purchased more often and used more intensely because they are each less burdening indeed – per kilometre driven – for both the environment and the environmental consciousness of transport users. Therefore, cleaner vehicles are necessary all right, but also reduced use of motor vehicles seems required.

Fourth, especially the OECD and the EU strongly bet on transport mode shifts as contributors to sustainable transport development. From road to rail and waterway transport, from short-haul air to high-speed rail transport, from urban car to bike trip, such shifts would diminish current problems of congestion, pollution, annoyance and safety risks from road traffic as well as around and between airports. Mode shifts, however, may not easily occur if the mode of origin is not significantly disqualified relative to the alternative, more "sustainable" mode. To beat the highly attractive private motor vehicle, alternatives like train, bus, light-rail and bicycle (in any inter-modal combination) must be made relatively more advantageous, financially and in several other respects (see also Chapter 17 by Gärling and Loukopoulos, this volume; Steg, 2003). Thus, "de-marketing" and "re-shaping" the car should be part of the story (Wright and Egan, 2000; Wright and Curtis, 2005). As long as freedom of mode choice is maintained, the promotion of alternative transport modes may prove to be a waste of public funds and it may well lead to an overall increase in mobility and transport volume.

Fifth, there exists a strong conviction that pricing policies are the best, market-compatible way of discouraging people to use "unsustainable" transport modes and shift to "sustainable" ones (see also Chapter 18 by Ubbels and Verhoef, this volume). Being sensitive to price changes, however, requires that other motivating factors are relatively less important. Price considerations in, for instance, mode-shift decisions would have to overrule such considerations as travel speed, flexibility, comfort, privacy and security. As already noted before, pricing is only one strategy for behaviour change. Therefore, a well-orchestrated mode-shift policy may best be based on a combination of several strategies simultaneously, so that other attraction factors would be affected as well. This would also reduce the probability of policy inconsistencies, as when, for instance, the kilometre price gives one signal, while infrastructure investments are giving another.

Sixth, official policy-designing bodies like the OECD, the European Commission and the WBCSD are operating at some distance of national governments and at an even greater distance of regular transport users. This may make it difficult to realise that the success of any policy measure crucially depends on its practicality-on-the-spot, cost-effectiveness and social acceptability – or "customer acceptance". Limitations in one or more of these implementation factors will also limit its benefits for sustainable transport development. An effective communication strategy, therefore, is indispensable. In it, the need for long term, well-coordinated changes should be continually emphasised.

Altogether so far, the strategic transport policies as envisioned by the various international bodies are problematic in various respects and may not soon enough lead to the needed reductions in pollution, congestion, accidents, CO_2 emissions and landscape degrading. At the same time, however, such reductions are broadly associated with the notion of sustainable transport (see further below). In terms of the commons dilemma model discussed next, international policymakers so far have done well to diagnose the problems and indicate possible solution directions, but they seem to be undecided as yet about *significant* long-term policy measures let alone their effective implementation. Banister (2000) also notes that policymakers have (only) reached the stage of contemplating limitations on car use and travel demand.

Greater attention to social and behavioural motives and mechanisms in using motor vehicles, however, is desirable if one wishes to strengthen the basis for effective policy-making. Let us, therefore, first consider the commons dilemma model and various general strategies for changing the behaviour of environment-burdening actors.

COMMONS DILEMMAS: A FOUR-STAGE MODEL OF COLLECTIVE RISK MANAGEMENT

The massive use of motor vehicles is a telling example of a large-scale commons (or common-resource) dilemma, reflecting a persistent conflict between many individual interests on the one hand, and a limited number of collective interests on the other (see Hardin, 1968; Edney, 1980; Vlek, 1996; Ostrom *et al.*, 2002). Table 4 provides a stepwise model approach revolving around the four components of problem diagnosis, policy decision-making, practical intervention and effectiveness evaluation. Together, these may be elaborated into twelve specific steps, as listed in Table 4.

Applied to collective transport problems this four-stage approach may be explicated as follows. In a well-tuned *problem diagnosis* one would analyse and describe the damage, risks and stress from motorised transport to human well-being, living environments and natural ecosystems. In addition, one would investigate the social and behavioural processes whereby damage is caused and risks are generated. Much of this has been well treated in Part 1 of this book, showing that scientific research plays a crucial role here. Moreover, in order to know how to best approach responsible actors one needs to familiarise oneself with their awareness of the problem, their perceptions of the damage and risks involved, and their background values and interests at stake.

Table 4: A Four-Stage Approach for Research and Policy-Making About Commons Dilemmas

I. Problem diagnosis
a. Analysis and assessment of collective risk, annoyance and stress
b. Analysis of socio-behavioural factors and processes underlying risk generation
c. Assessing problem awareness, risk appraisal and actors' values and benefits
II. Policy decision-making
d. Weighing of collective risk against total individual benefits ("need for change?")
e. If "risk unacceptable": setting environmental and/or social risk-reduction objectives
f. Translation of risk-reduction objectives into individual-behaviour goals
III. Practical intervention
g. Specifying feasible behaviour alternatives, selecting policy instruments
h. Zooming in on target groups, considering conditions for policy effectiveness
i. Full implementation and control of behaviour-change programme
IV. Effectiveness evaluation
j. Designing a monitoring and evaluation programme
k. Comprehensive evaluation of effects and side effects
l. Intermittent and *post hoc* feedback about intervention effects

Source: Adapted from Vlek (1996); see further Steg and Vlek (2007).

Under *policy decision-making*, the problem manager – that is, some policy-making body – would determine the extent to which the collective damage and risks outweigh the often large and varied collection of actors' individual benefits. Clearly, this involves scientific assessments as well as political judgments. If damage and risks are deemed unacceptable, then one would need to indicate how much and in which respects these should be reduced. This is to yield environmental and/or social risk reduction objectives that subsequently need to be translated into behavioural performance goals: what should actors actually do to help achieve the environmental and/or social goals adopted?

Practical intervention – Part 3 of this book – is aimed at initiating and supporting the cooperation of problem-causing actors. The latter need clear specifications of the behaviour changes required. Problem managers need effective policy instruments for approaching actors and encouraging them to cooperate; see next section. Subjects of intervention policies always differ in problem perceptions, knowledge of possible solutions, basic needs and values, and (thus) their capacity and motivation to cooperate. This means that interventions should be carefully tuned towards relevant characteristics of target groups, taking account of their different requirements for effective cooperation.

Effectiveness evaluation is an often-underestimated component of good policy-making. If one has a well-designed programme of collective damage reduction and risk management, and if one succeeds in applying it to the relevant groups of actors, then one would like to know as well the extent to which intervention policies are effective, in which respects, and for whom. Systematic evaluation of an intervention programme is important for policymakers' own feedback ("did it work?"), but it also enables one to provide feedback to targeted subjects about actual reductions in environmental and/or social damage and risks achieved ("did it help?").

General Strategies for Changing (Transport) Behaviour

Commons dilemmas can be "resolved" only by achieving a safer, sustainable *balance* between individual and collective benefits and risks. The nature and effectiveness of various strategies towards this end have been investigated in a great number of laboratory experiments as well as field studies. Most of these strategies can be categorised under seven general headings. The seven strategies can be briefly characterised as follows (description adapted from Vlek, 1996; see also Steg and Vlek, 2007).

1. *Provision of physical and/or technical alternatives and (re)arrangements* may change the basic structure of a commons dilemma by offering individual actors new or alternative behavioural options – or deleting options, thereby increasing or decreasing their freedom of choice. Through judicial application of some of the other strategies outlined below, these options may gradually become accepted as preferable to current, environmentally harmful options. Examples include setting up transport waste recycling facilities, closing town centres to motorised traffic and introduction of energy-efficient vehicles. The underlying assumption here is that individual behaviour is shaped in part by technological artefacts and systems and by the physical environment in which it takes place.

2. *Regulation-and-enforcement* generally serves to direct individual actors' choices. This strategy is based on laws, regulations and standards issued by the political leadership. Violations of these rules and norms usually meet with some kind of punishment, fine or disapproval. The underlying assumption is that official laws, regulations and standards are eventually internalised under threat of punishment. Such strategies imply a need for adequate organisational capacity for supervision, monitoring and enforcement.

3. *Financial-economic incentives* typically alter the pay-off structure of any commons dilemma and, in doing so, may change its very nature. Financial-economic strategies are designed to reward people monetarily for choosing behaviours that do not harm the collective, or to punish them for behaving in a collectively burdensome fashion. The relevant policy instruments are subsidies, discounts, levies, taxes, fines, tolls and the like. The basic underlying assumption is that the behaviour of individual actors is subject to the workings of the "price mechanism" and that the demand-price elasticities involved are reasonably high.

4. *Information, education and communication* may involve providing information, education, arguments, prompts and admonitions aimed at heightening actors' awareness (e.g., of environmental effects), increasing their knowledge (e.g., of policy goals and

behaviour alternatives), influencing their attitudes, increasing their trust in others' cooperation and (thereby) strengthening their inclination and ability to adopt other, less harmful behaviours. This strategy is based on the assumption that the behaviour in question is "reasoned" and that influencing the cognitive processes underlying behavioural choices will also affect those choices themselves.

5. *Social modelling and support* is based on the notion that collectively risky behaviour (and the cognitive structures underlying it) is strongly determined by social factors such as social norms and customs, processes of social comparison (in particular, status- and power-seeking), notions of individuality versus collectiveness, and the public examples set by prominent members of society. This strategy exploits these factors, for example by organizing family, company or community support for altering attitudes, preferences and habits, by offering people respected role models and by boosting their trust in others to cooperate in achieving common goals.

6. *Organisational change* aims to modify and adapt the structure and functioning of institutions, organisations and lifestyles, to make these more amenable to sustainable activities and behaviours. Changes may amount to regionalisation and downsizing of production-consumption systems, adopting a recycling system for enhancing materials' efficiency, or introducing company transport management for commuting personnel. The basic assumption here is that individual behaviour is embedded in and significantly directed by the institutions, organisations and generally adopted lifestyles that characterise the collective as a whole.

7. *Changing values and morality* involves appeals to the conscience of relevant actors, attempts to enhance their "altruism" towards other actors and towards future generations, and measures to increase mutual trust among those involved in the commons dilemma. The strategy may also seek to change people's conceptions of "quality of life" and particularly the relative importance they attach to environmental quality for sustainable development. The basic assumption here is that people's values and morality form the roots of fundamental attitudes that guide and support lifestyles and specific behaviours.

An eighth strategy may be labelled: "Do nothing; wait and see". This solution-by-default is nicely expressed in a Dutch saying that translates as "The quay will turn the ship". Such resignation will eventually necessitate adaptive behavioural responses, but these can never be anything but inappropriate, too little and too late. Making such environmental disaster scenarios palpable – for example, through "business-as-usual" studies – may allow individual actors to be (better) able to undertake due preventive action.

Tradable exploitation rights. One particular combination of government regulation (strategy 2 above), payment according to resource use (strategy 3), information provision (strategy 4) and privatisation (strategy 6), may prove to be an effective as well as efficient approach towards managing common goods, if applied to a suitably defined territory. This is the management system known as tradable emission permits or, more generally, tradable exploitation rights (see Bruneau, 2005, for a critical discussion). Theoretically, this is an almost ideal system of environmental risk management, because it seeks to achieve the necessary *balance* between individual and collective interests, between "government" and "market", between carrot and stick, and (thus) between individual freedom and social equality. In practical terms, however, for instance when implemented as a system of tradable car-kilometre allowances, it means making a number of serious political choices with regard to

- The total ceiling on exploitation: Which total number of car kilometres is acceptable for this country per year?
- The initial distribution of rights: Should the government, to begin with, allocate to all citizens an equal number of annual (or monthly) car kilometres?
- Government supervision: How could each citizen's permitted number of car kilometres per year be validly monitored and assessed?
- Government sanctioning: Which type and level of sanction – positive or negative – is most effective in making citizens' car use remain within the annually permitted number of kilometres?
- Securing full market operation: How could market imperfections and/or a "black market" in car-kilometre rights (yielding social inequities) be prevented?

Second-Order Management Approaches

A recurring question about commons dilemmas is: *who* should or may best establish diagnostic judgment, decide about management goals and strategies, undertake practical interventions, and evaluate their effects? This question refers to second-order strategies for managing commons dilemmas (see Yamagishi, 1995). Three different kinds of environmental resource management may be usefully distinguished, respectively: representative government, free-market economy, and stakeholder partnerships. Detailed discussion of these three management approaches is outside the scope of this chapter. It may be emphasised, however, that – given the existence of various government as well as market failures – new institutional arrangements are being designed and tested. The basic idea is that an optimal or acceptable management of common goods (*in casu:* sustainable transportation) may best be determined through relevant stakeholder dialogues and in multiparty decision-making (see Agrawal, 2002; Van Tatenhove and Leroy, 2003, for proposals and examples). Management through stakeholder dialogues requires a strong organisation, agreed procedures and open information exchange. Often, stakeholder dialogues are forced upon initiators of transport development projects (see, e.g., Flyvbjerg *et al.*, 2003) because some of the parties concerned have been left out of the process. However, especially in controversial cases, much time, effort and money can be saved by drawing in representatives from various stakeholder groups and following a joint procedure of information exchange, evaluation and decision-making.

Conditions for Subjects' Behaviour Change

The barriers to "sustainable" behaviour change are manifold. In fact, like solutions, barriers can also be listed under the several strategies discussed above. Examples include: absence of physical or technical alternatives (for strategy 1), inadequate and/or ineffective law enforcement (strategy 2), weakness and/or inconsistency of financial incentives (strategy 3), unawareness of one's own causal role and possible contribution to solutions (strategy 4), absence or invisibility of model behaviours by opinion leaders (strategy 5), organisational goals biased towards short-term "survival" (strategy 6) and importance of social status in consumer activities (strategy 7). Gatersleben (see Chapter 12, this volume) elaborates on the importance of affective and symbolic motives for car use.

In response to this, one may infer a number of basic conditions for effective behaviour change. Many of these also show up in research on social cooperation (e.g., Messick and Brewer, 1983; Kopelman *et al.*, 2002; Steg and Vlek, 2007). Let us go by Table 4 as well as the various strategies again, but now *from the subjects' point of view*. Under "problem diagnosis", subjects' own (long-term) problem awareness and their risk appraisal are important for understanding the need for policy formation. Under "policy decision-making" subjects need insights in policy goals and their own behaviour alternatives to possibly help achieve goals: "Where are we heading?" Under "practical intervention" subjects need to appreciate the feasibility, for themselves, of available behaviour options, and they must accept that policies are (probably) effective, both through their own behaviour change ("self-efficacy") and through their trust in the cooperation of others ("response efficacy"). Finally, under "effectiveness evaluation" subjects need to see where policies are leading and whether their cooperation strengthens the common good. Sustainable-transport policies, too, should preferably be developed (and perceived) such that they enhance, not diminish subjects' overall quality of life (see Chapter 3 by Gifford and Steg, this volume).

Commons-Dilemma Thinking

The commons dilemma model (Table 4) may be a useful basis for thinking about transport developments and transport policy-making. First, the model clearly holds that collective problems arise from individual behaviour patterns, that is, from the accumulation of negative external effects of individual behaviours (such as emitting exhaust gases). Thus, problem solving depends on individual behaviour change. Second, clear policy goals and decisions are needed about the directions in which the collective problems may best be resolved. Third, there is a spectrum of strategies for accomplishing actors' behaviour change. Each strategy has its own strengths and weaknesses. Thus no single strategy, be it technological innovation, spatial (re-)design, regulation or pricing, can be effective enough by itself. In many cases a well-designed combination of strategies may therefore be optimal, whereby "tradable exploitation rights" (see above) deserves special attention. Fourth, any campaign towards reducing environmental damage and risks from motorised transport should be well tuned towards the targeted subjects' basic characteristics, and its environmental, social and economic effects must be systematically evaluated.

DEFINING "SUSTAINABLE TRANSPORT DEVELOPMENT"

From its renewed formulation by the Brundtland Commission (WCED, 1987), the concept of sustainable development has been elaborated in three dimensions, viz. economic, social and environmental sustainability (e.g., Munasinghe, 1993). Gudmundsson and Höjer (1996) consider transport development on the basis of four sustainability principles: (1) safeguarding the natural resource base, (2) maintaining the option value of a productive capital base for future generations, (3) improving the quality of life for individuals and (4) realising an equitable distribution of quality of life. Following the authors (p. 280): ". . . the mobility patterns, which dominate western society today, are not in accordance with these sustainable development goals".

As quoted previously, the OECD (1997, p. 56) defined sustainable transportation in terms of meeting various access needs without permanent environmental harm and social inequity. The WBCSD (2004) says: "Sustainable Mobility is the ability to meet the needs of society to move freely, gain access, communicate, trade, and establish relationships without sacrificing other essential human or ecological values, today or in the future". The essence of both the OECD's and the WBCSD's definition is that "meeting mobility needs" cannot be sustainable when it involves irreparable harm to important social and environmental goods, now or later.

In 2001, the transport ministers of the then 15 EU countries acknowledged the economic, social and environmental dimensions of sustainable development and agreed to a more elaborate definition, as dissected in Table 5. While the definition from the OECD's Vancouver conference (1996) is fairly clear in setting boundary conditions ("no permanent harm . . ."), the 2001 circumscription by the EU ministers reflects a multi-objective goal orientation leaving much room for trade-offs and further consideration of key terms, for instance, "optimally using renewables", "minimizing noise" and "minimizing land use".

If the specifications of Table 5 are viewed against the "principal measures" listed in Table 1 and making up the EU's common transport policy for 2010, one may easily note the commons dilemma between transport benefits and environmental quality. Also notable is the dominance of economic and social development goals, whose achievement undoubtedly yields consistent transport growth, in all modalities but especially by road and air. The elaboration of environmental quality goals in both Tables 1 and 5 suggests

Table 5: Specifications of "Sustainable Transport" in EU Ministers' Definition (2001a)

Social	• Meeting basic access needs of individuals, companies and societies . . .
	• Safe, healthy and environmentally friendly . . .
	• Equitable within and between generations . . .
Economic	• Affordable, fairly and efficiently operated . . .
	• Supportive of a competitive economy . . .
	• Allowing balanced regional development . . .
Environmental	• Limiting emissions and waste within natural absorption capacity
	• Optimally using renewable energy, materials . . .
	• Controlled use of non-renewables . . .
	• Minimising land use . . .
	• Minimising noise generation

that policymakers believe they can both have their cake, that is, safeguarding environmental sustainability, and eat it, that is, letting motorised transport grow on.

This implicit policy tendency, however, goes against the logic of the IPAT-formula (Ehrlich and Holdren, 1971) in which total environmental impact (I) is viewed as the product of population size (P, the number of transport users), per-capita affluence (A, the amount of transport use per person) and the environmental resource intensity of the technology used for one unit of affluence (T, the environmental burden per transport kilometre): $I = P \times A \times T$. With a growing population of transport users, more transport per person (also indirectly, through the distribution of consumer goods and services) and greater environmental impact of motor vehicles (e.g., bigger cars, more powerful vans and trucks), the total environmental burden of transport can only increase, unless significant policies are directed at all three IPAT-components. As already indicated at this chapter's beginning, such a multi-level policy would be aimed at a reduction in the number of (motorised) transport users, in the amount of motorised transport per person, and in the environmental resource intensity of motorised transport per unit of distance covered.

In these respects the EU seems to be moving rather cautiously. In its communication about the notion of sustainable transport (EU, 2003, p. 13), the Commission unfolds "a strategy for sustainability" which emphasises the promotion of rail (instead of road and short air) transport, more direct user payment for infrastructure, elimination of cross-border bottlenecks, and reduction of transport's environmental burden.

This strategy actually is a partial elaboration of the EU's (2001b) "principal measures" summarised in Table 1. Apparently, greater emphasis is now laid on shifts between transport modes and on behavioural changes of transport users. However, the Commission's 2003 communication does not specify any sustainability objectives to be realised in due time. It may be more supportive for the EU to construct several long-term

Table 6: A Three-Dimensional Structure of Guiding Points and Indicators for Sustainable Transport Development

	Economic	*Social*	*Environmental*
General goal variables	• Production, consumption • Transport, trade • Employment, income • Profits, investments • Low-tax burden • Stable development	• Health, safety • Freedom, autonomy • Social equity • Self-esteem, status • Comfort, convenience • Variety, excitement • Community liveability	• Natural resources (energy, materials) • Natural sinks (regenerative capacity) • Natural habitats, ecosystems • Climate stability
Specific transport objectives and quality indicators	• Accessibility • Efficiency • Safety, security • Worker conditions • International standards • Infrastructure costs • Transport costs for users	• Sufficient mobility • Efficient mobility • Equal opportunity • User payability • User-friendliness • Healthy environment • Urban space use	• Land/space use • Air pollution (various) • CO_2 emissions • Soil, water pollution (waste, road run-off, fuel spills) • Noise • Hydrological impacts

transport policy scenarios, much like the German FEA (1998) and the OECD (2002; see also Geurs and Van Wee, 2003) have done, and to investigate how important sustainability objectives (cf. Table 6) could be met through one or the other scenario. Thus far at least, the overall EU policy for sustainable transport seems to be ineffective.

SUSTAINABLE-TRANSPORT POLICY-MAKING

In view of the above, let us attempt to elaborate what sustainable transport development may involve and how it might be achieved. Table 6 provides an ordering of general goals and specific objectives and quality indicators, aimed at defining and promoting sustainable transport development in the economic, social and environmental domains.

The difference between sustainable and unsustainable transport development may be marked via a two-stage approach. In stage one, minimum sustainability standards may be set, for example, with respect to accessibility and transport speed (economic quality), health and safety (social quality) and harmful emissions and land use (environmental quality). Minimum quality standards are necessary to prevent one dimension, for example, economic sustainability, to overrule either one or both of the other two; this would evidently not be sustainable in the overruled dimension(s). Secondly, one may identify and make trade-offs between conflicting sustainability objectives, for example, transport speed versus harmful emissions, or accessibility versus space use, or health versus comfort.

Thus, a sustainable-transport policy document may well cover the items in Table 6 and contain specifications of minimum standards for important economic, social and environmental quality objectives (see also Litman, 2005). It may also explicate the trade-offs proposed between mutually conflicting objectives, especially between economic objectives on the one hand and social and environmental objectives on the other.

CONVERGING PRINCIPLES FOR SUSTAINABLE TRANSPORTATION

Unfolding a three-dimensional structure for transport policy-making is one thing. In addition, one also needs principal directions for sustainable transport development. Given that the major problem of unsustainability is the continuing proliferation of private, fossil-fuelled motor vehicles (cars, vans, and trucks as well as motorised two- and three-wheelers) and the related road infrastructure, what seems necessary is a significant shift of emphasis away from that transport mode and towards modes that, in their collective environmental impacts, meet the basic requirements of sustainability. Five policy principles may be worth considering, as follows.

First, for all transport users alike, *freedom of mobility and transport is inversely proportional to the negative external effects (per person- or ton-kilometre) of the chosen transport mode.* This principle requires that the spectrum of negative external effects is well known, so that people may realise what their transport is bringing about for others and the environment. The principle further implies that there is a realistic choice of transport modes so that, if necessary, one may go for the least burdening one. When certain (e.g., long distance) trips are only possible by means of an environment-burdening mode of transport, such trips may have to be discouraged[3].

Second, *sustainable transport behaviours are guided by spatial design, infrastructure, vehicle technology and traffic regulation that indicate, invite or induce such behaviours.* This principle links up with the idea that a behaviouristic approach – involving external-environmental opportunities, incentives and other guidance – will be more effective than a cognitivistic approach aimed at changing users' internal attitudes, norms and values (within a transport system that invites them to behave otherwise). Needed mentality changes will be strongly conditioned by changes in people's external physical and technical environment.

Third, *the transport user pays for all costs of any trip made.* This principle requires the valid and reliable assessment of "all costs", especially the external costs of negative social and environmental impacts. This should include costs of infrastructure, traffic

[3] Thus, in contrast to relevant statements in its White Paper (EU, 2001a), the EU may well have to recommend restrictions on environmentally harmful transport of persons and goods.

regulation, mitigating (effects of) air pollution, noise abatement and accident prevention. Such assessment may also be helpful for valuing the total negative external effects of different transport modes, as required to follow the first principle above.

Fourth, *people's awareness of transport problems, risks and dilemmas for the future is continually promoted and improved.* This principle expresses the essential role of societal problem awareness for policy acceptance and cooperation. Promoting problem awareness obviously starts with adequate expert analyses, and it requires credible public communication.

Fifth, *public authorities, such as leaders of government, industry, schools, universities and the media, demonstrate a sense of urgency, while setting examples in their own transport attitudes, decisions and behaviours.* This principle expresses the key role of societal leadership, role models and social support in acknowledging the problems of unsustainable transport development and in indicating the course towards long-term sustainability.

SUBSTANTIVE TRANSPORT POLICIES

Which concrete policies would bring society closer to sustainable transport development? Table 7 presents a collection of feasible policies aimed at restoring the balance between motorised transport and environmental (and social) quality. Some of the items in Table 7 have already been mentioned by ISIRT, OECD, EU or WBCSD, as noted above. Other items may be recognised from various other literature (e.g., Lowe, 1990a, 1990b, 1992; Vlek and Michon, 1992; Banister and Button, 1993; Rees, 2003; Hensher and Button, 2003).

Table 7: Substantive Policies for Sustainable Transport Development

- Promoting compact urban design (and preventing urban sprawl) to reduce transport needs
- Developing, maintaining high-quality urban public transport, including light-rail
- Further developing, promoting infrastructure, facilities for convenient bicycling in urban areas
- Improving infrastructure and organisation for linking walking and cycling to public transport
- Shifting passenger transport from private motor vehicle to collective vehicle on road or rail
- Extending, improving transport services for young, disabled, handicapped and elderly people
- Expanding, improving regional, national and international railways
- Shifting short-haul passenger transport (<500 km) from air to rail
- Shifting freight transport from road to rail and inland waterway
- Adopting/maintaining great reluctance in developing new road infrastructure
- Largely banning the private motor vehicle from the centres of cities and towns
- Producing cleaner, quieter motor vehicles, making them use less fossil fuel
- Facilitating and promoting inter-modality between road, rail and waterway transport
- Making the use of private motor vehicles generally more difficult and less attractive
- Discouraging the "motor vehicle culture" of racing events, media advertising, clubs and media

From the OECD's multi-annual project on Environmentally Sustainable Transport (EST), a best-practices document has been published (OECD, 2000) which describes and honours sixteen "outstanding contributions" all over Europe. Among others, there are projects about car-sharing in Germany and in Switzerland, natural gas-fuelled city buses in Hungary, a "traffic-saving community" in Langenlois, Austria, non-motorised city transport in historic Spoleto, Italy, a transport-chain emissions profiling programme in Sweden, a school car-trip reduction programme in Victoria, British Columbia, and a "soft-mobility network" of bicycle paths in Wallonia, Belgium. These are, of course, well-intentioned projects marking inspiring beginnings with sustainable transport development. However, more-encompassing system changes and far-reaching field experiments seem necessary to eventually realise large-scale societal transformations of unsustainable transport systems.

CONCLUDING WISDOMS FROM A BEHAVIOURAL SCIENCE PERSPECTIVE

This chapter started with an overview of the worldwide growth in motorised transport and of the actual problem analyses and policy intentions formulated by the OECD (1997, 2002), UNEP (1999), EU (2001a, 2001b, 2003) and WBCSD (2004), among others. It was proposed that commons-dilemma thinking (see Table 4) can be supportive of systematic analysis and management planning, following the four components of problem diagnosis, policy decision-making, practical intervention and effectiveness evaluation. In view of the concerns and opinions expressed by the OECD, EU, WBCSD and others, it may be concluded that, policy-wise, much more seems needed with regard to practical interventions as well as effectiveness evaluation. To support this, five policy principles and a variety of substantive policies for sustainable transport development were proposed above. In addition, the following observations and suggestions may be useful.

Motorised transport is a large-scale commons dilemma involving many different individual users as diffuse sources whose negative impacts are each very small in proportion to the overall problem accumulation. In comparison to large point sources of environmental pollution, for instance, electric power plants or petrochemical refineries, motor vehicle users are less easily approached as responsible polluters and they can more easily find justification in the behaviours of their numerous road-traffic companions. Therefore, a sophisticated policy approach is required if one wants to achieve sufficient social understanding, acceptance and cooperation for effective policy implementation. A crucial element of this approach is the need to make popular but unsustainable transport modes less appealing, in order to achieve increased use of sustainable alternatives.

For scientific experts and policymakers outside the social and behavioural sciences it may be useful to stress the "forgotten" conditions for human behaviour change. Physical and technical alternatives, regulation and pricing are insufficient drivers of change by themselves. Besides these, problem awareness, self-efficacy, trust in others' cooperation and positive expectations about future quality of life are crucial factors that may make or break sufficient policy acceptance and cooperation.

In their ecological-economic analysis, Gudmundsson and Höjer (1996) identify two major difficulties for sustainable transport policies. On the one hand they doubt whether mobility is an undeniable good and not forced upon people through social and physical structures. On the other hand, they realise that different sectors of modern society are highly integrated with one another, so that transforming the transport sector would have effects spreading into other activity domains. These considerations give a deeper meaning to the issue of coercive versus non-coercive, or push versus pull (and persuasion) strategies for transport behaviour change (see other chapters in Part 3 of this volume). If the private motor vehicle would not be so attractive socially and economically, it would not be so hard to bring its use within the boundaries of sustainability. So it would seem that, particularly in urban areas, some individual freedom, comfort, pleasure and status have to be subordinated to the improvement and security of long-term environmental and social qualities for all.

From a behaviouristic perspective it is essential to include the supply side of "transport consumption" in the design of sustainable-transport policies. This means that the motor vehicle industry, fuel producers, road-building companies and connected organisations such as special clubs and media should be involved so as to gradually adopt new roles, develop new products and adapt their way of approaching potential customers, with a view on sustainable transport development.

Reflecting the scope and urgency of the sustainable-transport problem, the foreword to the WBCSD (2004) report states: "A clear message of 'Mobility 2030' is that if we are to achieve sustainable mobility it will require contributions from every part of society throughout the world". This is exactly what sustainable transport development seems to demand. Fundamental and far-reaching transformations in current transport systems and habits require solid scientific research, effective deliberations and well-organised interactions among government, industry and citizen representatives. Diagnosis, decision, intervention and evaluation are the four components of an overall approach aimed at finding common routes to sustainability.

REFERENCES

Agrawal, A. (2002). Common Resources and Institutional Sustainability. In: *The Drama of the Commons* (E. Ostrom, Th. Dietz, N. Dolšak, P.C. Stern, S. Stonich, and E.U. Weber, eds), National Academies Press, Washington, DC, 41–85.

Alcott, B. (2005). Jevons' Paradox, *Ecological Economics*, **54**, 9–21.

Banister, D. (2000). Sustainable Urban Development and Transport – A Eurovision for 2020, *Transport Reviews*, **20**, 113–130.

Banister, D. and K. Button (eds) (1993). *Transport, the Environment and Sustainable Development*, E & F Spon, London.

Berkhout, P.H.G., J.C. Muskens, and J.W. Velthuijsen (2000). Defining the Rebound Effect, *Energy Policy*, **28**, 425–432.

Bruneau, J.F. (2005). Inefficient Environmental Instruments and the Gains from Trade, *Journal of Environmental Economics and Management*, **49**, 536–546.

Button, K. and W. Rothengatter (1993). Global Environmental Degradation: The Role of Transport. In: *Transport, the Environment and Sustainable Development* (D. Banister and K. Button, eds), E & F Spon, London, 19–52.

ECMT (European Conference of Ministers of Transport). (1990). *Transport Policy and the Environment*, ECMT & OECD Publications, Paris.

ECMT (1993). *Transport Policy and Global Warming*. In: *Proceedings of an ECMT Seminar* (J. Short, ed.), OECD Publications ECMT Series 75 93 10 1, Paris Cedex 16, France.

ECMT (2004). *National Cycling Policies for Sustainable Urban Transport. Policy Note and Declaration*, ECMT, CEMT/CM(2004)11, Paris, April (www.cemt.org).

Edney, J.J. (1980). The Commons Problem: Alternative Perspectives, *American Psychologist*, **35**, 131–150.

EEA (European Environment Agency). (1999). *Environment in the European Union at the Turn of the Century*, EEA, and Luxembourg Office for Official Publications of the European Communities, Copenhagen (see also www.eea.eu.int).

EEA (2004). *Ten Key Transport and Environment Issues for Policy-Makers*, EEA, and Luxembourg Office for Official Publications of the European Communities, Copenhagen (see also www.eea.eu.int).

EEA (2006). *Transport and Environment: Facing a Dilemma – TERM 2005*, EEA Report No. 3/2006 (http://reports.eea.eu.int/eea_report_2006_3/en).

Ehrlich, P.R. and J.P. Holdren (1971). Impact of Population Growth, *Science*, **171**, 1212–1217.

EU (2001a). Definition of Sustainable Transport Adopted at Luxembourg Meeting of EU Ministers of Transport and Communication (www.health.fgov.be/ WH13/ krant/ krantarch2001/ kranttekstmei1/010509m05eu.htm).

EU (2001b). *White Paper on "European Transport Policy for 2010: Time to Decide"*, European Commission, Brussels (http://europa.eu.int/scadplus/leg/en/lvb/l24007/htm).

EU (2003). *Europe at a Crossroads; the Need for Sustainable Transport*, Luxembourg Office for Official Publications of the European Communities, p. 22 (see also europa.eu.int/scadplus/leg/en/lvb/l24007/htm).

FEA (Federal Environment Agency). (1998). *Sustainable Development in Germany: Progress and Prospects*, Umwelbundesamt, P.O. Box 330022, D-14191 Berlin, Berlin, Germany.

Flyvbjerg, B., N. Bruzelius, and W. Rothengatter (2003). *Megaprojects and Risk: An Anatomy of Ambition*, Cambridge University Press, Cambridge, UK/New York.

Geurs, K.T. and G.P. Van Wee (2000). *Environmentally Sustainable Transport: Implementation and Impacts for the Netherlands for 2030*, National Institute for Public Health and the Environment RIVM, Report 773002013, Bilthoven (NL), p. 144.

Geurs K.T. and G.P. Van Wee (2003). The Role of Non-Motorized Modes in an Environmentally Sustainable Transport System. In: *Sustainable Transport: Planning for Walking and Cycling in Urban Environments* (R. Tolley, ed.), Woodhead, Abington/Cambridge, UK, 32–47.

Gudmundsson, H. and M. Höjer (1996). Sustainable Development Principles and Their Implications for Transport, *Ecological Economics*, **19**, 269–282.

Hardin, G. (1968). The Tragedy of the Commons, *Science*, **162**, 1243–1248.

Hensher, D.A. and K.J. Button (eds) (2003). *Handbook of Transport and Environment*, Elsevier/Pergamon, Amsterdam.

ISIRT (International Scientific Initiatives on Road Traffic). (1992). *Special Issue on Road Transport and Environment*, *IATSS Research*, **15** (2), 5–151.

Kopelman, S., J.M. Weber, and D.M. Messick (2002). Factors Influencing Cooperation in Commons Dilemmas: A Review of Experimental Psychological Research. In: *The Drama of the Commons* (E. Ostrom, Th. Dietz, N. Dolšak, P.C. Stern, S. Stonich, and E.U. Weber, eds), National Academies Press, Washington, DC, 113–156.

Litman, T. (2005). *Well Measured; Developing Indicators for Comprehensive and Sustainable Transport Planning*, Victoria Transport Policy Institute, Victoria, BC (www.vtpi.org).

Lowe, M.D. (1990a). *Alternatives to the Automobile: Transport for Livable Cities*, Worldwatch Institute Paper no. 98, Washington, DC, p. 49.

Lowe, M.D. (1990b). Cycling into the Future. In: *State of the World 1990. A Worldwatch Institute Report on Progress Toward a Sustainable Society* (L.R. Brown, ed.), W.W. Norton, New York, and Unwin Hyman Ltd, London, 119–134.

Lowe, M.D. (1992). Shaping Cities. In: *State of the World 1992. A Worldwatch Institute Report on Progress Toward a Sustainable Society* (L.R. Brown *et al.*, eds), W.W. Norton, New York/London, 119–137.

Messick, D.M. and M.B. Brewer (1983). Solving Social Dilemmas: A Review. In: *Review of Personality and Social Psychology* (L. Wheeler and P. Shaver, eds), Vol. 4, Sage, Beverly Hills, CA.

Munasinghe, M. (1993). *Environmental Economics and Sustainable Development*, World Bank, Washington, DC.

OECD (Organisation for Economic Cooperation and Development). (1997). *Towards Sustainable Transportation*, Summary Proceedings of an International Conference held in Vancouver, Environment Canada, Hull (Quebec), March.

OECD (2000). The *EST!* Best Practice Competition (www.oecd.org).

OECD (2002). *Environmentally Sustainable Transport (EST): Futures, Strategies and Best Practice*, Report on the OECD Conference (www.oecd.org).

Ostrom, E., Th. Dietz, N. Dolšak, P.C. Stern, S. Stonich, and E.U. Weber (eds) (2002). *The Drama of the Commons*, National Academies Press, Washington, DC.

Pucher, J., N. Korattyswaropam, N. Mittal, and N. Ittyerah (2005). Urban Transport Crisis in India, *Transport Policy*, **12**, 185–198.

Rees, W. (2003). Ecological Footprints and Urban Transportation. In: *Sustainable Transport: Planning for Walking and Cycling in Urban Environments* (R. Tolley, ed.), Woodhead, Abington/Cambridge, UK, 3–19.

Steg, L. (2003). Can Public Transport Compete with the Private Car? *IATSS Research*, **27**, 27–35.

Steg, L. and C. Vlek (2007). Understanding and Managing Environmental Resource Use: A Behavioural Science Perspective. In: *Principles of Environmental Science* (J.J. Boersema and L. Reijnders, eds), Kluwer Academic Publishers, Dordrecht/Boston.

Tiwari, G. (1999). Planning for Non-Motorized Traffic – A Prerequisite for Sustainable Transport System, *IATSS Research*, **23**, 70–77.

UNEP (United Nations Environment Program). (1999). *GEO 2000: Global Environment Outlook*, Earthscan, London.

Van Tatenhove, J.P.M. and P. Leroy (2003). Environment and Participation in a Context of Political Modernisation, *Environmental Values*, **12**, 155–174.

Vlek, C.A.J. (1996). Collective Risk Generation and Risk Management: The Unexploited Potential of the Social Dilemmas Paradigm. In: *Frontiers in Social Dilemmas Research* (W.B.G. Liebrand and D.M. Messick, eds), Springer, Heidelberg/New York, 11–38.

Vlek, C. and J.A. Michon (1992). Why We Should and How We Could Decrease the Use of Motor Vehicles in the Near Future, *IATSS Research*, **15** (2), 82–93.

WBCSD (World Business Council for Sustainable Development). (2004). *Mobility 2030: Meeting the Challenges to Sustainability*, WBCSD Sustainable Mobility Project, Geneva, CH (www. wbcsd.ch).

WCED (World Commission on Environment and Development). (1987). *Our Common Future*, Oxford University Press, New York.

Wright, C. and J. Egan (2000). De-Marketing the Car, *Transport Policy*, **7**, 287–294.

Wright, C. and B. Curtis (2005). Reshaping the Motor Car, *Transport Policy*, **12**, 11–22.

Yamagishi, T. (1995). Social Dilemmas. In: : *Sociological Perspectives on Social Psychology* (K.S. Cook, G.A. Fine, and J.S. House, eds), Allyn & Bacon, Boston, 311–335.

Zacharias, J. (2003). Choosing Sustainability: The Persistence of Non-Motorized Transport in Chinese Cities. In: *Culture, Environmental Action and Sustainability* (R. García Mira, J.M. Sabucedo Cameselle, and J.R. Martinez, eds), Hogrefe & Huber, Göttingen, Germany, 219–229.

AUTHOR INDEX

Page numbers in roman type indicate that the listed author is cited on that page; page numbers in italic denote the page where the literature citation is given.

A

B

H

I

J

T

Subject Index

www.ingramcontent.com/pod-product-compliance
Lightning Source LLC
LaVergne TN
LVHW010554300926
849162LV00002B/326